Hydrophobic Interactions

Hydrophobic Interactions

Arieh Ben-Naim

Department of Physical Chemistry
The Hebrew University of Jerusalem
Jerusalem, Israel

Plenum Press · New York and London

Library of Congress Cataloging in Publication Data

Ben-Na'im, Aryeh.
  Hydrophobic interactions.

  Includes index.
  1. Surface chemistry. 2. Solution (Chemistry) I. Title.
QD506.B45   541'.39   79-510
ISBN 0-306-40222-X

© 1980 Plenum Press, New York
A Division of Plenum Publishing Corporation
227 West 17th Street, New York, N.Y. 10011

Printed in the United States of America

To my sons

ODED
RAANAN
and
YUVAL

Preface

My personal involvement with the problem of hydrophobic interactions (HI) began about ten years ago. At that time I was asked to write a review article on the properties of aqueous solutions of nonpolar solutes. While surveying the literature on this subject I found numerous discussions of the concept of HI. My interest in these interactions increased especially after reading the now classical review of W. Kauzmann (1959), in which the importance of the HI to biochemical processes is stressed.

Yet, in spite of having read quite extensively on the various aspects of the subject, I acquired only a very vague idea of what people actually had in mind when referring to HI. In fact, it became quite clear that the term HI was applied by different authors to describe and interpret quite different phenomena occurring in aqueous solutions. Thus, even the most fundamental question of the very definition of the concept of HI remained unanswered. But other questions followed, e.g.: Are HI really a well-established experimental fact? Is there any relation between HI and the peculiar properties of water? Is the phenomenon really unique to aqueous solutions? Finally, perhaps the most crucial question I sought to answer was whether or not there exists hard evidence that HI are really important —as often claimed—in biological processes.

It is not uncommon to find statements in many textbooks on biochemistry referring to HI as "a major driving force in folding of macromolecules, the binding of substrate to enzymes and most other molecular interactions in biology" (Stryer, 1975).

Such statements may be correct. However, in spite of my researches in this field over almost ten years, I cannot confirm that there is at present either theoretical or experimental evidence that unequivocally demonstrates the relative importance of HI over other types of interactions in aqueous solutions. This does not mean that the whole subject is of no interest. On the contrary, even if the whole field should ultimately prove to be totally irrelevant to biochemistry, it still deserves careful attention within the general framework of the study of the properties of aqueous solutions *per se*. It is also my belief that HI are, and will prove to be, of central importance to the understanding of complex biological processes.

The initial aim of this book was to try to answer some of these questions, but the reader who is looking for definitive answers will surely find this book disappointing. The reason is very simple: Our present knowledge and understanding of the HI are only in their very elementary stages. Much more work must be done on both the experimental and the theoretical frontiers before we are sufficiently well equipped to attempt to answer these basic questions. Thus, my aims in writing the book have shifted from the initial ones to less ambitious ones; namely, I have concentrated on surveying what is presently known rather than on what is completely understood in this field.

The style of the book is mostly descriptive. Theory is used either as a means for processing and interpreting experimental data, or for suggesting new and relevant experiments. Recently, some new theories have been devised to predict the strength of the HI. These are too advanced, however, to be included in this, rather elementary, exposition of the subject.

All chapters in the book deal with aqueous solutions, rather than with pure liquid water. Our understanding of HI will ultimately depend on a fuller understanding of the peculiar properties of liquid water, but it is not possible to establish such a link at present. However, in Chapter 5, some notions such as the "structure of water" and "structural changes in the solvent" are discussed in connection with the phenomena of HI.

Chapter 1 introduces the basic definitions of the concepts that are used throughout the book. The level of presentation is very elementary. Although some of the statements are based on statistical mechanical arguments, the latter are collected in the Appendixes and are not presented in the main text.

Chapter 2 is devoted to the properties of ideal dilute solutions. These solutions are useful in the study of the so-called solute–solvent interactions. These interactions are related, both directly and indirectly, to the main topic of the book, namely the solute–solute interactions.

Chapters 3 and 4 form the main body of the work. Here, we first survey the experimental data on the *pairwise* solute–solute interactions and then proceed to the more general case, i.e., interactions among many solute particles. In the last two sections of Chapter 4 we present a very short review of an immense literature dealing with the role of HI in micellar and biological systems. The prevalent approach in these two chapters is to present those experimental methods that provide information on the nature of the HI. In most of the biochemical literature, the concept of HI is, by contrast, usually used to explain some peculiar and complex processes. In other words, the concept of HI is used as input rather than as output. I believe that the former approach, though very popular, is unjustified because it applies a poorly understood concept to explain other, more complicated phenomena. Thus, to a certain extent, we diverge in Sections 4.8 and 4.9 from our main approach, and present some examples that are clearly relevant to the subject of the book, although they may not as yet be used to provide information that will enhance our knowledge of the HI.

Chapter 5 deals with the temperature and pressure dependence of the HI. Here, the concept of the "structure of water" is used to interpret some of the peculiarities of the entropy and the enthalpy changes associated with the process of HI. In order to define and use the concept of the structure of water we felt it necessary to appeal to some elementary concepts in statistical mechanics. We also use statistical mechanics in the Appendixes to provide a better background for some statements made throughout the book.

In a rapidly expanding field of research, it is likely that some results, either of experimental or of theoretical nature, may become obsolete during the time lag from the date of submission to actual publication. Therefore I have stressed, in most parts, the general principles rather than specific results. This approach has particularly influenced my decision to review only very briefly the recent progress made in the theory of HI.

Tables of data are used for illustration only. It is not claimed that a particular set of data is the most complete or most accurate of the kind. As the reader will immediately notice, the data themselves are very scant and fragmentary. I have also exercised the author's privilege and have added a few personal comments at the end of some sections. Here I have tried to express my personal view on the merits and potentialities of the particular topic discussed in that section and, in some cases, tried to suggest further experiments that should complement and improve our knowledge of that field.

Acknowledgments

I am very much indebted to many friends and colleagues who have read parts of the manuscript and have made useful comments. I am especially grateful to Dr. Carmel Jolicoeur for writing the first draft of Section 3.10 on the application of NMR and ESR to the problem of hydrophobic interactions. This section has been further revised with the kind help of Dr. Haim Levanon and Dr. Gerhard Hertz. I am also grateful for helpful comments from: Tarique Andrea, Rubin Battino, Kulbir Birdi, John Edsall, Zeev Elkoshi, Walter Kauzmann, Eddie Morild, David Oakenfull, Zvi Rappoport, Harold Scheraga, and Jacob Wilf.

The book was typed by Ms. Doris Ganeor and the art work of most of the figures was done by Ms. Rachel Behrend. Their help and patience are gratefully acknowledged.

Arieh Ben-Naim

Jerusalem, Israel

Contents

Chapter 4. Hydrophobic Interaction among Many Solute Particles

Chapter 5. Temperature and Pressure Dependence of the Hydrophobic Interactions

Appendixes

Chapter 1

Introduction and Fundamental Equations

1.1. WHAT ARE HYDROPHOBIC INTERACTIONS?

The term "hydrophobic interaction" (HI) has been used in connection with several closely related phenomena. It is difficult to trace the origin of this term [or of its predecessor—the "hydrophobic bond"—apparently first coined by Kauzmann (1954, 1959)], but it is quite clear that this concept has been used in the biochemical literature to describe a variety of biochemical processes, including conformational changes of a biopolymer, the binding of a substrate to an enzyme, the association of subunits to form a multisubunit enzyme, and processes involving high levels of aggregation such as the formation of biological membranes and the organization of biological molecules to form a functional unit in a living system.

All the processes mentioned above are very complex in the sense that many factors combine to determine their rate and equilibrium constants. At present it is impossible to assess the relative importance of the various factors involved, such as the van der Waals forces, hydrogen bonds, charge–charge interactions, or hydrophobic interactions. Instead, it has now become fashionable to study each of these factors separately in model systems. Such a step is indeed necessary in order to gain information and to characterize each of the factors involved. It might not be sufficient, because the combined effect of all these factors may not be *additive*, i.e., when two or more of these factors operate simultaneously, nonadditive or cooperative

1

effects might be significant. Hence, even a complete knowledge of each of these factors may not be sufficient to understand the way they combine in the more complex processes.

This book is devoted to one of these factors, the hydrophobic interaction. We shall soon see that even on this restricted topic we know very little, and much has to be done before a complete knowledge and characterization of these interactions can be achieved.

Consider a conformational change of a biopolymer, as depicted schematically in Figure 1.1. Ignoring, for the moment, all other factors and focusing our attention on the HI only, the immediate question that arises is: In what model system can the HI be studied in isolation? There appears to be a kind of hierarchy of elementary processes, the study of which might contribute to our understanding of the hydrophobic interaction. Some of these are depicted in Figure 1.2.

The simplest process is the transfer of a simple solute, such as argon or methane, from pure water into a nonpolar liquid such as hexane. The study of this process has been advocated by many authors [in particular by Kauzmann (1959) and later by Tanford (1973)]. The idea is that the free energy of transferring methane from water into a nonpolar solvent is a good measure of the free energy change for the transfer of a —CH_3 group from the random-coil conformation to the interior of the polymer in the native configuration. In this way we may identify and study in isolation one factor that contributes to the total standard free energy change of the complete conformational change as depicted in Figure 1.1. Of course, this is not exactly so; the —CH_3 group is not entirely surrounded by water molecules in the random-coil conformation and is not really in a pure nonpolar environment in the compact configuration. Nevertheless, using

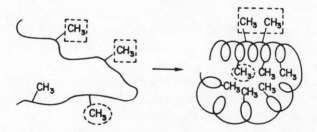

Figure 1.1. A schematic description of a conformational change of a biopolymer. One methyl group (circled) is transferred from an essentially aqueous environment to the interior of the polymer. Two methyl groups (in squares) approach each other and form a "dimer."

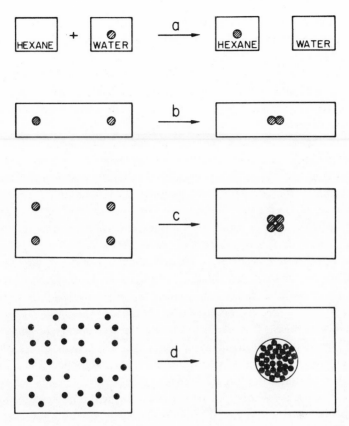

Figure 1.2. Various "model processes" that have been used in the study of the hydrophobic interaction. (a) Transfer of a simple solute (⊘) from water into hexane; (b) dimerization of two simple solutes in an aqueous solution; (c) formation of a tetramer; (d) aggregation of a large number of solute molecules.

this as an idealization of such a transfer process, we shall devote Chapter 2 to the study of the standard free energy of transferring simple solutes from water to nonaqueous solvents. [There is also the important question of which standard free energy change one must choose to represent this process; this topic is discussed in Section 2.6. A more detailed analysis of this question is given by Ben-Naim (1978a).]

Clearly, there are other, less extreme "transfer processes" involving nonpolar groups in the complicated process depicted in Figure 1.1. One other example shown in Figure 1.1 involves two methyl groups that are far apart in one configuration and are brought close together in another

configuration. This process may be studied in a model system as shown
in Figure 1.2(b). Chapter 3 is devoted to the study of the thermodynamics
of such "dimerization" processes.

It should be noted now that the thermodynamics of the two processes
illustrated in Figures 1.2(a) and 1.2(b) furnish essentially different information on the HI problem. In Chapter 3 we shall present some numerical
examples to demonstrate this point.

From the dimerization process one can easily generalize to higher-
order aggregation processes as depicted for tetramers in Figure 1.2(c) and
for very large aggregates in Figure 1.2(d). Here the new feature that may
reveal itself is the pairwise (or higher-order) nonadditivity effect. This
topic will be discussed in Chapter 4. In the limit of very large aggregates
[Figure 1.2(d)], we are effectively closing a cycle and returning to the
transfer process of Figure 1.2(a). The reason is the following: Let a large
number m of solute particles be transferred from a purely aqueous environment to a compact (say, spherical) aggregate. Let m_s of these be found
on the periphery and m_i in the interior of the aggregate. The latter are
completely surrounded by a nonpolar environment. Thus if m is very large,
m_s becomes negligible compared to m_i and the free energy of aggregation
will be essentially determined by m times the free energy of transferring
a single monomer from water to a nonpolar solvent, which is the process
with which we started in Figure 1.2(a).

This book is devoted mainly to the study of the thermodynamic aspects
of the processes listed above. This information is valuable for characterization of the nature of the HI and its dependence on temperature, pressure,
or the variation of the composition of the solvent.

Once having acquired an experimental knowledge of the thermodynamics of these processes, there still remains the interesting but difficult
question of "Why?" That is, can we explain the phenomenon of HI in terms
of what we know about the structure and properties of liquid water as a
solvent? This question is still largely unanswered. In the earlier literature
it was commonplace to discuss HI in terms of van der Waals forces between
the solute molecules, but clearly this could not be the whole story. The
van der Waals forces between, say, two methane molecules in water are
nearly the same as in any other solvent or in vacuum, and the peculiarities
of the aqueous environment cannot be accounted for by using only van der
Waals forces between the solute molecules. It was probably Kirkwood
(1954) who first stressed the importance of the role of the solvent in producing a new "force" between solute molecules other than the direct
van der Waals forces. This very important point was not completely ap-

preciated until very recently. It is true, though, that practitioners in the molecular theory of liquids have long recognized the solvent-induced forces which consist of a part of what is commonly referred to as the potential of average force. [As an example, two hard spheres do not exert attractive forces on each other. However, when they are placed in a solvent, which may also consist of hard spheres, one gets attractive forces which originate from the environment of these two particles—for more details see Ben-Naim (1974), Chapter 2.]

The average chemist or biochemist who is not familiar with the theory of liquids may find the HI a somewhat mysterious phenomenon. The reason is that we have been accustomed to the fact that a force between two particles is a property of the particles *themselves*, e.g., Coulombic forces arise from charges situated *in* the particles, van der Waals forces are due to fluctuations in the locations of the electric charges on the electrons of the two particles, and the strong repulsive forces are a result of the property of the impenetrable "volume" of the particles. When dealing with HI we, surprisingly, claim that the forces are mainly dependent on the properties of the *solvent*, and not on the *solutes*. This causes some difficulties in the appreciation of the origin of these forces. As in the case of hard spheres, where the solvent-induced forces can be attractive, we also obtain some new features that do not necessarily exist in the direct interaction. For example, the temperature dependence of the HI arises mainly from the solvent-induced forces and to a good approximation is nonexistent in the direct, say, van der Waals forces. [Further discussion on this topic may be found in Chapter 5].

Another property that could arise from the solvent-induced forces is the nonadditivity effect [even when we assume that the direct forces are strictly additive]. This will be further discussed in Chapter 4.

The above comments apply to any solvent. In particular, one might expect some unusual effects in water that may be related to the peculiarities of its structure. Some possible relations between HI and the properties of liquid water are discussed in Chapter 5.

We shall use the following nomenclature: The first process depicted in Figure 1.2(a) is described by the so-called *standard thermodynamics of transfer*. In particular, the standard free energy of transfer measures the relative preference of the solute for the two solvents. When one of the solvents is water, we shall use the term "hydrophobic hydration," since we are dealing here exclusively with the solute–solvent interactions. Once we have two or more interacting solute molecules in water we shall use the term "hydrophobic interaction." We can talk about pairwise HI, tripletwise

HI, etc., but this is usually unnecessary since the number of interacting solutes in a specific problem is known in advance. For convenience, we shall split the total free energy change of association into two terms, the direct interaction, which is presumed to be solvent independent, and an *indirect* part of the interaction, which carries the properties of the solvent. The latter will often be referred to as the HI part of the total free energy change. In a similar fashion one may split other thermodynamic quantities into these two corresponding contributions.

Comment

The above definitions of hydrophobic hydration and hydrophobic interaction are not yet universally accepted. The reader should be aware of the fact that in some publications the term HI is used for what we have called hydrophobic hydration, and sometimes the two terms are used for the same phenomenon. The literature also contains many reports of measurements that appear to be relevant, but not directly relevant, to the problem of HI. For example, free energies of dimerization of carboxylic acids in water certainly include contributions from HI, but it is almost impossible to extract information on the HI from such data (for more details see Section 3.7). Therefore the reader is urged to read articles on this subject with great care; this includes the pages of this book.

1.2. THE FUNDAMENTAL EXPRESSION FOR THE CHEMICAL POTENTIAL

We present here a brief survey of the most fundamental equations that will serve us throughout the entire text. These consist of the expressions for the chemical potential and the pseudo-chemical-potential. The content of the various terms is explained here in a qualitative fashion: the detailed derivations based on statistical mechanics are given in Appendix A.1. The reader should be aware of the fact that the thermodynamics alone is not sufficient to gain insight into the meaning of the various terms that appear in the expressions for the chemical potential. This is rendered possible only through statistical mechanical arguments.

Consider first a pure liquid, consisting of N_S molecules of S, at a specified temperature T and pressure P. The chemical potential of S in this system is defined by

$$\mu_S = \left(\frac{\partial G}{\partial N_S}\right)_{T,P} = \lim_{dN_S \to 0}\left[\frac{G(T, P, N_S + dN_S) - G(T, P, N_S)}{dN_S}\right]$$

$$= \lim_{dN_S \to 0} [G(T, P, (N_S/dN_S) + 1) - G(T, P, N_S/dN_S)]$$

$$= \lim_{M_S \to \infty} [G(T, P, M_S + 1) - G(T, P, M_S)] \qquad (1.1)$$

In the first two equalities we have used the formal definition of the chemical potential as a derivative of the Gibbs free energy with respect to the number of molecules (or moles, which is more commonly used in thermodynamics). In the third equality we have used the fact that G is an extensive thermodynamic quantity, i.e., for any real number $\alpha > 0$, we can write $G(T, P, \alpha N) = \alpha G(T, P, N)$. Thus for the choice of $\alpha = (dN_S)^{-1}$ we get the last form on the right-hand side of (1.1), where we replaced N_S/dN_S by M_S and replaced the limit $dN_S \to 0$ by $M_S \to \infty$.

From now on we shall always assume that the system is macroscopically large and that the chemical potential may be obtained from the difference

$$\mu_S = G(T, P, N_S + 1) - G(T, P, N_S) \qquad (1.2)$$

Classical statistical mechanics provides a useful expression for the chemical potential in terms of the properties of the single molecules and the local interaction of a single molecule with its entire environment. This relation has the form

$$\mu_S = kT \ln(\varrho_S \Lambda_S^3 q_S^{-1}) + W(S \mid S) \qquad (1.3)$$

where k is the Boltzmann constant; $\varrho_S = N_S/V$ is the number density (N_S and V could be either fixed quantities, as in the case of a closed system with a fixed volume, or average quantities, as in the case of an open system or system at a fixed pressure); Λ_S^3 is the momentum partition function, originating from integration over all possible momenta (or velocities) of the molecules in the system; and q_S is the internal partition function of a single molecule, including the rotational, vibrational, electronic, etc. partition functions of a molecule. We shall never need the explicit form of these functions, but we note here that we always assume that q_S is independent of the type of environment in which a molecule may find itself.

Perhaps the most important quantity that concerns us here is the one denoted by $W(S \mid S)$, which is the *coupling work* of S against an environment consisting of S molecules. We shall use the convention that in $W(i \mid j)$, i is the species for which the chemical potential is written, and j is the "solvent," i.e., the entire environment surrounding the i molecule.

It is also useful to cite the expression for $W(S \mid S)$ that can be obtained by statistical mechanical arguments. This relation is derived in Appendix A.1. It reads

$$W(S \mid S) = -kT \ln\langle\exp(-\beta B_S)\rangle \qquad (1.4)$$

where $\beta = (kT)^{-1}$ and B_S is referred to as the "binding energy" of one S molecule to the rest of the system, and the average is taken over all configurations of all the molecules in the system, excluding the one that has been added (which, in this particular case, is of the same species as all other molecules in the system).

We next introduce an important auxiliary quantity which will prove useful in later applications. This is the pseudo-chemical-potential, which is defined similarly to (1.2), but with the additional restriction that the added molecule be placed at a fixed position, say $\mathbf{R}_S$, in the system. Thus we define

$$\tilde{\mu}_S = G(T, P, N_S + 1, \mathbf{R}_S) - G(T, P, N_S) \qquad (1.5)$$

and we cite the corresponding statistical mechanical expression:

$$\tilde{\mu}_S = kT \ln q_S^{-1} + W(S \mid S) \qquad (1.6)$$

Note that in the above notation we have not included $\mathbf{R}_S$ explicitly. The reason is that in a macroscopic and homogeneous fluid all points in the system are equivalent (except for a small region near the surface of the system, which we can neglect). The tilde on $\tilde{\mu}_S$ should suffice to remind us that the added molecule has been placed at (some) fixed position in the system. Note also that for nonspherical molecules we may require that the added molecule be placed at a fixed position and orientation.

Combining equations (1.3) and (1.6) we get the following useful relation:

$$\mu_S = \tilde{\mu}_S + kT \ln \varrho_S \Lambda_S^3 \qquad (1.7)$$

This has a very simple interpretation (see also Figure 1.3). The free energy change for introducing a single molecule to the system, μ_S, is split into two parts, corresponding to two consecutive steps: first we place the molecule at a fixed position, and the corresponding change in free energy is $\tilde{\mu}_S$; secondly, we release the constraint of a fixed position, and this results in an additional change of free energy $kT \ln \varrho_S \Lambda_S^3$. [We note here briefly

Figure 1.3. The chemical potential μ is the change of free energy resulting from adding one particle to the system. This process may be split into two steps: first we introduce the particle at a fixed position, then we release the particle to wander in the entire volume. The corresponding two contributions to the free energy change are the two terms on the right-hand side of (1.7).

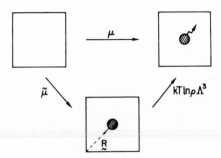

that the origin of the second term has three components: the momentum partition function $\Lambda_S{}^3$, resulting from the motion of the molecule; N_S, resulting from the fact that after releasing the particle from a fixed position it is no longer distinguishable from other particles in the system; and V, which results from the accessibility of the entire volume V to the released molecule. The second term on the right-hand side of (1.7) has been referred to as the liberation free energy (Ben-Naim, 1978a)].

Next we turn to the case of extremely dilute solutions of S in some solvent which we designate by W. The chemical potential of S in this case may be obtained either by adding dN_S molecules to a system of N_W solvent molecules, or simply adding one solute S to a pure solvent (but macroscopically large):

$$\mu_S = G(T, P, N_W, N_S = 1) - G(T, P, N_W) \qquad (1.8)$$

The statistical mechanical analogs of (1.3) and (1.4) are

$$\begin{aligned}
\mu_S &= kT \ln(\varrho_S \Lambda_S{}^3 q_S{}^{-1}) + W(S \mid W) \\
&= kT \ln(\varrho_S \Lambda_S{}^3 q_S{}^{-1}) - kT \ln\langle \exp(-\beta B_S) \rangle_0
\end{aligned} \qquad (1.9)$$

Most of the symbols in (1.9) are the same as in (1.3) and (1.4), with two important differences. In the first place $\varrho_S = 1/V$ since $N_S = 1$ in this case. Second, the coupling work $W(S \mid W)$ now contains the total binding energy of a solute S to an environment that consists *entirely* of solvent molecules W. (See also Appendix A.1.) The symbol $\langle \ \rangle_0$ stands for an average over all configurations of the *solvent* molecules.

The analog of the pseudo-chemical-potential is defined similarly to (1.5), which reads for this case

$$\begin{aligned}
\tilde{\mu}_S &= G(T, P, N_W, N_S = 1, \mathbf{R}_S) - G(T, P, N_W) \\
&= kT \ln q_S{}^{-1} + W(S \mid W)
\end{aligned} \qquad (1.10)$$

and similar to (1.7) we have

$$\mu_S = \tilde{\mu}_S + kT \ln \varrho_S \Lambda_S^3 \tag{1.11}$$

with the understanding that $\varrho_S = 1/V$ and that $\tilde{\mu}_S$ is the change of free energy corresponding to the process of inserting a simple solute S at a fixed position in a pure solvent W. In later applications we shall use special symbols such as μ_S^α to specify the phase into which the solute is inserted. This was not done here since we have described verbally the precise condition under which the process is carried out.

We have considered above two extreme cases where S was introduced to either pure S or pure W. Of course, a generalization of this process is possible. The solvent could be any mixture of S and W, or of any other components that may be present in the system. The general result is the following. The chemical potential of S in any solvent will have the form (1.11) with the understanding that $\varrho_S = N_S/V$ is the number density of S in the system, and that $\tilde{\mu}_S$ includes, besides $kT \ln q_S^{-1}$, the coupling work of S against whichever environment surrounds the added solute S.

With this brief survey of the basic definition of the chemical potential, we can now proceed to deal with the more important experimentally measurable quantity—the standard free energy of transferring a molecule S from one phase to another. The phases are denoted by α and β and they may include any number of components. The temperature T and pressure P are presumed to be fixed in this section.

The chemical potential of S in the two solvents is written as

$$\mu_S^\alpha = \tilde{\mu}_S^\alpha + kT \ln \varrho_S^\alpha \Lambda_S^3 \tag{1.12}$$

$$\mu_S^\beta = \tilde{\mu}_S^\beta + kT \ln \varrho_S^\beta \Lambda_S^3 \tag{1.13}$$

where Λ_S^3 is the same quantity in both phases (in classical statistical mechanics the momentum partition function depends only on the mass of the particles and on the temperature of the system); ϱ_S^α and ϱ_S^β are the number densities of S in the phases α and β, respectively—and so far we have imposed no restrictions on the magnitude of these densities. (Note also that since Λ_S has the dimensions of length, the product $\varrho_S \Lambda_S^3$ is a dimensionless quantity.)

We now consider several differences in the chemical potential of S in the two phases. The reader is urged to follow very carefully the precise conditions under which a given process is carried out. Failing to do so is a major source of misunderstandings and misleading conclusions.

Consider first the process of transferring one S molecule from α to β at given T and P, but with no other restrictions imposed on the composition of the phases. The free energy change for this process is

$$\Delta G\left(\begin{matrix} \alpha \to \beta \\ \varrho_S{}^\alpha, \varrho_S{}^\beta \end{matrix}\right) = \mu_S{}^\beta - \mu_S{}^\alpha = \tilde{\mu}_S{}^\beta - \tilde{\mu}_S{}^\alpha + kT \ln(\varrho_S{}^\beta/\varrho_S{}^\alpha) \qquad (1.14)$$

It is important to emphasize that the free energy change for transferring dN_S molecules from α to β is given by $\Delta G(\alpha \to \beta) = (\mu_S{}^\beta - \mu_S{}^\alpha)\, dN_S$, where here we have chosen $dN_S = 1$. The important point is that in order to compute $\Delta G(\alpha \to \beta)$ we *must* use the *chemical potentials* of S in the two phases. It is sometimes true that differences in *chemical potentials* are equal to differences in *standard chemical potentials* (see below), but this is *not always* true, and we shall see a specific example in Section 2.6. The starting point for computing $\Delta G(\alpha \to \beta)$ must involve the proper chemical potentials of S in the two phases.

Next we specialize relation (1.14) for one particular case—namely, we now require that the number densities $\varrho_S{}^\alpha$ and $\varrho_S{}^\beta$ be equal. Note, however, that the values of $\varrho_S{}^\alpha = \varrho_S{}^\beta$ are not required to be very small (as is common in the literature). Here we only require that the densities be the same in the two phases.

For this special case we obtain from (1.14) and (1.10)

$$\Delta G\left(\begin{matrix} \alpha \to \beta \\ \varrho_S{}^\alpha = \varrho_S{}^\beta \end{matrix}\right) = \tilde{\mu}_S{}^\beta - \tilde{\mu}_S{}^\alpha = W(S \mid \beta) - W(S \mid \alpha) \qquad (1.15)$$

This is an important equality. It relates the free energy change of two *different* processes. On the left-hand side we are dealing with a process where one S molecule is transferred from $\alpha \to \beta$ (in which we know that $\varrho_S{}^\alpha = \varrho_S{}^\beta$), and on the right-hand side the process involves the transfer of S from a *fixed* position in α to a *fixed* position in β (the other conditions of T, P, and $\varrho_S{}^\alpha = \varrho_S{}^\beta$ are the same in the two processes). Clearly, by forming the difference in (1.15) the internal and the momentum partition functions cancel out, as do the densities of the solute in the two phases, and we are left only with the difference in the coupling work of S in the two phases.

The next question concerns the measurability of the difference defined in (1.15). We consider an experimental setup from which we can obtain a connection with measurable quantities. The system is shown in Figure 1.4, where we have two phases α and β at the same P and T, separated by a partition that is permeable to S but not to any other component. In this case we can no longer control the densities of S in each phase, since these

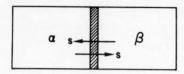

Figure 1.4. Two phases α and β, at the same temperature and pressure, are separated by a membrane which is permeable to a solute S. No restrictions on the magnitude of the densities $\varrho_S{}^\alpha$ and $\varrho_S{}^\beta$ in the two phases are preimposed.

are determined by the condition of chemical equilibrium, i.e.,

$$\mu_S{}^\alpha = \mu_S{}^\beta \tag{1.16}$$

Using this condition we form the difference between (1.12) and (1.13) to get

$$0 = [\mu_S{}^\beta - \mu_S{}^\alpha]_{\text{eq}} = [\tilde{\mu}_S{}^\beta - \tilde{\mu}_S{}^\alpha]_{\text{eq}} + kT \ln(\varrho_S{}^\beta/\varrho_S{}^\alpha)_{\text{eq}} \tag{1.17}$$

Note that here the internal and momentum partition functions of S cancel out, but the densities of S in the two phases appear in the form of a "partition coefficient" $(\varrho_S{}^\beta/\varrho_S{}^\alpha)_{\text{eq}}$, which is a measurable quantity. Thus, by measuring the partition coefficient of S in the experimental setup of Figure 1.4 we can evaluate the quantity $[\tilde{\mu}_S{}^\beta - \tilde{\mu}_S{}^\alpha]_{\text{eq}}$, which is the free energy change for transferring a single S from a *fixed* position in α to a *fixed* position in β, at the condition of equilibrium with respect to diffusion of S between the two phases. This in turn is equal to the difference in the coupling work of S to the two phases, respectively.

Obviously since we did not impose any condition on the magnitude of either $\varrho_S{}^\alpha$ or $\varrho_S{}^\beta$, the quantity $[\tilde{\mu}_S{}^\beta - \tilde{\mu}_S{}^\alpha]_{\text{eq}}$ depends on the densities of S in the two phases. This fact does not change the meaning or the importance of this quantity. For this reason we have recently referred to this quantity as the *generalized standard free energy of transfer* of S from α to β (Ben-Naim, 1978a).

There is one important limiting case of (1.17) that is most commonly referred to in the literature as the standard free energy of transfer of S from α to β. This is obtained when the densities $\varrho_S{}^\alpha$ and $\varrho_S{}^\beta$ are extremely low in α and β, in which case we know that $\tilde{\mu}_S{}^\alpha$ and $\tilde{\mu}_S{}^\beta$ become independent of $\varrho_S{}^\alpha$ and $\varrho_S{}^\beta$, respectively. We now define the standard chemical potentials of S in the two phases by

$$\mu_S{}^{\circ\alpha}(T, P, \mathbf{x}^\alpha) = \lim_{\varrho_S{}^\alpha \to 0} [\tilde{\mu}_S{}^\alpha(T, P, \mathbf{x}^\alpha, \varrho_S{}^\alpha)] + kT \ln \Lambda_S{}^3$$
$$= W(S \mid \alpha) + kT \ln q_S{}^{-1} \tag{1.18}$$

$$\mu_S{}^{\circ\beta}(T, P, \mathbf{x}^\beta) = \lim_{\varrho_S{}^\beta \to 0} [\tilde{\mu}_S{}^\beta(T, P, \mathbf{x}^\beta, \varrho_S{}^\beta)] + kT \ln \Lambda_S{}^3$$
$$= W(S \mid \beta) + kT \ln q_S{}^{-1} \tag{1.19}$$

where $\mathbf{x}^\alpha$ and $\mathbf{x}^\beta$ stand for the compositions of the phases α and β, respectively, excluding the densities of S.

Clearly if we form the difference as in (1.17) for the system at equilibrium (Figure 1.4) we obtain

$$\Delta\mu_{\mathrm{tr}}^\circ(\alpha \to \beta) \equiv \mu_S^{\circ\beta} - \mu_S^{\circ\alpha} = \lim_{\varrho_S \to 0} [\tilde{\mu}_S^\beta - \tilde{\mu}_S^\alpha]_{\mathrm{eq}}$$

$$= W(S \mid \beta) - W(S \mid \alpha)$$

$$= \lim_{\varrho_S \to 0} [-kT \ln(\varrho_S^\beta/\varrho_S^\alpha)_{\mathrm{eq}}] \qquad (1.20)$$

Thus the *conventional* standard free energy of transfer, defined in (1.20), is independent of the densities of S in the two phases; it is equal to the difference in the coupling work of S against pure phases of α and β (here "pure" means a solvent that does not contain S, but otherwise it can contain any number of other components). Note that in the limiting process in (1.20) we take only one density (either ϱ_S^α or ϱ_S^β) and let it decrease to zero; because of the condition of equilibrium the second density must tend to zero as well.

To summarize, $\Delta\mu_{\mathrm{tr}}^\circ(\alpha \to \beta)$ is *defined* as the free energy of transferring a single S from a fixed position in pure α to a fixed position in pure β (keeping T and P constant). It is also equal to the difference in the coupling work of S against the pure phases α and β; this, being a local property of S, is a measure of the difference in the solvation properties of the two phases with respect to the solute S. It is *measured* experimentally through the partition coefficient of S in the two phases, provided we let the densities of S be small enough to ensure the validity of the limit of ideal dilute solutions.

Note that in defining $\Delta\mu_{\mathrm{tr}}^\circ(\alpha \to \beta) = \mu_S^{\circ\beta} - \mu_S^{\circ\alpha}$ in (1.20) we did not use the subscript eq (for equilibrium) as we have done for $[\tilde{\mu}_S^\beta - \tilde{\mu}_S^\alpha]_{\mathrm{eq}}$. The reason is that once we get into the realm of ideal dilute solutions, both $\mu_S^{\circ\alpha}$ and $\mu_S^{\circ\beta}$ become independent of the density of S, and from the formal point of view they are characterized by the properties of a pure phase α (or β) having *one* solute S at some fixed position.

The quantity $\Delta\mu_{\mathrm{tr}}^\circ(\alpha \to \beta)$ is by far the most commonly used in the literature of aqueous solutions. However, we stress here that the more general quantity $[\tilde{\mu}_S^\beta - \tilde{\mu}_S^\alpha]_{\mathrm{eq}}$ defined in (1.17) has the same significance as $\Delta\mu_{\mathrm{tr}}^\circ(\alpha \to \beta)$, but unfortunately this has eluded the attention of most workers in this field. The probable reason for this is that in order to obtain relation (1.17) one must use arguments from statistical mechanics, whereas relation (1.20) is obtained by purely thermodynamic reasoning, supplemented by the knowledge of the existence of Henry's law (i.e., dilute ideal solutions).

We now present an important application of relation (1.17) that is of interest in the field of aqueous solutions, and that *may not* be obtained from purely thermodynamical arguments. Consider the same experimental setup as in Figure 1.4, but let the phase α be pure S and the phase β be pure water W (a concrete example might be benzene for S). At equilibrium conditions we obtain from (1.17)

$$[\tilde{\mu}_S{}^W - \tilde{\mu}_S{}^S]_{\text{eq}} = -kT \ln(\varrho_S{}^W/\varrho_S{}^S)_{\text{eq}} \tag{1.21}$$

Here $\varrho_S{}^S$ would simply be the number density of S in pure S, whereas $\varrho_S{}^W$ is the equilibrium density of S in W under these conditions. Clearly $[\tilde{\mu}_S{}^W - \tilde{\mu}_S{}^S]_{\text{eq}}$ measures the difference in the coupling work of S against pure S and against (almost) pure W. A slightly different situation arises if we let the two phases S and W be in direct contact (i.e., removing the partition that separates the two phases). In this case water penetrates into the benzene phase and slightly changes its properties, and in particular the density $\varrho_S{}^S$ changes. Relation (1.17) when applied to this case leads almost to the same relation as in (1.21) with the understanding that $\varrho_S{}^S$ is no longer the density of *pure S*, but the density of S in this phase after W has also reached its equilibrium between the two phases. Similarly $\varrho_S{}^W$ might be slightly different from $\varrho_S{}^W$ in (1.21) for the same reason. These differences become negligibly small if the two components S and W are very slightly miscible in each other.

Comment

The general expression for the chemical potential, along with the interpretation of the two terms on the right-hand side of equation (1.11) is considered to be a cornerstone relation of this book. Note that the significance of the pseudo-chemical-potential, as well as the meaning assigned to the standard free energy of transfer, cannot be perceived within the realm of pure thermodynamics. One must also use some elementary arguments from statistical mechanics. The serious reader is therefore urged to study carefully Appendix A.1. The effort invested in doing so is well justified, even if the only reward is the avoidance of plunging into the confusion that exists in the literature on this topic.

1.3. DEFINITION OF HYDROPHOBIC INTERACTION (HI)

In this section we present a definition of the concept of hydrophobic interaction (HI) between pairs of simple solute molecules in a solvent. In Chapter 4 the definition is extended to any number of solute molecules. It is true, though, that the latter case will ultimately be the most important one for the understanding of biological systems. Nevertheless, pairwise HI should be regarded as a necessary first step in the study of this field, and as a matter of fact almost all the experimental data that provide direct information on HI are concerned with interaction between pairs of molecules.

All the quantities that we shall define in this section may be used for any solvent. However, for the sake of simplicity we shall be using the term HI even when dealing with nonaqueous systems (where a term such as "solvophobic interaction" might be more appropriate).

The concept of hydrophobic interaction is introduced here by using "thermodynamic manipulations." But this is a somewhat deceptive statement, since the validity of the arguments that are employed are to be established through statistical mechanical ideas.

Consider two simple spherical solute molecules S in any solvent (which may be a pure component or a mixture of any number of components). We start with the two solutes at fixed positions R_1 and R_2 but at infinite separation $R_{12} = |R_2 - R_1| = \infty$ (by infinity we actually mean that R_{12} is large enough so that there is no correlation between the locations of the two molecules). From this initial state we bring the two solutes close together, say to $R_{12} = \sigma$, where σ is the molecular diameter of the solute molecules. The process is carried out within the solvent, keeping T and P constant. This will be referred to as the *HI process*.

The Gibbs free energy change that corresponds to this process is

$$\Delta G(\infty \to \sigma) = G(T, P, \text{solvent}, R_{12} = \sigma) - G(T, P, \text{solvent}, R_{12} = \infty) \quad (1.22)$$

where $\Delta G(\infty \to \sigma)$ is written explicitly in terms of the difference in free energy between the initial and the final states. The argument "solvent" stands for the composition of the solvent that consists of our system excluding the two solutes S. Note that since the solvent is homogeneous and isotropic we need specify only the scalar distance between the centers of the two solute molecules. Furthermore, since our initial state will always be $R_{12} = \infty$, we shall simplify the notation for $\Delta G(\infty \to \sigma)$ and use the shorter one $\Delta G(\sigma)$.

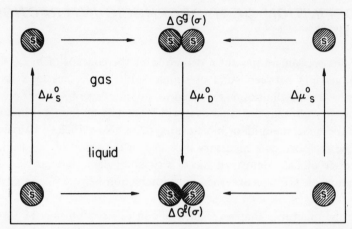

Figure 1.5. A cyclic process leading to equation (1.23). Two solutes S are brought from infinite separation to a small separation σ within the liquid. The corresponding free energy change is $\Delta G^l(\sigma)$. The second route is first to transfer the two solutes to the gaseous phase, then bring them to the separation σ, and finally introduce the "dimer" into the liquid.

We next seek to relate $\Delta G(\sigma)$ to experimentally measurable quantities. To this end consider the cyclic process depicted in Figure 1.5. Instead of carrying out the process as described above, we take an alternative route. First we transfer the two separate solutes from fixed positions in the solvent to fixed positions in a gaseous phase (essentially in a vacuum). Then we bring the two solutes to the required distance $R_{12} = \sigma$ in the gaseous phase. Denoting the pair of solutes at $R_{12} = \sigma$ by D, for "dimer," we finally transfer the "dimer" from the gaseous phase into the solvent. Since the change of free energy along the two routes must be the same, we have the equality

$$\Delta G(\sigma) = \Delta \tilde{\mu}_D - 2\Delta \tilde{\mu}_S + U_{SS}(R_{12} = \sigma) \qquad (1.23)$$

where $\Delta \tilde{\mu}_S$ is the generalized standard free energy of solution of S, $\Delta \tilde{\mu}_D$ has a similar significance when the "dimer" is viewed as a single molecular entity, and $U_{SS}(R_{12} = \sigma)$ is the direct intermolecular potential for the solute molecules at the specified separation $R_{12} = \sigma$. This quantity is the same as the $\Delta G^g(\sigma)$ of Figure 1.5.

A particular limiting case of relation (1.23), which is by far the most useful, occurs when the density of S is very small in the solvent. From the formal point of view it is sufficient to consider *only* two S molecules in the solvent, in which case we obtain, from (1.23),

$$\Delta G(\sigma) = \Delta \mu_D{}^\circ - 2\Delta \mu_S{}^\circ + U_{SS}(R_{12} = \sigma) \qquad (1.24)$$

where now we have used the conventional standard free energies of solution of the solute monomer S, and the "dimer" D, i.e., $\Delta\mu_D{}^\circ$ and $\Delta\mu_S{}^\circ$ are independent of ϱ_S.

Two comments are now in order. We have used a thermodynamic cycle to get the relations (1.23) and (1.24), but the reasoning underlying these relations is of statistical mechanical origin: the very usage of the pseudo-chemical-potential is extrinsic to classical thermodynamics. Second, the above relation connects the free energy of the HI process with some more familiar quantities: the direct pair potential U_{SS} and the standard free energies of solution of S and D. This is indeed a formal step forward. However, there still remains one obstacle in the application of either (1.23) or (1.24). That is, though $\Delta\mu_S{}^\circ$ and $\Delta\tilde{\mu}_S$ are measurable quantities, this is not the case for $\Delta\mu_D{}^\circ$ or $\Delta\tilde{\mu}_D$ since D is not a molecular entity for which we could measure the partition coefficient in the two phases. Nevertheless, we shall later see how these relations may be useful in suggesting various approximations to $\Delta G(\sigma)$.

It is worthwhile noting that the HI process described above is not a process that one could carry out in a laboratory. It was used here essentially as a convenient tool to obtain relations (1.23) and (1.24) through the thermodynamic cycle. In the next section we shall see how $\Delta G(\sigma)$ is related to the probability of finding two solute molecules S at a separation $R_{12} = \sigma$ in a *real* system, i.e., in a system where there are no restrictions on fixed locations for the solutes. In this sense we shall bring $\Delta G(\sigma)$ closer to the current concept of HI as a measure of the tendency of the two solute molecules to adhere to each other (particularly in aqueous fluids).

The general form of the function $\Delta G(R)$ [i.e., $\Delta G(\infty \to R)$] is shown in Figure 1.6 for simple solutes S in a simple solvent. One can show that at very low solvent densities $\Delta G(R) = U_{SS}(R)$, i.e., all the free energy change comes from the direct pair potential between the two solute molecules. As the density of the solvent increases, the minimum of the function $\Delta G(R)$ at $R \approx \sigma$ becomes deeper, and new maxima and minima develop at larger distances.

Statistical mechanics provides a very important relation between the

Figure 1.6. Schematic description of the function $\Delta G(R)$ for simple solutes in a solvent. The dotted curve represents the same function for the two solutes in vacuum. This function is identical with the pair potential $U_{SS}(R)$ for these solutes.

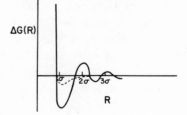

gradient of $\Delta G(R)$, with respect to R, and the average force operating between the two solute molecules. For this reason, this function is also referred to as the potential of average force. The magnitude of this force is given by

$$F_{SS}(R) = -\frac{\partial \Delta G(R)}{\partial R} \tag{1.25}$$

We now define, for any distance R, the quantity $\delta G^{\mathrm{HI}}(R)$ by

$$\Delta G(R) = U_{SS}(R) + \delta G^{\mathrm{HI}}(R) \tag{1.26}$$

Thus if we take the gradient of $\Delta G(R)$ we see that it consists of two parts, the direct force due to the direct solute–solute interaction and an *indirect* part that originates from the presence of a solvent.

Clearly, if we are interested in forces operating between two solute molecules, or between two nonpolar side-chain groups on a biopolymer, we must ultimately turn to $\Delta G(R)$ or its gradient. However, we see from (1.26) that $\Delta G(R)$ has one ingredient, $U_{SS}(R)$, which is independent of the type of solvent surrounding the solutes. Therefore, if we are searching for any special effects of a specific solvent, say water, we must focus our attention on the indirect part, denoted by $\delta G^{\mathrm{HI}}(R)$, of the free energy change, and we shall refer to this part as the HI between the two solute molecules at the distance R in the specified solvent.

It is important to remember that the total average force operating between the two solute molecules, or the probability of finding these two molecules at a distance R from each other, is governed by the function $\Delta G(R)$ (see Section 1.4). However, the *contribution* of the *solvent* to this force is given by $\delta G^{\mathrm{HI}}(R)$. And since we are mainly interested in the property of water as a medium in which nonpolar molecules attract each other (presumably more than in other solvents), it is justifiable to call this part simply the hydrophobic interaction.

One can often find, especially in textbooks on biochemistry, a description of HI in terms of van der Waals forces. This is partially correct, since $U_{SS}(R)$ is always present in the *total* force operating between the two solute molecules. However, the story cannot be ended at this stage, since $U_{SS}(R)$ contributes the same force in any other solvent. Therefore, any special features of aqueous fluids must be searched for in the indirect, or "hydrophobic," part of the interaction.

So far we have introduced the formal definition of the concept of HI. We now turn to formulating the problems that should be studied in connection with this concept.

The first and by far the most important question is concerned with the comparison of the form of the function $\delta G^{HI}(R)$ in water and in other liquids. The main question is whether water is really outstanding as a solvent for simple nonpolar solutes such as methane, argon, and the like. More specifically, the question may be stated as follows: Is there any specific distance R for which $\Delta G(R)$ has a particularly low minimum (or the probability function has a high maximum; see Section 1.4) in water as compared with other solvents?

If we find a positive answer to this question, and determine the distance at which such a minimum exists, we can claim that the hydrophobic interaction hypothesis has been validated. Unfortunately, a full answer to this question is not available at present, but research in this field is quite intense, and one may hope that at least partial answers will be found in the near future.

Other related questions are the temperature, pressure, and solvent composition dependence of the HI. All these properties are extremely important for a full understanding of the role of the aqueous environment in biological processes, and some of the partial answers are reviewed in the following sections.

1.4. PROBABILITY INTERPRETATION

We now present a brief but important alternate definition of HI that is equivalent to the one presented in Section 1.3. The proof of the equivalency is not given here (but may be found in Ben-Naim, 1974, Chapter 3), and knowledge of it is not essential for understanding the subject matter in the following sections.

As in Section (1.3), consider a system of any composition at a given pressure P and temperature T. Let the volume of this system be V. (More precisely we should refer to the average volume $\langle V \rangle$, since the volume is a fluctuating quantity. However, we disregard the distinction between V and $\langle V \rangle$ in the present elementary treatment.) The system is assumed to contain at least two solute molecules S, which for convenience we denote by 1 and 2. There are no restrictions on their locations as were imposed in Section 1.3. The probability of finding molecule 1 in some element of volume $d\mathbf{R}_1$ at $\mathbf{R}_1$ is denoted by $P_S^{(1)}(\mathbf{R}_1)\,d\mathbf{R}_1$. Clearly the probability of finding this particular molecule at any point in the system is unity, i.e.,

$$\int_V P_S^{(1)}(\mathbf{R}_1)\,d\mathbf{R}_1 = 1 \qquad (1.27)$$

In a homogeneous and macroscopically large system all points of the system $\mathbf{R}_1$ are equivalent, so $P_S^{(1)}(\mathbf{R}_1)$ is actually independent of $\mathbf{R}_1$. Hence, from (1.27) we obtain

$$P_S^{(1)} \int_V d\mathbf{R}_1 = 1$$

or

$$P_S^{(1)} = 1/V \tag{1.28}$$

If there are N_S molecules of solute S in the system, then the probability of finding any *specific* molecule in $d\mathbf{R}_1$ at $\mathbf{R}_1$ is $1/V$. If $d\mathbf{R}_1$ is an element of volume small enough so that no two molecules can occupy it at any given time, then the probability of finding any one of the N_S molecules in $d\mathbf{R}_1$ is simply $N_S P_1^{(1)}(\mathbf{R}_1)\, d\mathbf{R}_1$ or $\varrho_S\, d\mathbf{R}_1$, where $\varrho_S = N_S/V$ is the number density of the solute S.

Next we consider the two specific molecules 1 and 2 and denote by $P_{SS}^{(2)}(\mathbf{R}_1, \mathbf{R}_2)\, d\mathbf{R}_1\, d\mathbf{R}_2$ the probability of finding molecule 1 in $d\mathbf{R}_1$ at R_1 and simultaneously molecule 2 in $d\mathbf{R}_2$ at $\mathbf{R}_2$.

In a homogeneous fluid $P_{SS}^{(2)}(\mathbf{R}_1, \mathbf{R}_2)$ depends only on the scalar separation $R_{12} = |\,\mathbf{R}_2 - \mathbf{R}_1\,|$ between the centers of the two molecules, and not on the precise location of each of the molecules. Therefore we may write $P_{SS}^{(2)}(\mathbf{R}_1, \mathbf{R}_2) = P_{SS}^{(2)}(R_{12})$. We now introduce a physically plausible argument. If the distance R_{12} is very large (compared with the typical molecular diameters of the molecules), then the two events "molecule 1 in $d\mathbf{R}_1$ and molecule 2 in $d\mathbf{R}_2$" become *independent* events, and the joint probability density $P_{SS}^{(2)}(\mathbf{R}_1, \mathbf{R}_2)$ factors into two parts:

$$P_{SS}^{(2)}(\mathbf{R}_1, \mathbf{R}_2) = P_S^{(1)}(\mathbf{R}_1) P_S^{(1)}(\mathbf{R}_2) = V^{-2} \tag{1.29}$$

For any finite separation R_{12}, we expect that there is a *correlation* between the locations of the two molecules. The correlation function $g_{SS}(R_{12})$ is now introduced as a correction to equation (1.29), namely,

$$P_{SS}^{(2)}(R_{12}) = V^{-2} g_{SS}(R_{12}) \tag{1.30}$$

where for $R_{12} \to \infty$ we get $g_{SS}(R_{12}) \to 1$.

So far we have introduced g_{SS} by considering two specific solute molecules 1 and 2. If these are the only solute molecules of species S in the system we can end up with relation (1.30). However, if there are N_S molecules of solute S, we can speak of the probability of finding *any* solute S in $d\mathbf{R}_1$ and any other solute S in $d\mathbf{R}_2$; this probability is denoted by

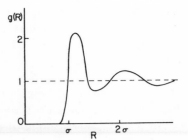

Figure 1.7. The general form of the radial distribution function $g(R)$ for a liquid consisting of simple spherical molecules of diameter σ.

$\varrho_{SS}^{(2)}(\mathbf{R}_1, \mathbf{R}_2)\, d\mathbf{R}_1\, d\mathbf{R}_2$, and by extension of the argument leading to (1.30) we get

$$\varrho_{SS}^{(2)}(R_{12}) = \varrho_S^2 g_{SS}(R_{12}) \qquad (1.31)$$

where $\varrho_S = N_S/V$ in (1.31) replaces $1/V$ in (1.30).

The function $g_{SS}(R)$ has been introduced here for a solute S in an unspecified solvent. This function, defined in a one-component liquid, is perhaps the most important function in the modern theory of liquids. It is often referred to as the pair correlation function, or the radial distribution function. The general form of this function for a simple fluid is shown in Figure 1.7. Note that $g(R) = 0$ for $R \lesssim \sigma$, since below $R = \sigma$ mutual penetration of two molecules is an extremely rare event. For $R \to \infty$, (which effectively means a few molecular diameters), we get $g(R) \to 1$, i.e., no correlation exists between the locations of the two molecules involved.

We now cite an important relation between $g_{SS}(R)$ and $\Delta G(R)$ of Section 1.3, which reads

$$g_{SS}(R) = \exp[-\Delta G(R)/kT] \qquad (1.32)$$

The proof of the validity of this relation may be found in Ben-Naim (1974, Chapter 3). It is very important to note that the two quantities $g_{SS}(R)$ and $\Delta G(R)$ refer to two different situations for the same system. When using $g_{SS}(R)$ we are concerned with solute molecules S which are free to wander in the entire volume of the system and we are interested in the probability of finding two such molecules at a certain distance from each other. When using $\Delta G(R)$, on the other hand, we are concerned with a process of bringing two molecules from *fixed* positions at infinite separation to a small separation R.

Relation (1.32) is valid for any concentration of solute S. A particular situation arises when S is very dilute in the solvent, in which case $g_{SS}(R)$, as well as $\Delta G(R)$, become independent of ϱ_S. Another limiting case of interest arises when the *total* density of the molecules ϱ_T in the system becomes very small, in which case one obtains

$$g_{SS}(R) \xrightarrow{\varrho_T \to 0} \exp[-U_{SS}(R)/kT] \tag{1.33}$$

which is essentially a Boltzmann distribution for the distances between pairs of S molecules. From this point of view, relation (1.32) may be regarded as a generalization of the Boltzmann distribution for systems at high densities, where the direct potential $U_{SS}(R)$ is replaced by the potential of average force. Thus, in the condensed phase a relatively high probability of finding two molecules at R is related to a low value of the free energy of the system when viewed as a function of the distance R.

Finally, we introduce another auxiliary function which is also of central importance in the general theory of fluids. Using relations (1.26) and (1.32) we obtain

$$\begin{aligned} g_{SS}(R) &= \exp\{-[U_{SS}(R) + \delta G^{\mathrm{HI}}(R)]/kT\} \\ &= y_{SS}(R) \exp[-U_{SS}(R)/kT] \end{aligned} \tag{1.34}$$

Thus the function $y_{SS}(R)$ is related to what we have called the HI part of the total work $\Delta G(R)$. This function is often more convenient to work with than $g_{SS}(R)$ because it is a smoother function of R at $R \lesssim \sigma$. This feature is exploited in Section 3.3 to obtain an approximate measure of HI.

1.5. INTEGRALS INVOLVING $g_{SS}(R)$

In the theory of liquids, integrals involving $g(R)$ are very common. These usually involve not only the pair correlation function, but also the pair potential itself. We shall be interested here in integrals with integrands that contain only $g_{SS}(R)$ and may be viewed as a kind of an average over all possible intermolecular separations.

For our particular interest in the properties of aqueous solutions of a nonpolar solute S, it would have been desirable to know the full pair correlation function $g_{SS}(R)$. Such a function would actually provide all the information we need on pairwise HI. Unfortunately, this information is still unavailable. Some attempts to obtain this function will be described in Section 3.2.

Acknowledging our inability to obtain the full function $g_{SS}(R)$, from either experimental or theoretical sources, we turn to ask what is the next best information that we can get from such sources. This leads us to the average value of $g_{SS}(R)$, which is a quantity accessible to experimental determination. The relevant quantity is defined by the integral

$$G_{SS} = \int_0^\infty [g_{SS}(R) - 1] 4\pi R^2 \, dR \qquad (1.35)$$

[For nonspherical solute molecules, $g_{SS}(R)$ is understood to be the average over all the possible orientations of the pair of solute molecules.]

To interpret G_{SS} it is useful to consider the quantity $\varrho_S G_{SS}$ rather than G_{SS} itself. In Section 1.4 we introduced the function $\varrho_{SS}^{(2)}(\mathbf{R}_1, \mathbf{R}_2)$ such that $\varrho_{SS}^{(2)}(\mathbf{R}_1, \mathbf{R}_2) \, d\mathbf{R}_1 \, d\mathbf{R}_2$ is the probability of finding *any* solute molecule in $d\mathbf{R}_1$ at $\mathbf{R}_1$ and any other solute molecule in $d\mathbf{R}_2$ at $\mathbf{R}_2$. We now introduce the conditional probability of finding a solute molecule in $d\mathbf{R}_2$ at $\mathbf{R}_2$ given a solute molecule at $\mathbf{R}_1$,

$$\varrho(\mathbf{R}_2/\mathbf{R}_1) \, d\mathbf{R}_2 = \frac{\varrho_{SS}^{(2)}(\mathbf{R}_1, \mathbf{R}_2) \, d\mathbf{R}_1 \, d\mathbf{R}_2}{\varrho_S^{(1)}(\mathbf{R}_1) \, d\mathbf{R}_1}$$

$$= \varrho_S g_{SS}(R) 4\pi R^2 \, dR \qquad (1.36)$$

where to obtain the last line of the right-hand side of (1.36) we have used relation (1.31) and also replaced $d\mathbf{R}_2$ by a spherical shell of radius R and width dR, R being the distance between $\mathbf{R}_1$ and $\mathbf{R}_2$. One can easily show (Ben-Naim, 1974, Chapter 2) that the quantity $\varrho_S g_{SS}(R) 4\pi R^2 \, dR$ is the average number of S molecules in a spherical shell of radius R around an S molecule fixed at the center of the sphere. Clearly $\varrho_S 4\pi R^2 \, dR$ is the average number of S molecules in such a spherical shell, the center of which has been chosen at random. Thus the quantity $\varrho_S[g_{SS}(R) - 1] 4\pi R^2 \, dR$ is a measure of the excess (or deficiency) in the average number of S molecules in a spherical shell at a distance R from a given S, fixed at its center, relative to the number obtained by eliminating the condition that an S molecule be placed at the center of the sphere. Taking the integral of this quantity over all distances R we get the average *total* excess (or deficiency) of S molecules in the entire surroundings of a given S molecule relative to a random point in the liquid. This quantity when divided by ϱ_S gives a measure, per unit density, of the average *affinity* of one S molecule toward another S molecule in the particular environment under consideration.

There are several experimental sources from which one can obtain the values of G_{SS}. The most direct one, which also provides all the values of G_{ij} (for any pair of species i and j present in the system), is by inversion of the Kirkwood–Buff theory of solution. This method is described in Section 3.6.

Another source that provides the low density limit of G_{SS},

$$G_{SS}^{\circ} = \lim_{\varrho_S \to 0} G_{SS} \tag{1.37}$$

is the density expansion of the osmotic pressure (and in fact any other quantity that measures deviations from the ideal dilute solution behavior), and which reads

$$\pi/kT = \varrho_S + B_2^* \varrho_S^2 + B_3^* \varrho_S^3 + \cdots \tag{1.38}$$

where π is the osmotic pressure of a system in which the solute density is ϱ_S. The coefficients B_K^* are the analogs of the virial coefficients in the density expansion of the pressure. Hence, these are also referred to as the virial coefficients of the osmotic pressure of a given solution. Statistical mechanics provides formal expressions for all the B_K^* in terms of the potentials of average force among K solute molecules. Of foremost importance is the second virial coefficient B_2^*, which is related to G_{SS}° by the equation

$$B_2^* = -G_{SS}^{\circ}/2 = -\tfrac{1}{2} \lim_{\varrho_S \to 0} \int_0^{\infty} [g_{SS}(R) - 1] 4\pi R^2 dR \tag{1.39}$$

It is important to note that this method provides only the low-density limit of G_{SS} as defined in (1.37). All the coefficients in the density expansion (1.38) are evaluated at $\varrho_S = 0$. One example of using this method for the study of HI is described in Section 3.5.

Chapter 2

Very Dilute Solutions and Hydrophobic Interaction

2.1. INTRODUCTION

In the previous chapter we noted that there are essentially two different concepts which are discussed in the literature under the same term, *hydrophobic interaction* (or hydrophobic bond). One is represented by the standard free energy of transfer of a solute S between water and some other nonaqueous solvent, and it measures the relative tendency of the solute S to prefer one environment over the other. The second measures the tendency of two (or more) solute molecules to aggregate in aqueous solutions. Both quantities have a common feature, i.e., they express some kind of "phobia" for the aqueous environment. However, they are quite different, and in general one cannot infer anything about one quantity from the other. In fact we shall see in Section 3.4 that in some cases they provide an apparently contradictory result as to the degree of "phobia" for two solvents.

This chapter is devoted to the first notion, which is essentially a measure of the "solute–solvent interaction." Although free energies of transfer do not provide direct information on HI, in the sense of the definition presented in Section 1.3, they are related to HI by various indirect relations, one of which has already been given in Section 1.3. But, more importantly, we shall see in Chapter 4 that free energies of transfer of single solute molecules

25

are important ingredients for the computation of the HI between a large number of solute molecules.

There is one further link between standard free energy of solution and free energy of the HI process. The standard free energy of solution of a hard-sphere particle in any solvent is equal to the negative of the HI between two such hard spheres at zero separation. This example has some theoretical value—it has been discussed in detail in Ben-Naim (1974) and will be briefly mentioned in Section 3.3.

2.2. STANDARD FREE ENERGIES OF TRANSFER

Consider the following experimental situation: pure water W and pure ethanol E, at the same temperature and pressure, are separated by a partition that is permeable to a solute S only. Let a solute S, say argon, be added to the system in minor quantities, so that ideal dilute solution behavior is expected, i.e.,

$$\mu_S{}^W = \mu_S{}^{\circ W} + kT \ln \varrho_S{}^W \tag{2.1}$$

$$\mu_S{}^E = \mu_S{}^{\circ E} + kT \ln \varrho_S{}^E \tag{2.2}$$

so that $\mu_S{}^{\circ W}$ and $\mu_S{}^{\circ E}$ are independent of the solute densities $\varrho_S{}^W$ and $\varrho_S{}^E$, respectively. At equilibrium, with respect to the transfer of the solute S, we have

$$0 = \mu_S{}^W - \mu_S{}^E = \mu_S{}^{\circ W} - \mu_S{}^{\circ E} + kT \ln(\varrho_S{}^W/\varrho_S{}^E)_{eq} \tag{2.3}$$

$$\Delta\mu_{tr}^{\circ}(E \to W) = -kT \ln(\varrho_S{}^W/\varrho_S{}^E)_{eq} \tag{2.4}$$

where "eq" stands for equilibrium, and note that the quantity $\Delta\mu_{tr}^{\circ}$ is essentially the difference in the coupling work of S against pure E and W, namely,

$$\Delta\mu_{tr}^{\circ}(E \to W) = W(S \mid W) - W(S \mid E) \tag{2.5}$$

Clearly if one phase is a rare gas (or vacuum) then $W(S \mid G) = 0$ (no coupling work) and the standard free energy of transfer turns into a standard free energy of solution, which in our notation will be written as $\Delta\mu_{tr}^{\circ}(G \to E)$ or $\Delta\mu_{tr}^{\circ}(G \to W)$.

Thus, we have also the relation

$$\Delta\mu_{tr}^{\circ}(E \to W) = \Delta\mu_{tr}^{\circ}(G \to W) - \Delta\mu_{tr}^{\circ}(G \to E) \tag{2.6}$$

As an explicit example consider the transfer of methane from ethanol to water (at $P = 1$ atm and $T = 283$ K):

$$\Delta\mu_{\text{tr}}^{\circ}(G \rightarrow E) = 325 \text{ cal/mol}, \qquad \Delta\mu_{\text{tr}}^{\circ}(G \rightarrow W) = 1747 \text{ cal/mol} \qquad (2.7)$$

and

$$\Delta\mu_{\text{tr}}^{\circ}(E \rightarrow W) = 1747 - 325 = 1422 \text{ cal/mol} \qquad (2.8)$$

Using (2.4) we can rewrite this result in the form

$$(\varrho_S^W/\varrho_S^E)_{\text{eq}} = \exp[-\Delta\mu_{\text{tr}}^{\circ}(E \rightarrow W)/RT]$$
$$= \exp(-1422/566) = 0.081 \qquad (2.9)$$

The numerical result obtained in this example can be interpreted in two ways: first, the work required to transfer one methane molecule from a *fixed* position in pure E into a *fixed* position in pure W (at the specified P and T) is a positive quantity and its value is $1422/N_0$ cal (N_0 being the Avogadro number, $N_0 = 6.023 \times 10^{23}$). A second interpretation results from (2.9)—namely, under the experimental conditions described above, the density of methane in ethanol would be about 12.3 times larger than in water. In other words, methane would, on the average, favor the ethanol environment over the aqueous environment. This is the prevailing interpretation of $\Delta\mu_{\text{tr}}^{\circ}(E \rightarrow W)$ in the literature in terms of the relative "phobia" of methane for the two liquids.

A slightly more elaborate analysis should be made when one of the phases is pure S. In Section 1.2 the thermodynamics of transfer for this case has been treated. Here we shall only add a possible interpretation in terms of the relative "phobia" of S with respect to the two solvents. As in Section 1.2, consider an experimental situation where pure S, say benzene, is in equilibrium with benzene in a dilute solution in water; at equilibrium we have

$$\mu_S^S = \mu_S^W$$

Hence

$$0 = \mu_S^W - \mu_S^S = \tilde{\mu}_S^W - \tilde{\mu}_S^S + kT \ln(\varrho_S^W/\varrho_S^S)_{\text{eq}} \qquad (2.10)$$

Here $\tilde{\mu}_S^W - \tilde{\mu}_S^S$ has exactly the same meaning as before, namely, this is the free energy of transferring an S from a fixed position in S into a fixed position in W. However, in contrast to equation (2.4), where both ϱ_S^W and ϱ_S^E were variables and only the ratio was characteristic of the two solvents, here ϱ_S^S is constant, i.e., the density of pure S, hence also ϱ_S^W, must be constant since the ratio $(\varrho_S{}^W/\varrho_S{}^S)_{eq}$ is constant. Clearly, this ratio cannot be assigned the meaning of the relative "phobia" of S with respect to the two "solvents" (one of which is pure S), as we have done for the quantity defined in (2.4). However, if we measure the "partition coefficient" in (2.10) for different solvents, say pure S against W, and pure S against E, then the ratio of these two quantities is

$$\frac{(\varrho_S{}^W/\varrho_S{}^S)_{eq}}{(\varrho_S{}^E/\varrho_S{}^S)_{eq}} = (\varrho_S{}^W/\varrho_S{}^E)_{eq} \tag{2.11}$$

and has the same significance as in (2.4). Finally we note again that the meaning assigned to $\tilde{\mu}_S{}^W - \tilde{\mu}_S{}^S$ in (2.10) is a result of statistical mechanical considerations, and cannot be obtained by purely thermodynamical arguments (see also Appendix A.1).

2.3. SOME NUMERICAL EXAMPLES OF $\Delta\mu_{tr}^{\circ}$ BETWEEN PURE PHASES

The Ostwald absorption coefficient of a solute S in a liquid W is defined as the ratio of the number densities of S in the liquid and in the gas at equilibrium

$$\gamma_S{}^W = (\varrho_S{}^W/\varrho_S{}^G)_{eq} = \exp[-\Delta\mu_{tr}^{\circ}(G \to W)/RT] \tag{2.12}$$

Relation (2.12) has been written for the case where both $\varrho_S{}^W$ and $\varrho_S{}^G$ are very small to ensure that each phase is an ideal dilute one (which in the gas, containing only S, is identical to an ideal gas). Otherwise, if the density of S in either phase is high, then a similar relation holds, but with the replacement of $\Delta\mu_{tr}^{\circ}$ by the generalized standard free energy of solution $\Delta\tilde{\mu}_{tr}(G \to W)$. In this section we present some numerical values of $\Delta\mu_{tr}^{\circ}$ per mole of solute, so we use the gas constant R in (2.12) rather than the Boltzmann constant k.

Table 2.1 provides some values of γ_S for methane and ethane in pure water and in pure ethanol at three temperatures. The fourth column contains the ratio of the two γ_S's at equilibrium with respect to S between the two solvents:

$$\eta \equiv (\gamma_S{}^E/\gamma_S{}^W)_{eq} = (\varrho_S{}^E/\varrho_S{}^W)_{eq} \tag{2.13}$$

This quantity may be interpreted as the equilibrium partition coefficient of S between the two phases E and W [as in Figure 2.1(a)] or between

Table 2.1

Ostwald Absorption Coefficients for Methane and Ethane in Water and in Ethanol at Three Temperatures[a]

t (°C)	$\gamma_S^W \times 10^3$	$\gamma_S^E \times 10^3$	η	$\Delta\mu_{tr}^{\circ}(E \to W)$ (kcal/mol)
		Methane		
10	44.8	561	12.5	1.42
20	37.0	540	14.5	1.57
30	31.9	519	16.3	1.69
		Ethane		
10	69.1	3361	56.0	2.20
20	51.4	2910	56.6	2.36
30	40.5	2580	63.7	2.52

[a] Data from Yaacobi and Ben-Naim (1973).

E and W through an intermediate gaseous phase [as in Figure 2.1(b)]. The presence of partitions that are permeable to S only is essential. Otherwise, the two pure liquids E and W might mix either directly or indirectly by distillation through the gaseous phase G. In such a case (2.13) may not be valid.

The fifth column presents the values of $\Delta\mu_{tr}^{\circ}(E \to W)$ in kcal/mol.

Note that the tendency to "avoid water" relative to ethanol, as expressed by the values of η, is larger for ethane than for methane. More interestingly, this tendency increases with increasing temperature. This

Figure 2.1. Three possible experimental setups for measuring partition coefficients of a solute S between two solvents: (a) The two solvents W and E are separated by a partition permeable to S *only* (dotted

partition). (b) The two solvents W and E are separated by an impermeable partition (dark). The solute S is, however, in equilibrium with the gas G through the permeable partitions (dotted). (c) The two solvents W and E are separated as in (b). Transference of all components can be achieved from one phase to the other through the gaseous phase. If E and W are almost totally immiscible (say benzene and water), then the partition coefficient of S measured in (c) will be very close to the one measured in (a) or (b). A different value is obtained in (c) if E and W are soluble in each other.

seems to contradict our intuitive expectation, which tells us that at high temperatures the peculiarities of water should diminish and therefore the ratio η should decrease with temperature. Indeed, if one examines the values of η as a function of temperature beyond, say, $t \approx 80°C$, one finds the expected behavior. However, at lower temperatures, the opposite behavior is observed. This phenomenon is an interesting one and is related to the effect of the solute on the structure of water, a subject that is dealt with in Chapter 5. (More detailed treatment may be found in Ben-Naim, 1974, Chapter 7.)

An instructive example of a standard free energy of transfer between two liquids is the case of light and heavy water. Table 2.2 presents some values of γ_S in the two liquids as well as the standard free energy of transfer from H_2O to D_2O (at three temperatures and 1 atm).

Note that all the values of γ_S for methane and ethane in D_2O are higher than the corresponding values in H_2O. This, according to the interpretation given above, would mean that the "phobia" for H_2O is greater than for D_2O, or equivalently, the standard free energy of transferring a solute molecule from H_2O to D_2O is negative. Before we continue to discuss these particular two solvents we stress again that we do not consider $\Delta\mu_{tr}^{\circ}$ as a quantity that conveys information on *hydrophobic interaction* in the sense of Section 1.3. It is, however, a legitimate measure of the relative "phobia" or affinity for the two solvents.

Table 2.2

Ostwald Absorption Coefficients of Methane and Ethane in H_2O and D_2O^a

t (°C)	$\gamma_S^{H_2O} \times 10^3$	$\gamma_S^{D_2O} \times 10^3$	$\Delta\mu_{tr}^{\circ}(H_2O \rightarrow D_2O)$ (cal/mol)
		Methane	
10	44.8	48.5	−45
20	37.0	39.3	−35
25	34.2	36.0	−30
		Ethane	
10	69.1	73.3	−33
20	51.4	53.9	−28
25	45.3	47.9	−33

^{a} Data from Ben-Naim *et al.* (1973).

Table 2.3

Standard Free Energies of Transfer from H_2O to D_2O^a

Solute	$\varrho_S \times 10^3$		$\Delta\mu_{tr}^{\circ}(H_2O \to D_2O)$ (cal/mol)
	H_2O	D_2O	
Propane	1.53	1.56	−11.6
Butane	1.26	1.29	−14.0
Glycine	313.0	324.0	−20.6
Norvaline	72.0	69.6	+19.3
Norleucine	82.4	79.6	+20.9
$CH_3CH_2CH_2-$	(conditional transfer, see text)		+39.9
$CH_3CH_2CH_2CH_2-$			+41.2

^{a} Data from Kresheck *et al.* (1965) at 25°C.

We now turn to a specific example which is instructive in demonstrating a typical source of confusion that exists in the literature.

Kresheck *et al.* (1965) reported values of $\Delta\mu_{tr}^{\circ}$ for propane and butane. Table 2.3 cites the solubilities of these two gases in H_2O and D_2O in molar concentrations ϱ_S at 25°C. From these values we computed the standard free energy of transfer, namely,

$$\Delta\mu_{tr}^{\circ}(H_2O \to D_2O) = -RT \ln(\varrho_S{}^{D_2O}/\varrho_S{}^{H_2O})_{eq} \qquad (2.14)$$

The experimental situation corresponds to the setup of Figure 2.1(b). Comparing the values obtained with those in Table 2.2 we see that $\Delta\mu_{tr}^{\circ}$ for propane and butane are somewhat smaller than the corresponding values for methane and ethane, but the sign is negative in all cases, and the conclusion is that for all these solutions, the "phobia" for H_2O is slightly *greater* than for D_2O.

Kresheck *et al.* also reported values of $\Delta\mu_{tr}^{\circ}$ for some amino acids. We cite the values for three of them in Table 2.3.[†] We note that $\Delta\mu_{tr}^{\circ}$ is negative for glycine and positive for norvaline and norleucine. This in itself means that glycine prefers D_2O over H_2O, whereas the contrary is

[†] The solubilites of the amino acids are quite large and therefore relation (2.14) may not be valid in this case. Here we present the processing of the data as given by Kresheck *et al.* (1965).

true for the so-called "hydrophobic" amino acids. This result seems curious at first glance. We have seen in Tables 2.2 and 2.3 that for simple hydro-carbon molecules, such as propane and ethane, $\Delta\mu_{tr}^{\circ}$ is *negative*. Here, on the other hand, we started with glycine, for which $\Delta\mu_{tr}^{\circ}$ is negative. We added a nonpolar group to form norvaline and we find that $\Delta\mu_{tr}^{\circ}$ becomes *positive*. These findings seem to be contradictory to our expectations. If we write the approximate relation

$$\Delta\mu_{tr}^{\circ}[R_n\text{---}CH(NH_2)COOH] \approx \Delta\mu_{tr}^{\circ}(R_n\text{---}) + \Delta\mu_{tr}^{\circ}[\text{---}CH(NH_2)COOH]$$

$$(2.15)$$

i.e., split $\Delta\mu_{tr}^{\circ}$ of an amino acid into two parts—one corresponding to the "hydrophobic" group, the other to the "hydrophylic" group—we can estimate that

$$\Delta\mu_{tr}^{\circ}(CH_3CH_2CH_2\text{---}) \approx \Delta\mu_{tr}^{\circ}(\text{norvaline}) - \Delta\mu_{tr}^{\circ}(\text{glycine}) \quad (2.16)$$

$$\Delta\mu_{tr}^{\circ}(CH_3CH_2CH_2CH_2\text{---}) \approx \Delta\mu_{tr}^{\circ}(\text{norleucine}) - \Delta\mu_{tr}^{\circ}(\text{glycine}) \quad (2.17)$$

and we find in the last two rows of Table 2.3 that $\Delta\mu_{tr}^{\circ}$ of the "hydrophobic" group is *positive*, in sharp contrast to the values of $\Delta\mu_{tr}^{\circ}$ for propane and butane. This apparent contradiction has caused some confusion in the literature (see, e.g., Franks, 1975, page 72). The reason for this confusion is subtle and arises from the fact that relation (2.15) is in general invalid, i.e., one cannot split the free energy of transfer of a molecule of the type A—B into the sum of the free energies of transferring A and B *separately*. This point is further elaborated upon in Appendix A.2; here we shall only cite the final result of that appendix. Namely, the standard free energy of transfer of a molecule $R_n\text{---}CH(NH_2)COOH$ can be approximated by the sum of two terms: the standard free energy of transfer of $\text{---}CH(NH_2)COOH$ (or approximately of glycine), and a *conditional* standard free energy of transfer of the nonpolar residue. The important point that should be stressed is that in the second step we are transferring the group R_n *not* from *pure* H_2O into *pure* D_2O, but from H_2O containing a polar group $\text{---}CH(NH_2)COOH$ into D_2O containing the same polar group. Since the polar group may have a considerable effect on the solvent around it, this "modified" solvent may not be considered to be pure H_2O or D_2O. (In a sense, the conditional standard free energy of transfer is more closely related to the process of transferring S from an H_2O ionic solution to a D_2O ionic solution. In both cases the presence of ions modifies the properties of the solvent around them.)

Table 2.4

Ostwald Absorption Coefficient of Argon in Water and in Some Other Solvents at 25°C and 1 atm[a]

Solvent	$\gamma_S \times 10^3$	$\Delta\mu_{tr}^{\circ}(G \to L)$ (cal/mol)
Water	34.1	2001
Methanol	269	778
Ethanol	260	798
n-Hexane	472	445
n-Heptane	415	521
n-Octane	367	594
n-Nonane	338	642
n-Decane	311	692
Benzene	240	845
Cyclohexane	334	649
Ethylene glycol	37.1	1951

[a] Data from Clever *et al.* (1957), Saylor and Battino (1958), and Ben-Naim (1968).

In conclusion, the relative "phobia" of the four hydrocarbons cited above for H_2O is *greater* than for D_2O. The presence of a polar group such as —$CH(NH_2)COOH$ has a considerable effect on the properties of the solvent around it, which in turn is reflected in the values of the *conditional* standard free energy of transfer. We repeat our warning again that all the results of this section are relevant to the question of the relative "phobia" of S for the two solvents, and not to the problem of hydrophobic interaction as defined in Section 1.3. [A different attitude is maintained, however, in the paper of Kresheck *et al.* (1965).]

We next turn to an extension of the data of Table 2.1. Table 2.4 shows values of γ_S and $\Delta\mu_{tr}^{\circ}(G \to L)$ at 25°C, for argon in water and in some other liquids. It is quite clear that in *most* cases $\Delta\mu_{tr}^{\circ}(G \to L)$ is positive and large in water as compared with other solvents—but note that in ethylene glycol the value of $\Delta\mu_{tr}^{\circ}(G \to L)$ is almost the same as in water, showing that argon "feels" almost equally comfortable in these two liquids [in fact, at a lower temperature, about 20°C, the values of γ_S of argon in the two solvents become identical to each other—for more details see Ben-Naim (1968)].

The conclusion of this section may be stated as follows: if we let a small quantity of an inert solute distribute itself between water and other non-aqueous solvents, then in most cases we shall observe that the solute prefers the nonaqueous environment. But exceptions to this rule exist; one of them is ethylene glycol, and it is almost certain that having richer experimental data we could find more and more exceptions to this rule.

2.4. SOME NUMERICAL EXAMPLES OF $\Delta\mu_{tr}^{\circ}$ BETWEEN MORE COMPLEX PHASES

We now extend the data presented in Section 2.3 to include the case where one of the phases is a mixture of at least two components. This is interesting for two reasons: In the first place, there exists a considerable amount of discussion in the literature on the effect of added solvents (electrolytes or nonelectrolytes) on the hydrophobic interaction (here we use this term in a loose sense; more strictly, all the effects studied in the present section are relevant to the question of the relative "phobia" of a solute for various solvents—not to hydrophobic interaction). Secondly, some of the data presented here are further discussed in subsequent sections as ingredients for the computation of hydrophobic interaction in the sense of Section 1.3.

Table 2.5 presents some values of γ_S and $\Delta\mu_{tr}^{\circ}(G \to L)$ for methane in aqueous solutions of simple electrolytes. All of these solutions are less "attractive" to methane than pure water, i.e., all values of γ_S in the ionic solutions are smaller than the corresponding value in pure water. This is

Table 2.5

Ostwald Absorption Coefficient for Methane in Various Ionic Solutions (at 25°C and 1 atm)[a]

Solution	$\gamma_S \times 10^3$	$\Delta\mu_{tr}^{\circ}(G \to L)$ (cal/mol)
Pure water	34.1	2001
NaCl 1 M	24.2	2204
NaBr 1 M	24.6	2195
NaI 1 M	25.3	2178
LiCl 1 M	26.6	2148
KCl 1 M	24.9	2188
CsCl 1 M	25.5	2173
NH$_4$Cl 1 M	27.9	2122

[a] Data from Ben-Naim and Yaacobi (1974).

Table 2.6

Values of the Ostwald Absorption Coefficient γ_S for Methane in Different Mixtures of Ethanol–Water[a]

x_{EtOH}	$\gamma_S \times 10^3$				
	10°C	15°C	20°C	25°C	30°C
0.00	44.80	40.51	37.00	34.19	31.92
0.02	47.46	42.73	39.27	36.78	35.09
0.03	47.60	43.55	40.30	37.68	35.64
0.045	47.70	43.77	40.90	38.99	37.83
0.06	46.50	43.59	41.10	38.90	37.01
0.09	45.60	43.00	41.30	40.39	40.23
0.12	45.10	43.70	43.00	43.38	44.38
0.15	47.90	48.05	48.20	48.34	48.55
0.20	60.40	61.62	62.70	62.72	64.53
0.25	80.00	80.92	81.90	83.27	84.77
0.40	155.4	156.7	158.0	160.0	161.9
0.60	274.0	271.6	271.0	273.0	277.2
0.80	404.0	400.6	397.0	394.4	392.4
1.00	561.0	550.7	540.0	529.3	518.6

[a] Data from Yaacobi and Ben-Naim (1973).

the well-known salting-out effect of ions in aqueous solutions. For non-electrolytes the effect may be to either increase or decrease the value of γ_S. Perhaps the best way to present the effect of a nonelectrolyte solute is to follow the change of γ_S, of say methane, in mixtures of water and ethanol and in mixtures of water and p-dioxane. Some values of γ_S in such mixtures are presented in Tables 2.6 and 2.7.

There is a clear-cut difference between the initial effect of ethanol and that of dioxane on γ_S of methane. At very small concentrations, ethanol increases γ_S relative to pure water (i.e., makes the solution more "comfortable" to methane), between $0.03 \lesssim x \lesssim 0.1$ there is an opposite effect of ethanol on γ_S, and beyond that γ_S increases quite steeply upon further additions of ethanol. This behavior is temperature dependent, and it is most pronounced at lower temperatures. At about 30°C we observe almost a monotonic increase of γ_S with increasing concentration of ethanol.

Addition of dioxane seems to have a different effect on γ_S of methane. Even at low temperatures we do not observe the initial increment of γ_S upon the addition of dioxane. At all the temperatures for which data are

Table 2.7

Values of the Ostwald Absorption Coefficient γ_S for Methane in Different Mixtures of Water–Dioxane[a]

x_{dioxane}	$\gamma_S \times 10^3$				
	10°C	15°C	20°C	25°C	30°C
0.00	44.80	40.51	37.01	34.19	31.92
0.015	45.63	41.50	38.45	36.27	34.78
0.03	45.16	42.11	39.49	37.24	35.30
0.06	45.97	43.95	42.32	41.02	40.00
0.09	48.90	46.86	45.98	46.16	47.34
0.12	53.76	53.00	52.70	52.82	53.35
0.15	60.78	60.38	60.62	61.47	62.93
0.20	75.66	75.84	76.63	78.02	80.01
0.40	161.2	162.4	163.7	164.8	166.0
0.70	262.3	265.4	267.2	268.0	267.6
1.00	384.1	387.6	390.7	393.3	395.6

[a] Data from Ben-Naim and Yaacobi (1975).

available, we observe either a minimum for γ_S as a function of x_{dioxane} or a monotonic behavior as in the case of ethanol at high temperatures.

It is perhaps worth noting that tetraalkylammonium salts exhibit an interesting effect on γ_S. For ions $(R_n)_4 N^+$ with small n the properties of the salt are similar to those of the simple ions, but as n increases we often observe an opposite effect of *salting-in*.

We have presented here a very small sample of results on the effect of a third component on γ_S of a solute S in aqueous solutions. A very thorough review of this subject has recently been published by Wilhelm *et al.* (1977). The effects of some more-complex additives, such as proteins, will be mentioned in Chapter 4.

2.5. SURVEY OF THEORETICAL METHODS OF ESTIMATING $\Delta\mu_{tr}^{\circ}(G \to W)$

This section presents a short survey of the theoretical methods that have been used to estimate $\Delta\mu_{tr}^{\circ}(G \to W)$, i.e., the standard free energy of solution of a simple solute in water. From Section 1.2 and Appendix A.1

it is recognized that the computation of the standard free energy of solution of a solute S in water W is equivalent to the computation of the coupling work associated with the introduction of a single solute S to some fixed position in the water, i.e.,

$$\Delta\mu_{tr}^{\circ} = W(S \mid W) \qquad (2.18)$$

[In this section we shall always refer to $\Delta\mu_{tr}^{\circ}(G \to W)$, which for short we denote by $\Delta\mu_{tr}^{\circ}$.]

There have been several attempts to compute these quantities by theoretical means. None of them are completely satisfactory at present. This situation is quite understandable since water is a complex fluid and at present there exists no efficient method of computing $W(S \mid W)$ even for the simplest solutes, such as inert gases, in a simple solvent. It is therefore no wonder that semitheoretical methods have taken the place of the standard molecular theory, based on the fundamentals of statistical mechanics. All the semitheoretical methods involve serious approximations, especially when they are applied to such a complex fluid as water.

In accordance with the general aims and style of this book we shall not endeavor to present details of the theories employed. Instead a qualitative description of the various attempts will be presented.

Perhaps the most successful approach that has been used quite extensively on this problem is the application of the scaled-particle theory (SPT). Since this theory is also used in a different way in Chapter 4, we shall present in Appendix A.4 some further details on the basic assumptions and results of this theory. For reviews of this theory we refer the reader to Reiss (1966) and Pierotti (1976).

As a first step towards the application of the SPT, we assume that the solute–solvent pair potential may be conveniently split into a repulsive and an attractive part. Following this separation we can also split $W(S \mid W)$ into two terms,

$$W(S \mid W) = W(\text{rep} \mid W) + W(\text{atr} \mid W) \qquad (2.19)$$

where the first term on the right-hand side of (2.19) corresponds to the coupling of the repulsive part of the solute–solvent interaction, and the second term is the work required to "turn on" the attractive part of the interaction. We note that the split of $W(S \mid W)$ may be achieved in a rigorous fashion by using a consecutive coupling procedure (Ben-Naim, 1974, Chapter 7). Clearly, if the repulsive part of the solute–solvent potential function is very steep, we may approximate the first term on the right-hand side of

(2.19) by the work required to introduce a hard-sphere (HS) solute to a fixed position in the liquid, i.e.,

$$W(\text{rep} \mid W) \approx W(\text{HS} \mid W) = W(\text{cav} \mid W) \tag{2.20}$$

The last equality on the right-hand side of (2.20) follows from the very definition of a HS solute, namely, a region in the solvent from which the centers of all the molecules are precluded. This is also precisely the definition of a "cavity" at some fixed location in the liquid.

The computation of $W(\text{HS} \mid W)$, which is the most important output of the SPT, may be viewed as either a first step in the process of computing $W(S \mid W)$ in (2.19), for a real solute S, or we can view $W(\text{HS} \mid W)$ as the standard free energy of solution of a HS solute, the latter being considered to be an idealization of the simplest solute for which a computation may be carried out.

A slightly different approach is to split the process of introducing S to a fixed position into two parts; first we create a cavity of suitable size to accommodate the solute, and then the solute is inserted into the cavity. In such a process the choice of the size of the cavity has an element of ambiguity, and any judicious choice must depend on the knowledge of the solute–solvent interaction.

In essence, the SPT provides an approximate expression for the work $W(\text{HS} \mid W)$. Let the cavity produced by the HS solute have a radius r; then if the solvent molecules are assigned a hard core diameter a, the work required to create such a cavity is given by a polynomial of the form

$$W(\text{HS} \mid W) = W(r) = K_0 + K_1 r + K_2 r^2 + K_3 r^3 \tag{2.21}$$

where the radius of the cavity r is related to the diameter a_{HS} of the HS and to the diameter a of the solvent molecules by

$$r = (a_{\text{HS}} + a)/2 \tag{2.22}$$

The coefficients of the polynomial in (2.21) are functions of the temperature T, the pressure P, the solvent density ϱ, and the molecular diameter of the solvent molecules a. (For an explicit expression for the K_i see Appendix A.4.)

It should be noted that the SPT is not a pure molecular theory in the following sense. In a proper molecular theory one should, in principle, be able to compute both $W(\text{HS} \mid W)$ and the average density ϱ from a knowledge of the molecular properties of the liquid at some specified T, P, and number of molecules N. Here, on the other hand, only the diameter of the solvent molecules is employed (which itself is not always a well-defined quantity, especially for complex molecules), and in addition to the specification of T and P, we must know the average density ϱ. The latter effectively introduces other characteristics of the solvent into the theory. Therefore, the theory has the character of a correlation between two macroscopically measurable quantities $W(\mathrm{HS} \mid W)$ and ϱ (the former is measurable only approximately through a partition coefficient of a simple solute; see Section 2.2). This feature of the theory automatically precludes the consideration of the structural changes in the solvent which may be induced by the dissolution process. On the other hand, the very fact that the solvent density is used as an input in the theory brings in implicitly structural features of the solvent.

Of course, an exact molecular theory should be capable of providing expressions for both $W(\mathrm{HS} \mid W)$ and ϱ, and perhaps also a functional relation between the two quantities. The SPT suggests such an approximate functional dependence between the two quantities as given by the polynomial in (2.21).

We have touched upon the notion of the "structural changes in the solvent"—without elaborating on its precise definition. However, a formal definition of this concept is available. We shall return to this topic in Chapter 5 when we shall discuss the temperature and the pressure dependence of the HI.

An interesting correlation between $\Delta\mu_{\mathrm{tr}}^{\circ}$ and the surface tension σ is shown in Figure 2.2. It should be noted that the point for liquid water falls almost on the same line as for the other solvents. This fact in itself suggests

Figure 2.2. Linear correlation between the standard free energy of solution of argon (at 25°C) in various solvents and the surface tension of the solvents. The solvents used, in order of increasing surface tension, are: n-hexane; isooctane; 2,3-dimethylhexane; n-heptane; 2,4-dimethylhexane; 3-methyl-heptane; n-octane; n-nonane; methycyclohexane; n-decane; cyclohexane; n-dodecane; carbon tetrachloride; n-tetradecane; fluorobenzene; toluene; p-xylene; benzene; chlorobenzene; carbon disulfide; bromobenzene; nitromethane; iodobenzene; nitrobenzene; water (Reproduced with changes from Battino and Clever, 1966.)

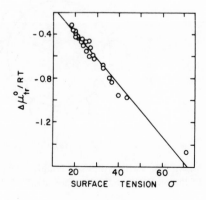

that there might exist a "universal" (at least for those solvents that appear in this figure) relation between $\Delta\mu_{tr}^{\circ}$ and the surface tension σ. Such a relation is certainly useful for correlating two macroscopic quantities. But even an exact relation between these two quantities cannot replace a molecular theory; the latter should in principle be able to provide both $\Delta\mu_{tr}^{\circ}$ and σ in terms of the molecular properties of the system.

The computation of $\Delta\mu_{tr}^{\circ}$ is often attempted by assuming that the most important contribution to the work of creating a cavity is the work required to extend the surface area of the solvent.

Thus, if the radius of the cavity is r then the new surface created is πr^2 and the associated work is $\pi r^2 \sigma$. This is certainly not the whole, or even the most important, contribution to $\Delta\mu_{tr}^{\circ}$. But even if we do not know precisely how to compute all the other contributions, we are facing a serious difficulty in the very definition of the surface area of a cavity of microscopic dimensions. To demonstrate this point, consider a solvent consisting of molecules with diameter a. Let us create two cavities in this solvent, one with a radius $r \gg a$ and one with a radius $r \approx a$. The situation is depicted in Figure 2.3. Clearly in the first case the surface area is πr^2, and the fact that this is not a purely geometrical surface but one made up of molecules has a negligible effect on the sharpness of our definition of the area. This, however, is not the case when the radius of the cavity becomes of the same order of magnitude as the diameter of the molecules. Here it is not clear at all how one should define the area of the cavity. This basic difficulty has already been pointed out by Kirkwood (1939). As is demonstrated in Figure 2.3, a cavity of radius $r \approx a$ may be "empty" in the formal sense of our definition of a cavity (i.e., no centers of solvent molecules are within that region), but the same cavity may be filled with particles, in the

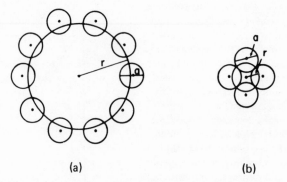

(a)   (b)

Figure 2.3. Two cavities in a liquid: (a) with a radius $r \gg a$ and (b) with a radius $r \approx a$. In the second case the "surface area" of the cavity is not a well-defined quantity.

ordinary sense of the word "filled." This is the reason why the surface area of such a cavity loses its sharp meaning.

Another relation between two macroscopic quantities which has a similar character as the relation between $\Delta\mu_{tr}^{\circ}$ and σ is the following: Consider the thermodynamic relation (for a closed, one-component system)

$$dG = -S\,dT + V\,dP \tag{2.23}$$

from which one obtains

$$\left(\frac{\partial G}{\partial V}\right)_T = V\left(\frac{\partial P}{\partial V}\right)_T = -\varkappa_T^{-1} \tag{2.24}$$

where $\varkappa_T$ is the isothermal compressibility. Integrating this relation we obtain

$$\Delta G = -\int \varkappa_T^{-1}\,dV \tag{2.25}$$

This relation strictly applies for macroscopic changes in the volume of the system. Oakenfull and Fenwick (Oakenfull, 1973; Oakenfull and Fenwick, 1974b) suggested applying this relation for the volume changes associated with the creation of a cavity. If one assumes that $\varkappa_T$ is approximately constant we may integrate (2.25) to yield

$$\Delta G(\text{cav} \mid W) \cong -\varkappa_T^{-1} V_{\text{cav}} \tag{2.26}$$

Again, such a relation has the character of a correlation between two macroscopic quantities. [In fact, both the creation of a volume and the extension of the surface contribute to $\Delta G(\text{cav} \mid W)$, and we have discussed above two separate correlations: between $\Delta\mu_{tr}^{\circ}$ of a HS and σ, on the one hand, and between $\Delta\mu_{tr}^{\circ}$ and V_{cav}, on the other hand.] But clearly the very meaning of the volume of the cavity that should be used in a relation such as (2.26) is not clear. The difficulty is much the same as we have noted above for the surface area of a microscopic cavity.

All the methods discussed above apply to the problem of calculating the standard free energy of solution of a hard-sphere solute. If this part is viewed only as a first step in computing $\Delta\mu_{tr}^{\circ}$ for a real solute, then we must turn to the next step and compute the contribution of the attractive part of the potential, i.e., the second term on the right-hand side of (2.19). This may be done by viewing the attractive part of the potential as a small perturbation to the strong repulsive potential of the hard spheres. Estimates along these lines have been made by several authors. For details on the theoretical arguments employed in these calculations the reader is referred to the original articles (Pierotti, 1963, 1965, 1976; Hermann, 1971, 1972, 1975; Klapper, 1973; Pratt and Chandler, 1977).

Another approach, which in principle may be used to estimate $\Delta\mu_{tr}^{\circ}$, is to apply the well-known simulating techniques, either the Monte Carlo or the molecular dynamics technique, to this problem. However, at present such attempts have not met with success. The main reason is that there are serious difficulties in calculating the free energy of a system by these simulation techniques. The standard free energy of solution is, in essence, a difference between free energies of two systems (before and after adding the solute). The difficulties encountered in such a computation have been discussed recently by Owicki and Scheraga (1977).

Comment

Progress in the computation of $\Delta\mu_{tr}^{\circ}$ from a purely molecular theory will certainly be difficult and will ultimately be dependent on our progress in the field of the molecular theory of liquid water. Effort expended in devising even approximate methods of estimating $\Delta\mu_{tr}^{\circ}$ should be rewarding, since such quantities are essential ingredients in the theory of hydrophobic interactions, as defined in Chapter 3.

2.6. SOME WORDS OF CAUTION

This chapter has been devoted to the characterization of the so-called solute–solvent interaction, or the hydrophobic hydration problem when the solvent is water. The motivation for studying this topic in the context of the present book has been given in Chapter 1—namely, the standard thermodynamic quantities of transfer may be used to estimate one of the driving forces in the conformational change of a biopolymer. However, such an application is not free of some serious difficulties. One question is to what extent a conformational change of a biopolymer—even in the most favorable cases where only HI are involved—may be reduced to a process of transfer of a simple solute between two liquids. The analysis of this question requires some statistical mechanical tools and we defer it to Appendix A.5. Here we shall address ourselves to a more technical problem. Suppose one can reduce the very complicated process of a conformational change to a simple process of transferring a solute from water to a nonpolar solvent. The question is: Which of the multitude of "standard" processes of transfer should be used for such a representation? This problem is sometimes referred to as being concerned with the mere choice of *units* of concentration, and it is argued that there is no fundamental preference for one unit over the other. We shall now demonstrate that this is not so and that the choice of correct standard quantities is crucial if we want to use the transfer process as a model for a conformational change process. Unfortunately, most articles and books that deal with this question advocate the choice of the wrong "units" or standard states. This will be demonstrated below. For more details on this subject, the reader is referred to Ben-Naim (1978a).

To begin with, we note that there are many standard quantities in use in the literature. We shall, for simplicity, compare only two of these, namely, the quantity $\Delta\mu_{\text{tr}}^{\circ\varrho}(E \to W)$, based on the number (or molar) density, and the quantity $\Delta\mu_{\text{tr}}^{\circ x}(E \to W)$, based on the mole-fraction scale. The latter is by far the most commonly used in the literature. We note also that our main criticism concerning $\Delta\mu_{\text{tr}}^{\circ x}$ will be applicable to other standard free energies of transfer, but we shall not discuss these here.

We recall that the chemical potential of a simple solute S in a very dilute solution in a solvent W may be written as

$$\mu_S{}^W = W(S \mid W) + kT \ln \varrho_S{}^W \Lambda_S{}^3 = \mu_S{}^{\circ\varrho} + kT \ln \varrho_S{}^W \qquad (2.27)$$

This relation may be used either for a constant volume and closed system where $\varrho_S = N_S/V$, or for a constant pressure system where V is to be replaced by the average volume. Here we also assume for simplicity that the solute molecules are structureless, but all our conclusions are valid for more general cases as well.

The standard free energy of transferring S from one solvent W to another E, based on the number density, is obtained from

$$\mu_S{}^E - \mu_S{}^W = \mu_S{}^{\circ\varrho E} - \mu_S{}^{\circ\varrho W} + kT \ln(\varrho_S{}^E/\varrho_S{}^W) \qquad (2.28)$$

and for the special case where $\varrho_S{}^E = \varrho_S{}^W$ we have

$$\Delta\mu_{\text{tr}}^{\circ\varrho} \equiv \Delta G\!\left(\begin{matrix} W \to E \\ \varrho_S{}^W = \varrho_S{}^E \end{matrix}\right) = \mu_S{}^{\circ\varrho E} - \mu_S{}^{\circ\varrho W} = W(S \mid E) - W(S \mid W) \qquad (2.29)$$

i.e., this is the free energy change for transferring a solute S from W to E, with P and T constant and the condition that $\varrho_S{}^W = \varrho_S{}^E$, provided that these are very small in order to secure the validity of the ideal dilute behavior of the solution. This may be referred to as the *thermodynamic* standard process. But the last equality shows that $\Delta\mu_{\text{tr}}^{\circ\varrho}$ is also the difference between the coupling work of S to the two solvents E and W. We consider this to be the more important interpretation for our purposes. It should be noted that the latter interpretation has a statistical mechanical origin and cannot be obtained from purely thermodynamic reasoning. We shall denote the process of transferring S from a fixed position in W to a fixed position in E by $\Delta G(W \to E; L)$. This quantity "senses" the difference in the solvation free energy of S in the two solvents, and it depends on the properties of the local (L) environment of S in the two phases. Thus we have the equality

$$\Delta G\begin{pmatrix} W \to E \\ \varrho_S{}^W = \varrho_S{}^E \end{pmatrix} = \Delta G(W \to E; L) \tag{2.30}$$

which means that the two *different* processes produce the *same* free energy change. This equality is unique for the free energy. If we consider any other thermodynamic function, say the entropy, then, in general the entropy changes associated with the two processes indicated in (2.30) will be different from each other. We shall return to this point in Chapter 5.

Next we turn to the mole-fraction scale. Since we are dealing with very dilute solutions we may write the transformation of variables as

$$\varrho_S{}^W = \varrho_W{}^W x_S{}^W, \qquad \varrho_S{}^E = \varrho_E{}^E x_S{}^E \tag{2.31}$$

where $\varrho_W{}^W$ and $\varrho_E{}^E$ are the number densities of the pure solvents W and E, respectively. The standard chemical potential of S, based on the mole fraction, is obtained from (2.27) as follows:

$$\begin{aligned} \mu_S{}^W &= W(S \mid W) + kT \ln \varrho_S{}^W \Lambda_S{}^3 \\ &= [W(S \mid W) + kT \ln \varrho_W{}^W \Lambda_S{}^3] + kT \ln x_S{}^W \\ &= \mu_S{}^{\circ x W} + kT \ln x_S{}^W \end{aligned} \tag{2.32}$$

The corresponding standard free energy of transfer is obtained by analogy to (2.28) and (2.29), i.e.,

$$\Delta\mu_{\mathrm{tr}}^{\circ x} \equiv \Delta G\begin{pmatrix} W \to E \\ x_S{}^W = x_S{}^E \end{pmatrix} = \mu_S{}^{\circ x E} - \mu_S{}^{\circ x W}$$
$$= W(S \mid E) - W(S \mid W) - kT \ln(\varrho_W{}^W / \varrho_E{}^E) \tag{2.33}$$

Thus $\Delta\mu_{\mathrm{tr}}^{\circ x}$ is the free energy change associated with the transfer of one S from W to E, in which we have the condition $x_S{}^W = x_S{}^E$ (P and T are constant as before). This process is different from the process indicated on the left-hand side of Equation (2.30). In contrast to $\Delta\mu_{\mathrm{tr}}^{\circ\varrho}$, where we have given a second interpretation in terms of the difference in the coupling work of S in the two phases, this cannot be done for $\Delta\mu_{\mathrm{tr}}^{o\alpha}$. We see from (2.33) that $\Delta\mu_{\mathrm{tr}}^{o\alpha}$ contains, in addition to $W(S\,|\,E) - W(S\,|\,W)$, a term that depends on the densities of the pure solvents. Clearly the expression for $\Delta\mu_{\mathrm{tr}}^{o\varrho}$ is simpler than that for $\Delta\mu_{\mathrm{tr}}^{o\alpha}$. But simplicity, although an important feature of $\Delta\mu_{\mathrm{tr}}^{o\varrho}$, cannot be used as the sole argument for favoring this particular standard quantity. We now present a more profound argument, which leads to the inevitable conclusion that whereas $\Delta\mu_{\mathrm{tr}}^{o\varrho}$ may be used as a measure of the difference of the solvation properties of the two phases, $\Delta\mu_{\mathrm{tr}}^{o\alpha}$ cannot serve that purpose.

Consider a schematic process of a conformational change of a bipolymer, as depicted in Figure 2.4. We follow a particular nonpolar group, say a $—CH_3$ group which is darkened in the figure. We also assume that this process effectively removes the nonpolar group from an aqueous environment, and introduces it into an essentially nonpolar environment, which consists of similar groups. Figure 2.4(b) shows a process in which a methane molecule is transferred from water to a nonpolar environment "similar" to the one in the interior of the polymer. The similarity is stressed by the dotted lines that represent the total interaction of the group under consideration with all of the molecules surrounding it.

Clearly, the total interaction, as well as the free energy of interaction, depends on the average density of the molecules in the surroundings. To take an extreme example, suppose that the darkened group in the interior

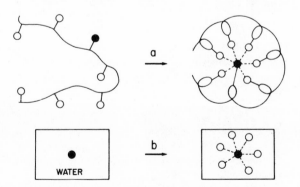

Figure 2.4. (a) A schematic description of a conformational change of a biopolymer. We follow a particular nonpolar group (dark circle) that is transferred from an (essentially) aqueous environment to an (essentially) nonpolar environment. (b) A model process representing the transfer of the nonpolar group in (a). The transfer is made from pure water into a nonpolar environment which is "similar" to the environment of the darkened circle in the interior of the polymer.

of the biopolymer interacts with very few other groups that happen to be around it. In such a case the appropriate nonpolar solvent in Figure 2.4(b) must also be chosen to have a similar low density of molecules. In the limit of very low density, the free energy of interaction of the darkened group with its surroundings must tend to zero. The standard free energy of transfer $\Delta\mu_{\text{tr}}^{\varrho\varrho}$ for the process in Figure 2.4(b) responds in an appropriate manner, namely,

$$\Delta\mu_{\text{tr}}^{\varrho\varrho}(W \to E) = W(S \mid E) - W(S \mid W)$$

$$\xrightarrow{\varrho_E \to 0} - W(S \mid W) \qquad (2.34)$$

It can be shown that $W(S \mid E)$ tends to zero linearly with ϱ_E as $\varrho_E \to 0$.

On the other hand the behavior of $\Delta\mu_{\text{tr}}^{\text{o}x}$ under the same limiting condition is

$$\Delta\mu_{\text{tr}}^{\text{o}x}(W \to E) = W(S \mid E) - W(S \mid W) - kT\ln(\varrho_W{}^W/\varrho_E{}^E)$$

$$\xrightarrow{\varrho_E \to 0} - \infty \qquad (2.35)$$

which is certainly inconsistent with our expectation. This is a very fundamental difference between the two standard quantities.[†]

Thus $\Delta\mu_{\text{tr}}^{\varrho\varrho}$ is the simplest quantity that conveys information on the solvation free energy of the solute in the two solvents. Moreover it is the only quantity that tends to zero as the solvent density tends to zero. Other standard free energies of transfer may diverge to infinity in this limit.

From Appendix A.1 it will be clear that when we construct $\Delta\mu_{\text{tr}}^{\varrho\varrho}$ we eliminate *all* the liberation free energy of the solute S, and we are left only with a quantity that "senses" the *local* environments of the two solvents. This is not the case with $\Delta\mu_{\text{tr}}^{\text{o}x}$, which, as shown in (2.33), still contains the volume of the phases, a residue of the liberation free energy of the solute.

[†] In deriving this result we have used the limiting relations $\varrho_S{}^W = \varrho_W{}^W x_S{}^W$ and $\varrho_S{}^E = \varrho_E{}^E x_S$. However, the exact relations are $\varrho_S{}^W = (\varrho_W{}^W + \varrho_S{}^W)x_S{}^W$ and $\varrho_S{}^E = (\varrho_E{}^E + \varrho_S{}^E)x_S{}^E$. Now if we do not first take the limit of infinite dilution with respect to S we get instead of (2.35)

$$\Delta\mu_{\text{tr}}^{\text{o}x}(W \to E) = \Delta G\binom{W \to E}{x_S{}^W = x_S{}^E} = W(S \mid E) - W(S \mid W) - kT\ln\left(\frac{\varrho_W{}^W + \varrho_S{}^W}{\varrho_E{}^E + \varrho_S{}^E}\right)$$

Note that for finite concentrations of S this does not diverge to $\pm\infty$ when one of the solvent densities goes to zero. However, in general, one takes first the limit $\varrho_S \to 0$ before the construction of the standard free energy of transfer.

We have discussed only free energy changes, but the same kind of analysis can be done for other thermodynamic quantities. *The conclusion is that the only standard quantities that have the property of sensing the local environment of the solute are those based on the number densities.* [For more details, see Ben-Naim (1978a).]

As we have seen, both $\Delta\mu_{\mathrm{tr}}^{\circ\varrho}$ and $\Delta\mu_{\mathrm{tr}}^{\circ x}$ can be given thermodynamic interpretations, in terms of transfer at constant ϱ_S or constant x_S, respectively. But thermodynamics alone could not be used to help us make the choice between the two for measuring the average free energy of interaction (or solvation) of S.

We now present another quantity which is also often used in the literature. It has a disadvantage similar to that of $\Delta\mu_{\mathrm{tr}}^{\circ x}$, but in contrast to $\Delta\mu_{\mathrm{tr}}^{\circ x}$ it does not have even a simple thermodynamic interpretation.

Consider a pure liquid S in equilibrium with pure water, and suppose for simplicity that the mutual solubility of these two solvents is very small, so that we can write for the chemical potential of S in the two phases

$$\mu_S{}^S = W(S \mid S) + kT \ln \varrho_S{}^S \varLambda_S{}^3 \tag{2.36}$$

$$\mu_S{}^W = W(S \mid W) + kT \ln \varrho_S{}^W \varLambda_S{}^3 \tag{2.37}$$

The standard free energy of transferring one S from a fixed position in W to a fixed position in S may be defined in complete similarity to (2.29) as

$$\Delta G(W \to S; L) = W(S \mid S) - W(S \mid W) \tag{2.38}$$

and may be measured by

$$\Delta G(W \to S; L) = kT \ln(\varrho_S{}^W/\varrho_S{}^S)_{\mathrm{eq}} \tag{2.39}$$

A different quantity has been used, especially by Tanford (1973), which is defined by

$$\mu_S{}^P - \mu_S{}^{\circ x W} = W(S \mid S) - W(S \mid W) + kT \ln(\varrho_S{}^S/\varrho_W{}^W) \tag{2.40}$$

where $\mu_S{}^P$ is the chemical potential of pure S, and $\varrho_S{}^S$, $\varrho_W{}^W$ are the number densities of the pure solvents S and W, respectively. Clearly this quantity also diverges to $+\infty$ or $-\infty$ when one of the liquid densities goes to zero, and therefore cannot be used as a proper measure for the transfer process of Figure 2.4(a). Moreover, since on the left-hand side of (2.40) we have a difference of a proper chemical potential $\mu_S{}^P$ and a standard chemical potential $\mu_S{}^{\circ x W}$, such a difference does not lend itself to a simple thermodynamic interpretation, as was the case for $\Delta\mu_{\mathrm{tr}}^{\circ x}$.

Comment

The title of this section should be taken very seriously. The literature is full of tables of quantities such as $\Delta\mu_u{}^\circ$, ΔG_{tr}°, ΔG°, etc. In some cases these are referred to as "standard free energy of transfer," but the process to which they refer is not specified. In other cases a meaning is assigned to these quantities without proper justification. It is difficult to overstate the confusion that exists in this field. Therefore the only wise advice that can be given to the reader is to be very very careful and critical when reading about this subject.

Chapter 3

Pairwise Hydrophobic Interaction (HI)

3.1. INTRODUCTION

This chapter forms the kernel of the whole book. Here we shall survey the available information on the most elementary, but fundamental, process to which the term hydrophobic interaction (HI) may be applied. This is essentially the dimerization process. The concept of "dimer" is used here in quite a broad sense. In fact, there always exists some ambiguity in the very definition of a dimer formed by two solute particles. Although we shall be discussing dimers formed by complex solute molecules, we shall restrict ourselves first to simple spherical solute molecules.

We recall from Sections 1.3 and 1.4 that the fundamental function, the gradient of which provides the average force between the solutes in a solvent, is $\Delta G(R)$. This is the Gibbs free energy change associated with the process of bringing two solutes from a *fixed* position at infinite separation to some distance R apart. This process is carried out within the solvent at some specified temperature T and pressure P.

In classical statistical mechanics one can always separate $G(R)$ into two terms

$$\Delta G(R) = U_{SS}(R) + \delta G^{HI}(R) \tag{3.1}$$

where $U_{SS}(R)$ is the *direct* solute–solute pair potential, i.e., this is the work required to carry out the same process as described above in vacuum. The function $\delta G^{HI}(R)$ will be referred to as the *indirect* part of the work associated with the above-mentioned process. This function exposes the characteristic features of the solvent in which the process is carried out. If the solvent is water we shall refer to this part as the HI part.

It is important to note that the sign of the *force* operating between two solute molecules depends on the sign of the gradient of $\Delta G(R)$, namely,

$$F_{SS}(R) = -\frac{\partial \Delta G(R)}{\partial R} \tag{3.2}$$

Attractive forces are obtained when the $F_{SS}(R)$ is negative, and repulsive forces when $F_{SS}(R)$ is positive. This is sometimes confused with the sign of the function $\Delta G(R)$ itself. We saw in Figure 1.6 that $\Delta G(R)$ oscillates as a function of R. The absolute magnitude of this function is of less significance than the sign of its slope at each point.[†]

In Section 1.4 we also noted that $\Delta G(R)$ is related to the solute–solute pair correlation function by

$$g_{SS}(R) = \exp[-\Delta G(R)/kT] \tag{3.3}$$

where $\varrho_S g_{SS}(R)4\pi R^2\,dR$ is the average number of solute particles in a spherical shell of width dR at a distance R from the center of a given solute S. Thus any information on $\Delta G(R)$ is completely equivalent to information on $g_{SS}(R)$.

Following the split of $\Delta G(R)$ in (3.1) we also defined the function $y_{SS}(R)$ as

$$y_{SS}(R) = g_{SS}(R)\exp[U_{SS}(R)/kT] = \exp[-\delta G^{\text{III}}(R)/kT] \tag{3.4}$$

The meaning of the function $y_{SS}(R)$ is somewhat less straightforward than $g_{SS}(R)$. We may think of two solute molecules for which the direct pair potential $U_{SS}(R)$ has been "switched-off"; in this case the meaning of $y_{SS}(R)$ coincided with the meaning of $g_{SS}(R)$ for such particles. A special case occurs for hard-sphere solutes. If we "switched-off" the direct pair potential between the two solutes, then what is left is indistinguishable from two cavities at a distance R from each other.

Since our main interest is focused on the properties of the solvent and not of the solute, we shall be mainly concerned with either $\delta G^{\text{III}}(R)$ or $y(R)$. These functions are somewhat more convenient for theoretical treatment as will be clear from the following sections.

[†] This is true if we are interested in *forces* acting on molecules. However, if we are interested in the probability of finding two solute particles a certain distance R apart, it will be $\Delta G(R)$ rather than its gradient which is of central importance.

The main question that is explored in this chapter is whether the function $\delta G^{\mathrm{HI}}(R)$ has any outstanding properties in water as compared with other solvents. This is sometimes referred to as the unique role of water in producing these HI. The presently available data are still scarce and fragmentary and it seems that any claim of uniqueness of the aqueous environment of simple solute must await more extensive data on a large number of solvents.

3.2. THE FULL PAIR CORRELATION FUNCTION $g_{SS}(R)$

For a one-component fluid there exist several methods of evaluating the pair correlation function $g(R)$. These range from experimental techniques, based on x-ray or neutron diffraction data, to various simulation techniques, such as the Monte Carlo and the molecular-dynamics methods, and to purely theoretical methods depending essentially on approximate integral equations for $g(R)$.

In proceeding to two-component systems, say of A and B, we face the problem of evaluating three functions $g_{AA}(R)$, $g_{AB}(R)$, and $g_{BB}(R)$ [we consider, for simplicity, two simple spherical molecules, where $g_{AB}(R)$ $= g_{BA}(R)$]. So far there is no experimental method for resolving these three functions from any experimental source. We are left, therefore, with either the computer–experimental or the purely theoretical methods. These methods are quite involved even for a mixture of simple spherical molecules for which all the pair correlation functions, $g_{\alpha\beta}(R)$, are functions of only the intermolecular distance R. In complex fluids, the pair correlation function depends on both the location and orientation of the pair of molecules. This poses tremendous difficulties in any theoretical method of evaluation of these functions.

We are interested in a two-component system: water and a simple solute S such as argon, methane, or the like. In particular, in the context of this book we are interested in the function $g_{SS}(R)$. (The other pair correlation functions g_{SW} and g_{WW} are also of interest in the general theory of aqueous solutions but will not concern us here.)

Besides the extreme technical difficulties that we encounter in any attempt to compute $g_{SS}(R)$ by theoretical means, there exists one more difficulty that should be borne in mind. This has to do with the unavailability of appropriate experimental data with which to compare the outcome of the theory. To appreciate the nature and extent of this difficulty we consider first a one-component simple fluid. Here there are several theoretical methods of computing $g(R)$. The ultimate test of the success of a theory is its ability to predict the form of the function $g(R)$ as close as possible to the experimental one. The main input in the theory of fluids is the pair potential $U(R)$, which has a relatively simple functional form for spherical nonpolar molecules. In the case of pure liquid water, we do have experimental data on $g(R)$. However, our knowledge of the pair potential for two water molecules is far from satisfactory. (There also exists the disturbing question of whether the pair potential for water is really the main ingredient for our theory. It is well known that higher-order potentials might be important, if not dominating, in determining the properties of this liquid.)

Therefore, one tries to guess the best form of the pair potential, use it in the theory, and compare the result obtained, say of $g(R)$, with the experimental data. The extent of agreement between the two results can serve as an index of our success in guessing a good pair potential and of the quality of the theoretical means employed in the course of the calculation.

The situation is far more difficult when we proceed to a two-component system, say of argon and water. Here we have to face not only all the difficulties involved in the choice of judicious pair potentials U_{SS}, U_{SW}, and U_{WW}, and the more lengthy and expensive computations, but at the end of the road, if we ever get there, we come out with a result, say of $g_{SS}(R)$, which we cannot compare with anything known in advance. Therefore, whatever the method we use, we cannot assess, even qualitatively, the extent of the success or failure of that particular theory.

The difficulties discussed above are quite serious and probably will not be overcome in the near future. It is for this reason that the main effort in the field of HI should be focused on the *experimental*, rather than on the theoretical side of the problem.

In the rest of this section we shall discuss very briefly some theoretical attempts to compute $g_{SS}(R)$. We shall present no details of the theories involved, since these would take us beyond the scope of the present book.

Perhaps the earliest attempt in this direction was the calculation of the cavity–cavity distribution function in water (Ben-Naim, 1969, 1972a). The calculations were based on Hill's (1958) original work on the problem of hole–hole distribution in simple fluids. We have seen in Section 3.1 that the function $y_{SS}(R)$ for two hard-sphere solutes in a solvent may be viewed also as the cavity–cavity pair correlation function. For cavities large enough to accommodate a real solute molecule, the computation of $y_{SS}(R)$ is extremely difficult. But there is one case in which such a computation becomes feasible, namely, for two cavities each of which has a radius equal to the radius of the solvent molecules. Such cavities formally correspond to a hard-sphere solute of zero diameter, and therefore bear no direct relation to any real solute molecules. Nevertheless, the result of this computation provides two rewards that may be relevant to larger cavities. In the first place the hole–hole distribution function in water has some features that are characteristic of the structure of liquid water. Second, this function may serve as a starting point for extensions to cavity–cavity distributions of more realistic radii.

Consider a fluid consisting of molecules having an effective hard-core radius r_0. We select a spherical cell of radius r_0 centered at some fixed point in the fluid. The probability of finding this cell occupied by the center of a molecule is

$$P_1 = \varrho \, \frac{4\pi r_0^3}{3} \tag{3.5}$$

Since a cell of this size may accommodate at most one center of a molecule at any given time, the probability of finding it empty is simply

$$P_0 = 1 - P_1 \tag{3.6}$$

Next we define the probability that two such cells at a given distance R from each other will be simultaneously empty. This probability is denoted by $P_{HH}(R)$, and the corresponding hole–hole correlation function is defined by

$$g_{HH}(R) = P_{HH}(R)/P_0^2 \tag{3.7}$$

For the particular radius r_0 of these cavities one can compute $g_{HH}(R)$ from the knowledge of the pair correlation function $g(R)$ of the pure liquid (Hill, 1958; Ben-Naim, 1969). The result obtained, though not relevant to any real solute particles, is not dependent on any modelistic assumptions on the pair potential for the liquid under consideration.

Figure 3.1 shows the function $g_{HH}(R^*)$ as a function of the reduced distance R^* (i.e., the distance divided by the diameter of the molecules, which was taken as 2.8 Å for water and 3.4 Å for argon).

The difference between the curve for argon (at 84.25 K and 0.71 bar) and for water (at 4°C and 1 bar) is quite striking. For argon we find two peaks at about $R^* \approx 1$ and $R^* \approx 2$, which correspond also to the locations of the peaks of $g(R)$ of argon. On the other hand in water we find one peak at about $R^* \sim 1.46$ (or $R = 4.1$ Å). This finding may indicate that because of the peculiarities of the structure of water, the first peak in the cavity–cavity distribution function, even for larger cavities, may be depen-

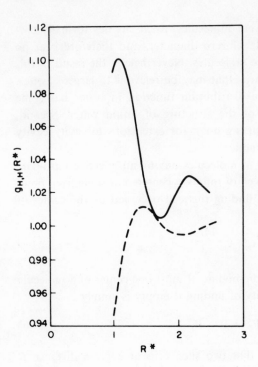

Figure 3.1. $g_{HH}(R^*)$ for water (dashed line) and for argon (solid line) as a function of the reduced distance $R^* = R/a$ (with $a = 2.8$ Å for water and $a = 3.4$ Å for argon).

dent more on the structure of the liquid than on the diameter of the solutes involved. This aspect of the problem deserves further careful study.

The next attempt to compute $g_{SS}(R)$ in a two-component mixture was carried out in a two-dimensional fluid (Ben-Naim, 1971a, 1972b). The basic idea was to see if $g_{SS}(R)$ had any peculiar features for a simple solute in waterlike particles in a two-dimensional system. The theoretical method used was a numerical solution of four integral equations for $g_{\alpha\beta}$ in this two-component system. Although the reduction of the problem to a two-dimensional system is a drastic simplification compared with the three-dimensional case, the computations are still prohibitively lengthy (and expensive). One interesting result that has been obtained from these calculations was that $g_{SS}(R)$ has indeed a high peak at $R^* \approx 1$. The variation of the function $g_{SS}(R)$ with the solute diameter was also studied. It was found that the peak at $R^* \approx 1$ is not a monotonic function of the molecular diameter of the solute (other parameters of the system being unchanged). There seems to be a higher peak for solute molecules that bears some relation to the size of the cavities that may be formed by the solvent molecules. These results are very tentative, however, not only because they were carried out in a two-dimensional system, but also because of many approximations that had to be introduced to make such a computation feasible.

Another attempt at such a computation in a three-dimensional system was reported by Dashevsky and Sarkisov (1974). Their water–water pair potential was constructed from atom–atom interactions (i.e., H–H, H–O, and O–O interactions situated on a pair of water molecules). They used the Monte Carlo technique to compute the free energy of the hydrophobic interaction [which is equivalent to the function $g_{SS}(R)$ or $y_{SS}(R)$; see Section 3.1]. As we have noted before, there are no experimental data with which one can compare the results from such a computation, but it is interesting to note that at least for one point, namely, for $R \approx 1.54$ Å, they obtained a result in good agreement with the experimental data, which we shall describe in Section 3.3.

A different approach to the problem has been suggested by Hermann (1974, 1975). The starting point for this theory is the first-order perturbation theory for liquids developed previously by Leonard et al. (1970). The main assumption is that the water–water pair potential may be described by a Lennard-Jones spherically symmetric potential function. This is quite a gross simplification since it is now well known that the strong angular dependence of the water–water pair potential has probably a dominant effect on the peculiarities of liquid water. Nevertheless with a judicious choice of the ε and σ parameters of the Lennard-Jones function a reasonably good agreement between computed and experimental values of the standard free energy of solution of hydrocarbons in water has been obtained. The calculations were also extended to include the function $\delta G^{\mathrm{HI}}(R)$ between two simple solutes in water. Unfortunately these results cannot be compared with experimental data, except for one value of $\delta G^{\mathrm{HI}}(R = 1.5$ Å$)$ (see Section 3.3) for which agreement with experimental results was achieved.

A somewhat more heuristic approach to the problem of HI has been suggested by Clark et al. (1977). This theory is based on the so-called Gurney model, which has been applied previously for ionic solutions by Friedman and Krishnan (1973).

Using the notation of Section 3.1 we write the free energy for the HI process as

$$\Delta G(R) = U(R) + \delta G^{\mathrm{HI}}(R) \tag{3.8}$$

where $\delta G^{\mathrm{HI}}(R)$ has been referred to as the indirect, or hydrophobic, part of the work. Within the present theory, the function $\delta G^{\mathrm{HI}}(R)$ is approximated by using the Gurney intuitive idea of cospheres of solvent molecules around each solute. These cospheres are not well-defined regions, but certainly there exists some region around the solute molecules in which the "structure" of the solvent is different from the structure of the pure solvent. When two such solutes approach each other, there is an overlap of the two cosphere regions assigned to the two solutes. This volume of overlap provides the basis for R dependence of the function $\delta G^{\mathrm{HI}}(R)$.

It is intuitively clear that in the cosphere region around the solute molecule the structure of the solvent is different from the structure of the solvent in the bulk. Furthermore, when two such cospheres overlap, the "structure" of the solvent in the overlapping region might again be different from the structure in each cosphere. The reason is that in this region the solvent molecules are affected by the combined field of force of the two solutes. (This statement applies to any possible definition of the concept of the structure of the solvent. We have intentionally avoided a discussion of this concept here, but we shall return to this topic in Chapter 5.)

The essential assumption of the theory is that the indirect part of the free energy change $\Delta G(R)$ may be related to the properties of the solvent in the overlapping region between the two cospheres of the solute molecules. However, because of our lack of knowledge of the properties of the solvent in this region, it is difficult to make a sound theory based on this approach. Furthermore, as in other heuristic approaches, there is no way of suggesting a systematic improvement on the "first-order trial"; it either succeeds or fails to reproduce the experimental results, a test that still has not been done.

Two more recent and promising approaches to the problem have been reported by Marčelja et al. (1977) and by Pratt and Chandler (1977). Both approaches start from well-founded elements of statistical mechanics. They differ markedly in the type of approximations that are introduced to proceed from the exact formalism to the end product. These are quite advanced methods and their description falls beyond the scope and level of this book.

Comment

This section has been addressed to a reader who is nonexpert in the theory of liquids. The main reason for including such a section in this book is to give the reader a general feeling for the extent of the complexity of the theoretical approaches to the problem of HI. The difficulties presented in this section will not be easily overcome in the foreseeable future. Therefore, the readers are advised to be very suspicious about the apparent success of one theory or another. All of them involve drastic approximations, and it is almost impossible at present to assess their validity or usefulness.

Perhaps with the development of a new generation of superfast computers one could overcome some of the technical difficulties involved in either the solution of integral equations or in executing a simulation of the behavior of aqueous solutions. One may also hope that new ingenious theoretical tools will be developed that will be appropriate to deal with the theory of hydrophobic interactions, and that are not a straightforward adaptation of theories that were developed for simple fluids. Until that end is reached, we still have a great deal of work to do on the experimental side. A survey of what is presently known is contained in the rest of this chapter.

3.3. VALUES OF δG^{HI} AT ONE SPECIFIC DISTANCE

We recall the fundamental relation between $g_{SS}(R)$ and $\Delta G(R)$,

$$g_{SS}(R) = \exp[-\Delta G(R)/kT] \tag{3.9}$$

and similarly the corresponding relation for the indirect part of the free energy change:

$$y_{SS}(R) = \exp[-\delta G^{HI}(R)/kT] \tag{3.10}$$

$$\delta G^{HI} = \Delta G(R) - U_{SS}(R) \tag{3.11}$$

As noted in Chapter 2, the ultimate function that should be computed in the HI problem is $\Delta G(R)$, since this is the function that governs the probability distribution for the solute–solute distances. However, the quantity $\delta G^{HI}(R)$ is the most important part of $\Delta G(R)$ since it is this part that reflects the properties of the solvent, while $U_{SS}(R)$ is presumed to be invariant to environmental changes (this is always assumed in the classical theory of fluids, without which little progress could have been achieved in this field). This does not necessarily mean that δG^{HI} is numerically the largest part of ΔG for those values of R that are of importance to us (i.e., $R \approx \sigma$, where σ is the molecular diameter of the solute molecule). To put it in other words, $U_{SS}(R)$ is presumed to be a relatively simple and fairly well-known function for simple molecules. A good description of this function may be obtained by a Lennard-Jones type of function which has the form

$$U_{SS}(R) = 4\varepsilon[(\sigma/R)^{12} - (\sigma/R)^6] \tag{3.12}$$

where ε is an energy parameter and σ can be identified with the hard-core diameter of the solute. Once we have agreed upon the form of $U_{SS}(R)$, there still remains the most difficult task of finding any means by which information on δG^{HI} may be obtained. It is for this reason that we shall now focus our attention on a single value of R for which $\delta G^{\mathrm{HI}}(R)$ may be estimated. It turns out that the very fact that we have eliminated the direct part U_{SS} from $\Delta G(R)$ leads to an interesting way of approximating δG^{HI}.

In Section (1.3) we derived a general relation between $\delta G^{\mathrm{HI}}(R)$ and the standard free energies of solution

$$\delta G^{\mathrm{HI}}(R) = \Delta\mu_R^\circ - 2\Delta\mu_S^\circ \tag{3.13}$$

where $\Delta\mu_S^\circ$ is the *experimental* standard free energy of solution of the solute S, and $\Delta\mu_R^\circ$ is formally the standard free energy of solution of a *pair of solutes* at a fixed separation R, which we have called a "dimer" in Section 1.3. Clearly, since such a "dimer" does not exist as a molecular entity, we cannot evaluate $\Delta\mu_R^\circ$ experimentally. It is at this point that statistical mechanics provides its important aid. We shall present here a rather brief and qualitative argument; more details are to be found in Ben-Naim (1974, Chapter 8).

We write the statistical mechanical expressions for $\Delta\mu_R^\circ$ and $\Delta\mu_S^\circ$ as

$$\Delta\mu_S^\circ = -kT\ln\langle\exp(-B_S/kT)\rangle_0 \tag{3.14}$$

$$\Delta\mu_R^\circ = -kT\ln\langle\exp[-B_{SS}(R)/kT]\rangle_0 \tag{3.15}$$

where B_S is the binding energy of a single solute to all the solvent molecules for some particular configuration, and the symbol $\langle\ \rangle_0$ stands for an average over all possible configurations of the solvent molecules. Similarly $B_{SS}(R)$ is the binding energy of a *pair of solutes* at a fixed separation R. More specifically we write

$$B_{SS}(R) = \sum_{i=1}^{N} [U(\mathbf{X}_i, \mathbf{R}_1) + U(\mathbf{X}_i, \mathbf{R}_2)] \tag{3.16}$$

where $R = |\mathbf{R}_1 - \mathbf{R}_2|$, $U(\mathbf{X}_i, \mathbf{R}_k)$ is the solute–solvent pair potential at the specified configuration, and the summation is over all the solvent molecules.

Combining (3.14) and (3.15) with (3.13), we obtain

$$\delta G^{\mathrm{HI}}(R) = -kT\ln\frac{\langle\exp[-B_{SS}(R)/kT]\rangle_0}{\langle\exp(-B_S/kT)\rangle_0^2} \tag{3.17}$$

To get an approximation for $\delta G^{\mathrm{HI}}(R)$ we have to find a way of relating the numerator of (3.17) to experimental quantities. This is rendered possible since the function $\delta G^{\mathrm{HI}}(R)$ is a smooth function of R even for values of R smaller than $R \approx \sigma$, i.e., whereas the function $\varDelta G(R)$ steeply increases for small values of R, it does so because of the strong repulsive forces originating from U_{SS}; the indirect part of $\varDelta G(R)$, however, remains finite up to $R = 0$. This argument is the basis for obtaining an approximate value of δG^{HI} at one distance. Note that once the direct solute–solute interaction has been eliminated in the definition of δG^{HI} we no longer have to worry about the question of impenetrability of the two solutes into each other. The simplest example of such a situation would be two hard-sphere solutes, for which the solute–solute interaction is of the form

$$U_{SS}(R) = \begin{cases} \infty & \text{for } R \le \sigma \\ 0 & \text{for } R > \sigma \end{cases} \tag{3.18}$$

Clearly two such solutes *cannot* approach each other closer than $R = \sigma$. However, in (3.17) this function does not play any role, and $B_{SS}(R)$ can be equivalently viewed as either the binding energy of two hard spheres or the binding energy of two cavities of suitable diameters (i.e., two regions that are impenetrable to the centers of the solvent molecules). Adopting the two-cavities point of view, we can bring their centers to any required distance, even up to $R = 0$, without violating any physical rules. In the case of real solutes, the notion of a "cavity" may be replaced by the notion of a "field of force" exerted by the solute molecules on the solvent. In this case the centers of the two "fields of force" originating from the two solute molecules may be moved to any required distance. With this brief description of one property of the numerator in (3.17) we now turn to a concrete example.

Consider two methane molecules, which for all our purposes may be viewed as spherical molecules, the centers of which are $\mathbf{R}_1$ and $\mathbf{R}_2$. Clearly, for two *real* solutes the work required to bring these two solutes to a separation R considerably smaller than the molecular diameter $\sigma \sim 3.82$ Å would be extremely large. However, since we are interested not in the direct interaction between the two solute molecules, but in the indirect part of the work, i.e., the part that reflects the properties of the solvent, we can choose any distance $R = |\mathbf{R}_2 - \mathbf{R}_1|$ for which relation (3.17) may be converted to a more convenient form. More specifically, we choose $R = \sigma_1 = 1.53$ Å, which is the C–C equilibrium distance in ethane. At this particular distance the binding energy of the *two methane* molecules is approximated by the binding energy of *one* ethane molecule. In mathematical terms the nature of our approximation is

$$B_{SS}(R = \sigma_1) \approx B_{\mathrm{Et}} \tag{3.19}$$

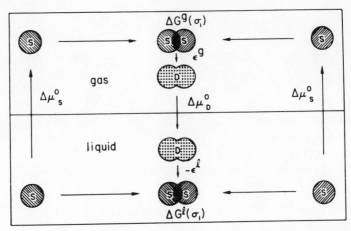

Figure 3.2. A modification of the thermodynamic cycle of Figure 1.5. Two methane particles S are brought from fixed positions and infinite separation to a close separation $R = \sigma_1$. A different route for the same process is to transfer the two solutes into the gaseous phase, bringing them to the separation $R = \sigma_1$. At this stage the "dimer" of solute particles is replaced by one ethane molecule. This molecule is transferred into the liquid and then replaced by the "dimer" of solute molecules at $R = \sigma_1$.

where B_{Et} is the binding energy of ethane to the entire solvent for a specific configuration. The situation is schematically illustrated in Figure 3.2. (A more detailed discussion on the nature of this approximation is given in Appendix A.6.)

Having replaced $B_{SS}(R = \sigma_1)$ by B_{Et} in (3.17) we can immediately identify the numerator as the standard free energy of solution of ethane, i.e.,

$$\varDelta\mu^{\circ}_{\mathrm{Et}} = -kT \ln\langle\exp(-B_{\mathrm{Et}}/kT)\rangle_0 \qquad (3.20)$$

and relation (3.17) is converted into the approximate relation

$$\delta G^{\mathrm{HI}}(\sigma_1) = \varDelta\mu^{\circ}_{\mathrm{Et}} - 2\varDelta\mu^{\circ}_{\mathrm{Me}} \qquad (3.21)$$

where the HI is expressed in terms of the standard free energies of solution of ethane and methane, which are experimentally determinable quantities.

To summarize, we have used an approximation in (3.19), which is a reasonable one for solutes having weak attractive interactions with the solvent. The outcome of this approximation is a useful relationship between an essentially molecular quantity, $\delta G^{\mathrm{HI}}(\sigma_1)$, and experimentally measurable quantities. Having a method for estimating $\delta G^{\mathrm{HI}}(\sigma_1)$ for two methane molecules, we can immediately turn to study the effect of various factors such as temperature, pressure, or addition of solutes on the HI. It is true, though, that we would have liked to know the effect of all these factors on $\delta G^{HI}(R \approx \sigma)$; but since we do not have any way of estimating this quantity, we shall, for the present, satisfy ourselves with studying the quantity $\delta G^{HI}(\sigma_1)$, hoping that its response to the variation of external conditions will be similar to the response of the quantity $\delta G^{HI}(\sigma)$.

We now turn to presenting some examples of $\delta G^{HI}(R = \sigma_1)$ in several solvents. All the quantities are computed from relation (3.21) using experimental data on the standard free energy of solution of methane and ethane. Figure (3.3) exhibits values of $\delta G^{HI}(\sigma_1)$ for water and some other non-aqueous solvents, for which the relevant experimental quantities are available. Two prominent differences between water and all the other liquids are clearly conspicuous. In the first place the absolute magnitude of $\delta G^{HI}(\sigma_1)$ is larger in water than in all the other solvents. (This, of course, does not

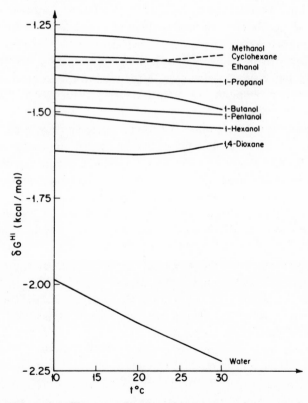

Figure 3.3. Values of $\delta G^{HI}(\sigma_1)$ as a function of temperature, for two methane molecules in various solvents.

necessarily mean that other solvents not represented in this figure would not exhibit an "anomalous" behavior similar to water. Our conclusions here are drawn only for the limited data that are available to us at present.) A typical difference, say between water and ethanol at 25°C, is about 0.5 kcal/mol, which is of the order of kT at this temperature. From this difference one can conclude that the indirect work of bringing two methane molecules from infinite separation to $R = \sigma_1$ is significantly more negative in water as compared to, say, ethanol. In probability terms, using relation (3.10) we get

$$\frac{y_{SS}(\sigma_1, \text{ in water})}{y_{SS}(\sigma_1, \text{ in ethanol})} = \exp(1.36) = 3.9 \qquad (3.22)$$

The ratio of $y_{SS}(R)$ in two solvents (and at infinite dilution with respect to S) is equal to the ratio of the probabilities of finding the two solutes at the distance R from each other. Hence the last result may be interpreted as follows: For two methane molecules, of an effective hard-core diameter $\sigma \sim 3.82$ Å, the probability of finding them at a distance $R \sim \sigma_1 = 1.53$ Å is almost zero. However, the ratio of such probabilities in two different solvents is equal to the ratio of the functions $y_{SS}(R)$, in which the direct pair potential has been canceled out. Hence we can assign meaning to the *ratio* of the two probabilities of finding the two solutes at $R = \sigma_1$ in the two solvents. The result (3.22) means that if we have two methane molecules in water and in ethanol and the direct solute–solute pair potential has been "switched-off," then the probability of finding the two molecules at $R = \sigma_1$ is about four times larger in water than in ethanol. Similar results would have been obtained for all the other solvents represented in Figure 3.3.

We stress, however, that from $\delta G^{\mathrm{HI}}(\sigma_1)$ we cannot infer anything about $\delta G^{\mathrm{HI}}(\sigma)$. But we believe that differences of $\delta G^{\mathrm{HI}}(\sigma_1)$'s reflect in a qualitative manner the differences of $\delta G^{\mathrm{HI}}(\sigma)$'s in the same pair of solvents.

The second important difference between water and all the other solvents in Figure 3.3 is the temperature dependence of $\delta G^{\mathrm{HI}}(\sigma_1)$. In all of the nonaqueous solvents the temperature dependence of $\delta G^{\mathrm{HI}}(\sigma_1)$ is very small, whereas in water we have a clear-cut large and negative slope of $\delta G^{\mathrm{HI}}(\sigma_1)$ as a function of temperature.

We believe that the two properties mentioned above are the most important characteristics of the HI. Though we certainly have to be careful with the interpretation of $\delta G^{\mathrm{HI}}(\sigma_1)$, we feel that it is quite safe to draw conclusions regarding the ratios of the values of $\delta G^{\mathrm{HI}}(\sigma_1)$ in two solvents as representing the ratio of δG^{HI} at some more realistic distances. We shall

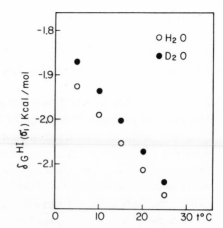

Figure 3.4. Values of $\delta G^{\mathrm{HI}}(\sigma_1)$ as a function of temperature in H_2O and D_2O [based on data from Ben-Naim *et al.* (1973)].

see in Chapter 4 that for the HI among more than two solute molecules we can improve the reliability of our estimates of the HI for more realistic configurations.

Figure 3.4 presents values of $\delta G^{\mathrm{HI}}(\sigma_1)$ for H_2O and D_2O. Clearly the values are larger (in absolute magnitude) in H_2O, which means that the tendency to "dimerize" is larger in H_2O as compared to D_2O, or the HI is larger in H_2O than in D_2O. This is consistent with the results we have obtained from the solubilities of methane in the two solvents, where we concluded (Section 2.3) that the "phobia" of methane for H_2O is larger than the "phobia" for D_2O. It should be stressed, however, that such a consistency should not be expected to hold in every case, and an example to the contrary is given in Section 3.4.

We conclude this section with a brief comment on a topic that is of some theoretical rather than practical value. The reader who is still wondering how we could dare to "fuse" two methane molecules to form an ethane molecule (which actually we did not do!) is warned of an even more profound shock. Instead of two methane molecules, we start with two hard spheres at $R = \infty$. Instead of bringing the two molecules to $R = 1.53$ Å, we bring the two hard spheres to *zero separation*, which is clearly an impossible process, since two hard spheres cannot penetrate one another. Yet relation (3.17) is still valid for any R, including $R = 0$. This is equivalent to the process of bringing two cavities of suitable size to zero separation. But two cavities at zero separation are indistinguishable from one cavity at a fixed position, i.e., for $S =$ hard sphere, we have

$$\langle \exp[-B_{SS}(R = 0)/kT] \rangle_0 = \langle \exp(-B_S/kT) \rangle_0 \qquad (3.23)$$

and relation (3.17) reduces to

$$\delta G^{\mathrm{HI}}(R=0) = -kT \ln \frac{\langle \exp(-B_S/kT) \rangle_0}{\langle \exp(-B_S/kT) \rangle_0{}^2}$$
$$= -\Delta\mu^{\circ}_{\mathrm{HS}} \tag{3.24}$$

where $\Delta\mu^{\circ}_{\mathrm{HS}}$ is the standard free energy of solution of a hard-sphere solute. The physical meaning of Equation (3.24) is simple: the work required to bring two identical cavities from infinite separation to zero separation is the same as the work required to eliminate one cavity, which, in turn, is the same as minus the standard free energy of solution of a hard-sphere solute of a suitable diameter. This result is one *exact* relation between HI in the sense of Section 1.3 and the measure of the "phobia" for the solvent in the sense of Section 2.2. One more relation of this kind will be discussed in Section 4.6.

3.4. EFFECT OF VARIATION OF THE SOLVENT ON $\delta G^{\mathrm{HI}}(\sigma_1)$

In Section 3.3 we developed an approximate measure of the strength of the HI. We stressed that though the approximation itself is a reasonable one, it does not provide information on HI at the distance $R \approx \sigma$ that is of most interest to us, but at a far shorter distance $R = \sigma_1$. However, having at present no other direct and reliable measure of the HI at the required distances, we resort to what is available and adopt the quantity $\delta G^{\mathrm{HI}}(\sigma_1)$ as a simple, convenient, and useful probe for characterizing the HI, and in particular its response to variations of the thermodynamic parameters such as temperature, pressure, and solvent composition. This section reports on some representative effects of added solutes on the HI. Temperature and pressure effects are discussed in Chapter 5.

Table 3.1 provides some values of $\delta G^{\mathrm{HI}}(\sigma_1)$ for aqueous solutions of simple electrolytes at one temperature. We also report the corresponding values of the entropy and the enthalpy of the HI; these are defined in the usual way by

$$\delta S^{\mathrm{HI}} = -\frac{\partial \delta G^{\mathrm{HI}}}{\partial T} \tag{3.25}$$

$$\delta H^{\mathrm{HI}} = \delta G^{\mathrm{HI}} + T\delta S^{\mathrm{HI}} \tag{3.26}$$

Note that since $U_{SS}(R)$ is presumed to be temperature independent, δS^{HI} is the entropy change for the HI process (i.e., there is no distinction

Table 3.1

Values of $\delta G^{\mathrm{HI}}(\sigma_1)$ (in cal/mol), δS^{HI} (in e.u.), and δH^{HI} (in cal/mol) for Various Aqueous Solutions of Simple Electrolytes, at $t = 10°C^a$

Solute (1 M)	δG^{HI}	δS^{HI}	δH^{HI}
Pure water	−1990	11	1100
NaCl	−2167	10	700
NaBr	−2184	11	900
NaI	−2190	12	1200
LiCl	−2113	12	1300
KCl	−2157	12	1200
CsCl	−2140	12	1300
NH$_4$Cl	−2095	11	1000

^{a} Based on data from Ben-Naim and Yaacobi (1974).

between the derivative of $\Delta G(R)$ and of $\delta G^{\mathrm{HI}}(R)$ with respect to the temperature). On the other hand, δH^{HI} is the enthalpy change corresponding to the indirect part of the HI. In order to obtain the enthalpy change for the HI process we have to add $U_{SS}(R)$ to $\delta H^{\mathrm{HI}}(R)$.

From Table 3.1 we see that all the electrolytes strengthen the HI relative to pure water (i.e., δG^{HI} becomes more negative). Although variation in the effect of the various salts is rather small, we note the order of these salts in regard to their effect on the HI:

$$NaCl > KCl > CsCl > LiCl > NH_4Cl$$

On the other hand, if we fix the cation and change the anion we get the following order:

$$NaCl < NaBr < NaI$$

Clearly there is a need for more detailed experimental work in order to establish the relation between variation of the size of the ions and their effect on HI. It is also worth noting that whereas some of the tetraalkylammonium salts have a *salting-in* effect on the solubility of methane and ethane, their effect on the HI seems to be similar to that of the simpler salts. Figure 3.5 shows some values of $\delta G^{\mathrm{HI}}(\sigma_1)$ for aqueous solutions of tetraalkylammonium salts as a function of the molarity of the salt (at 30°C).

Figure 3.6 also shows the dependence of $\delta G^{\mathrm{HI}}(\sigma_1)$ on the concentration of NaCl and urea. In both cases it is clear that the strength of the HI

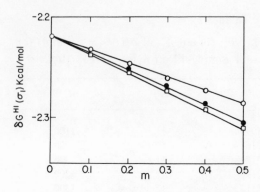

Figure 3.5. Values of $\delta G^{HI}(\sigma_1)$ as a function of the molarity m of the salts at 30°C. ●, $(C_4H_9)_4NBr$; ○, $(C_3H_7)_4NBr$; □, $(OHCH_2CH_3)_4NBr$ [from Tenne and Ben-Naim (1976)].

increases (i.e., δG^{HI} becomes more negative) almost linearly as the solute concentration increases. The data for urea are of particular importance since its effect on various biochemical processes such as aggregation of biopolymers or conformational changes has been studied quite extensively. It is important to remember, though, that we are using one particular measure for the HI, namely, $\delta G^{HI}(\sigma_1)$, and it is certainly possible that other results could have been obtained by using different measures of HI.

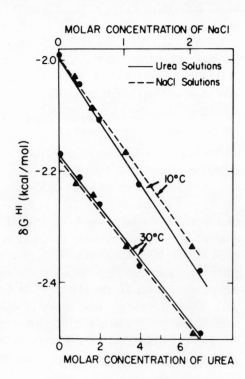

Figure 3.6. Values of $\delta G^{HI}(\sigma_1)$ as a function of the molarity of urea (lower scale) and the molarity of NaCl (upper scale) for aqueous solutions at two temperatures [based on data from Ben-Naim and Yaacobi (1974)].

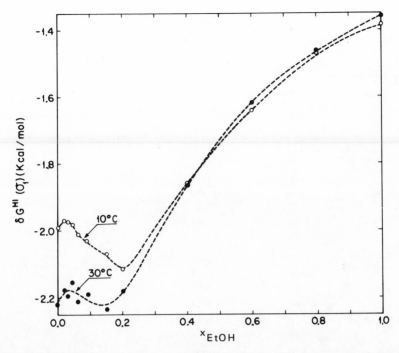

Figure 3.7. Variation of the strength of the HI with the mole fraction of ethanol in water–ethanol mixtures at two temperatures [based on data from Yaacobi and Ben-Naim (1973)].

The best way of illustrating the effect of a nonelectrolyte "solute" on the HI is to follow the change of δG^{HI} upon changing the mole fraction of an organic component in a two-component solvent system. Figure 3.7 shows the variation of δG^{HI} with the mole fraction of ethanol at two temperatures. The corresponding values of δG^{HI}, δS^{HI}, and δH^{HI} for the water–ethanol system are presented in Table 3.2. At the lower temperature (10°C) there is a clear-cut positive initial slope to the curve of δG^{HI} near $x = 0$. We may conclude that addition of *very* small quantities of ethanol causes a *weakening* of the HI. This effect is restricted to the region of $0 \leq x_{EtOH} \lesssim 0.03$; at higher concentrations of alcohol there is a pronounced increase in the strength of the HI, which persists up to about $x_{EtOH} \approx 0.2$; beyond that the curve turns upwards and changes monotonically until it reaches the relatively weak HI in pure alcohol.

We also note that the temperature dependence is large and negative in the water-rich region ($0 \simeq x_{EtOH} \lesssim 0.2$), whereas in all of the remaining region it is small and changes sign at about $x \approx 0.4$.

Table 3.2

Values of $\delta G^{HI}(\sigma_1)$ (in cal/mol), δS^{HI} (in e.u.), and δH^{HI} (in cal/mol) for Various Mixtures of Water and Ethanol

x_{EtOH}	t (°C)	δG^{HI}	δS^{HI} (15°C)	δH^{HI} (15°C)
0.00	10	−1990	11	1500
	30	−2220		
0.02	10	−1971	13	1700
	30	−2174		
0.03	10	−1974	12	1300
	30	−2197		
0.045	10	−1979	11	1200
	30	−2157		
0.06	10	−2010	9	800
	30	−2213		
0.09	10	−2029	10	1000
	30	−2193		
0.12	10	−2083	8	100
	30	−2186		
0.15	10	−2073	5	−800
	30	−2235		
0.20	10	−2115	1	−1600
	30	−2182		
0.40	10	−1862	1	−1700
	30	−1862		
0.60	10	−1640	0	−1400
	30	−1618		
0.80	10	−1472	1	−1400
	30	−1470		
1.00	10	−1387	0	−1200
	30	−1362		

The general form of the curve of $\delta G^{HI}(\sigma_1)$ is similar to the curve of $\Delta\mu^\circ_{tr}$ for transferring methane from pure water into a mixture of water and ethanol (see Section 2.4). However, a close examination shows that that the conclusion drawn from the two quantities regarding HI may be different.

As a specific example let us first consider ethanol and water as the two phases. The standard free energy of transferring a methane molecule from water to ethanol at $t = 10°C$ is

$$\Delta\mu^{\circ}_{\mathrm{tr}}(W \to E) = -1.42 \text{ kcal/mol} \tag{3.27}$$

This result was interpreted in Section 2.3 in terms of the greater "phobia" of methane for water than for ethanol. If we look at values of $\delta G^{\mathrm{HI}}(\sigma_1)$ in these two solvents, we get a result that is consistent with the above conclusion, i.e., we find '

$$\delta G^{\mathrm{HI}}(\sigma_1, \text{in water}) = -1.99 \text{ kcal/mol}$$
$$\delta G^{\mathrm{HI}}(\sigma_1, \text{in ethanol}) = -1.39 \text{ kcal/mol} \tag{3.28}$$

Thus the tendency to form "dimers" is greater in water than in ethanol. Both results (3.27) and (3.28) thus indicate that the methane molecule tends to avoid the aqueous environment more than it does the ethanol environment.

The point we would like to demonstrate now is that the two quantities $\Delta\mu^{\circ}_{\mathrm{tr}}$ and δG^{HI} do not *necessarily* convey consistent information. As an example, consider pure water (W) and a mixture of water and ethanol with a mole fraction of ethanol $x = 0.2$, which we refer to as the mixture (M). We have the following experimental results:

$$\Delta\mu^{\circ}_{\mathrm{tr}}(W \to M) = -0.17 \text{ kcal/mol} \tag{3.29}$$

and

$$\delta G^{\mathrm{HI}}(\sigma_1, \text{in water}) = -1.99 \text{ kcal/mol}$$
$$\delta G^{\mathrm{HI}}(\sigma_1, \text{in mixture}) = -2.11 \text{ kcal/mol} \tag{3.30}$$

Thus, (3.29) means that the "phobia" of methane is greater for water than for the mixture. But from (3.30) we are tempted to conclude that the tendency to avoid the solvent is greater in the mixture than in pure water. This paradox can be easily resolved by noting that $\Delta\mu^{\circ}_{\mathrm{tr}}(W \to M)$ reflects the change of the *entire* surroundings of methane from W to M. On the other hand, $\delta G^{\mathrm{HI}}(\sigma_1, \text{in water})$ (or any standard free energy of dimerization) reflects only a *partial* change in the environment of the two methane molecules. Each methane molecule is initially surrounded by pure water, and finally only a part of its surface is exposed to water and a part to the second methane molecule. Thus the two quantities $\Delta\mu^{\circ}_{\mathrm{tr}}$ and δG^{HI} inherently measure two different processes, and in general they are not expected to provide the same kind of information on the solvation properties of the system.

Table 3.3

Values of $\delta G^{HI}(\sigma_1)$ (in cal/mol), δS^{HI} (in e.u.), and δH^{HI} (in cal/mol) for Various Mixtures of Water and p-Dioxane

x_{dioxane}	t (°C)	δG^{HI}	δS^{HI} (15°C)	δH^{HI} (15°C)
0.00	10	−1990.2	11	1500
	25	−2167.7		
0.015	10	−2009.0	12	1300
	25	−2177.0		
0.03	10	−2046.0	10	800
	25	−2200.0		
0.06	10	−2098.6	8	1100
	25	−2211.0		
0.09	10	−2128.4	10	400
	25	−2219.4		
0.12	10	−2163.0	5	−800
	25	−2233.0		
0.15	10	−2147.0	6	−800
	25	−2206.6		
0.20	10	−2136.0	2	−1300
	25	−2169.2		
0.40	10	−1929.4	1	−1800
	25	−1945.5		
0.70	10	−1768.7	0	−2100
	25	−1763.8		
1.00	10	−1612.0	1	−1400
	25	−1611.6		

Table 3.3 reports the values of δG^{HI}, δS^{HI}, and δH^{HI} for the water–dioxane system. The general dependence of δG^{HI} on the mole fraction of the organic component is very similar to the case of the water–ethanol system. The only difference is in the very-low-concentration region. This is demonstrated in Figure 3.8, where the quantity of δG^{HI} (in mixture) − δG^{HI} (in water) is drawn as a function of the mole fraction of the organic component. Clearly a very small quantity of ethanol weakens the HI, whereas dioxane has an opposite effect in the same concentration range.

We summarize the results reported in this section with a qualitative, and somewhat risky, conclusion. Suppose we take pure water at room temperature as our reference solvent and make a small variation in its "structure," by either changing the temperature, replacing it by D_2O, or adding solutes. We presume here that we have in mind some definition of the concept of the "structure of water," and also that we "know" how this structure changes upon changing the temperature, replacement by D_2O, or adding solute. Thus, if one assumes that increasing the temperature leads to a *decrease* in structure, replacement of H_2O by D_2O *increases* the structure, addition of ionic solutes *decreases* the structure, and addition of ethanol *increases* the structure, then we may conclude that whenever we *increase* the "structure of the water" we *weaken* the strength of the HI and vice versa. The above conclusion is certainly very qualitative since we have not defined the structure of water and we have not proved that the

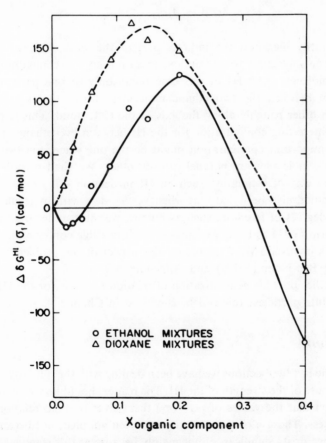

Figure 3.8. Values of $\Delta\delta G(\sigma_1) = \delta G(\sigma_1$, in mixture$) - \delta G(\sigma_1$, in water$)$ as a function of the mole fraction of the organic component, for water–ethanol and water–dioxane mixtures at 10°C.

above changes indeed occur. Nevertheless, what we have noted above is in qualitative conformity with what is believed to be true by many researchers in the field, and even if the details have not been supported by any proof, we find it worthwhile noting it for further thought and perhaps as a challenge for new research. Some discussion of the concept of the structure of water and its relevance to the problem of HI is to be found in Chapter 5.

We have discussed in the last two sections (Sections 3.3 and 3.4) one measure of the HI between two solute particles. It is clear that the same type of approximate measure may be used for other solutes to obtain other measures of the HI. For example, taking two ethane molecules and a butane molecule we may define the quantity

$$\delta G^{\mathrm{HI}} = \Delta\mu_{\mathrm{Bu}}^{\circ} - 2\Delta\mu_{\mathrm{Et}}^{\circ} \tag{3.31}$$

This quantity measures the indirect part of the work required to bring two ethane molecules from infinite separation to the configuration of a butane molecule. The HI in this case refers only to one particular configuration between the two ethane molecules: where the two molecules face each other roughly along the axis of the C–C bond. This is probably the less interesting configuration for the HI between two ethane molecules. A more important configuration might be the one where the two axes of the ethane molecules are parallel to each other. We shall describe an approximate way of estimating such an HI in Section 4.7.

It should also be noted that whenever we deal with molecules having internal degrees of freedom, such as butane, we must understand relations of the form (3.31) as being averages over all possible conformations of the molecules involved. More details on this aspect of the problem are to be found in Ben-Naim (1972a) and Appendix A.7.

Another possible generalization of relation (3.31) is for the HI among many solute particles; this will be discussed in Chapter 4.

Comment

In the last two sections we have been dealing with the quantity $\delta G^{\mathrm{HI}}(\sigma_1)$ as a measure of the strength of the HI. The reader should be aware, however, of the fact that the values of $\delta G^{\mathrm{HI}}(\sigma_1)$ themselves bear no relation to any real process. The use of this quantity has been adopted, not because of its importance or its significance, but mainly because at this stage we have no better measure of the HI between two simple solutes in a solvent. The application of this quantity would certainly be obsolete once we can devise a better one to replace it. For the moment the emphasis should be focused not on the absolute values of $\delta G^{\mathrm{HI}}(\sigma_1)$ but on differences between these quantities in two media. This difference is a more significant quantity for comparing the properties of two solvents.

Finally, a comment on the apparent uniqueness of water as a medium for HI is in order. We have seen that values of $\delta G^{\mathrm{HI}}(\sigma_1)$ are relatively more negative in water than in some other solvents. But the number of solvents for which the relevant data are available is quite small. It is very possible that HI might be large, or even larger than in water, in other liquids such as hydrofluoric acid, sulfuric acid, hydrogen peroxide, hydrazine, glycerol, etc. These are some liquids in which the values of $\delta G^{\mathrm{HI}}(\sigma_1)$ (or any other measure of HI) might be measured before we can conclude that HI is really a unique property of aqueous fluids.

3.5. SOLUTE–SOLUTE AFFINITY AT INFINITE DILUTION

The term "solute–solute affinity" was introduced in Section 1.5. Here we present one aspect of this concept that is relevant to the problem of HI.

We start by citing the well-known virial expansion of the pressure of a real gas in a power series in the (number) density of the molecules

$$P/kT = \varrho + B_2 \varrho^2 + B_3 \varrho^3 + \cdots \qquad (3.32)$$

where B_K is referred to as the Kth virial coefficient of the gas. One of the most elegant results of statistical mechanics is that it provides general expressions for the B_K in terms of the properties of exactly K molecules in a given volume V and at a temperature T. In particular, and of foremost importance, the second virial coefficient $B_2(T)$ of a system of spherical molecules S is given by

$$B_2(T) = -\frac{1}{2} \int_0^\infty \{\exp[-U_{SS}(R)/kT] - 1\} 4\pi R^2 \, dR \qquad (3.33)$$

where $U_{SS}(R)$ is the pair potential operating between a pair of S molecules. Thus knowledge of $U_{SS}(R)$ is sufficient for the computation of $B_2(T)$, and hence the first-order deviation from the ideal gas law. The inverse problem is a more difficult one, i.e., even if we have full information on $B_2(T)$ as a function of temperature, it is not possible to determine $U_{SS}(R)$ uniquely (and certainly not so for nonspherical molecules). Nevertheless, information on $B_2(T)$ is still valuable as a sort of overall measure of the strength of the intermolecular forces operating between pairs of molecules.

By analogy with the expansion of the pressure of real gases in (3.32) there exists a formally similar expansion for the osmotic pressure of a solution. For simplicity we shall be discussing a two-component system of a solvent W (say water) and a simple spherical solute S, the number density of which is $\varrho_S = N_S/V$. The so-called virial expansion of the osmotic pressure reads

$$\pi/kT = \varrho_S + B_2{}^*\varrho_S{}^2 + B_3{}^*\varrho_S{}^3 + \cdots \qquad (3.34)$$

where $B_K{}^*$ is referred to as the Kth virial coefficient of the osmotic pressure. In a two-component system $B_2{}^*$ is determined by the properties of a system of pure solvent containing exactly two solute molecules S. The formal analogy between the expansion (3.32) and (3.34) has been shown by McMillan and Mayer (1945). The statistical mechanical expression for $B_2{}^*$ is

$$B_2{}^* = -\frac{1}{2}\int_0^\infty [g_{SS}^\circ(R) - 1]4\pi R^2\,dR$$

$$= -\frac{1}{2}\int_0^\infty \{\exp[-\varDelta G(R)/kT] - 1\}4\pi R^2\,dR$$

$$= -\frac{1}{2}G_{SS}^\circ \qquad (3.35)$$

where $g_{SS}^\circ(R)$ is the solute–solute pair correlation function at infinite dilution defined by

$$g_{SS}^\circ(R) = \lim_{\varrho_S\to 0} g_{SS}(R) \qquad (3.36)$$

Clearly $B_2{}^*$ may be viewed as a kind of an average of the overall solute–solute correlation—which we prefer to refer to as the average affinity between the solute molecules as discussed in Chapter 1. [In the literature one often finds the term "solute–solute interaction" in a more general sense, i.e., in the context of any manifestation of deviations from the ideal dilute behavior. In this book we reserve the term "solute–solute interaction" for the *direct* pair potential $U_{SS}(R)$ only. Clearly in $B_2{}^*$ (or in G_{SS}°) there is more than just the direct solute–solute interaction.]

Experimentally, $B_2{}^*$ may be determined from the limiting slope of the function

$$F(\varrho_S) = (\pi/kT - \varrho_S)/\varrho_S \qquad (3.37)$$

i.e., if we plot $F(\varrho_S)$ as a function of ϱ_S and measure the initial slope at $\varrho_S = 0$ we obtain

$$B_2^* = \lim_{\varrho_S \to 0} \left[\frac{\partial}{\partial \varrho_S} F(\varrho_S) \right] \tag{3.38}$$

This can be simply verified by substituting (3.34) into (3.37), performing the derivative, and taking the limit $\varrho_S \to 0$.

The second virial coefficient of the osmotic pressure may also be determined by any other thermodynamic quantity that measures deviations from the ideal dilute solution behavior. Perhaps the most general quantity that is relevant to the present context is the activity coefficient of the solute, which we denote by γ_S^D. Using the thermodynamic identity

$$\left[\frac{\partial(\mu_S/kT)}{\partial \varrho_S} \right]_{T,\mu_W} = \frac{1}{\varrho_S} \left[\frac{\partial(\pi/kT)}{\partial \varrho_S} \right]_{T,\mu_W} \tag{3.39}$$

we obtain from (3.34)

$$\left[\frac{\partial(\mu_S/kT)}{\partial \varrho_S} \right]_{T,\mu_W} = \frac{1}{\varrho_S} + 2B_2^* + 3B_3^* \varrho_S + \cdots \tag{3.40}$$

which upon integration (at a fixed T and μ_W) yields

$$\mu_S/kT = \mu_S^\circ/kT + \ln \varrho_S + 2B_2^* \varrho_S + \tfrac{3}{2} B_3^* \varrho_S^2 + \cdots \tag{3.41}$$

where μ_S° is an integration constant, or the standard chemical potential. We now define the activity coefficient of the solute S by

$$\ln \gamma_S^D = 2B_2^* \varrho_S + \tfrac{3}{2} B_3^* \varrho_S^2 + \cdots \tag{3.42}$$

Hence B_2^* may be determined from the following limiting slope:

$$\lim_{\varrho_S \to 0} \left(\frac{\partial \ln \gamma_S^D}{\partial \varrho_S} \right) = 2B_2^* \tag{3.43}$$

A word of warning is in order here. We have emphasized above that information on B_2^* may be obtained from deviations from *ideal dilute solutions* as conveyed by the quantity γ_S^D. In some first-order theories of solutions, such as the lattice theory of regular solutions, one uses an activity coefficient that measures the extent of *dissimilarity* of the two components. A typical result of this kind, for a mixture of two components A and B

with mole fraction x_B, is

$$\ln \gamma_A{}^{\text{sym}} = \frac{C x_B{}^2 (2 W_{AB} - W_{AA} - W_{BB})}{2kT} \tag{3.44}$$

where the W_{ij} are the interaction energies between species i and j situated on adjacent lattice points, and C is the number of nearest neighbors of each lattice point. Here $\gamma_A{}^{\text{sym}}$ measures the deviation from the *symmetrical-ideal* behavior, i.e., this is a first-order term in the parameter $(2W_{AB} - W_{AA} - W_{BB})$, *not* in the solute density. Clearly this quantity does not convey information on the solute–solute affinity at infinite dilution, as does $\gamma_S{}^D$ in (3.42). Failure to use different notations for the various activity coefficients is a common source of error in the literature. [This subject is treated in greater detail in Ben-Naim (1974), Chapter 4.]

Thus by measuring the limiting slope given by either (3.38) or (3.43), one obtains the solute–solute affinity at infinite dilution. It is important to realize that this method always gives $B_2{}^*$ (or G_{SS}°) in that limit though the extrapolation is carried out from finite concentrations. Measurement of deviations from ideal dilute solutions at *finite* concentration does not *yield* G_{SS} at that concentration, but a combination of G_{SS}° and higher-order virial coefficients. The latter are more difficult to obtain experimentally and much more difficult to interpret theoretically, so we shall not discuss these any further. In Section 3.6 we shall present a different way of evaluating G_{SS} at any required concentration.

Suppose that we have experimental means of obtaining $B_2{}^*$. For spherical solutes with an effective hard-core diameter σ we may split the integral (3.35) into two terms:

$$
\begin{aligned}
B_2{}^* &= -\frac{1}{2} \int_0^\sigma (-4\pi R^2)\, dR - \frac{1}{2} \int_\sigma^\infty [g_{SS}^\circ(R) - 1] 4\pi R^2\, dR \\
&= \frac{1}{2} \left(\frac{4\pi\sigma^3}{3} \right) - A_2{}^* = 4v_S - A_2{}^*
\end{aligned} \tag{3.45}
$$

In this way we have split $B_2{}^*$ into two contributions. The first contribution results from the strongly repulsive region $0 \le R \le \sigma$, where one may assume that $g_{SS}(R) = 0$, and is proportional to the volume of the solute S. The second is a measure of the *essentially attractive* region $\sigma \lesssim R < \infty$. This last statement is obviously imprecise since g_{SS}° is an oscillating function of R in this region so there are alternating regions of attraction and repulsion between the pairs of solutes.

On the basis of the above theoretical analysis Kozak *et al.* (1968) have processed a large quantity of experimental data on aqueous solutions of alcohols, amines, carboxylic acids, etc.

The first step in their work is to estimate the contribution of the "volume" of the solutes to B_2^*. This is not easy to do for the case of non-spherical molecules for which a simple split of B_2^* [as in (3.45)] does not exist. Noting, however, that for spherical molecules this contribution to B_2^* is $4v_S$, where v_S is the volume of the solute S, they suggested replacing v_S by $\bar{v}_S^\circ$, the partial molar volume of S at infinite dilution, to obtain an estimate of the analog of the first term on the right-hand side of (3.45) for nonspherical molecules. [This is quite a doubtful procedure: in the first place because $\bar{v}_S^\circ$ is determined not only by the "volume" of the solute but by any type of interaction that may exist between the solute and the solvent; second, and perhaps more important, is the fact that $\bar{v}_S^\circ$ gets contributions from "structural changes in the solvent," a quantity which depends primarily on the properties of the *solvent* rather than on the solute. An explicit expression for $\bar{v}_S^\circ$ is given in Ben-Naim (1974), Section 7.7.] Using such an estimate of the contribution from the repulsive part of the solute–solute interaction and using the experimental values of B_2^*, one can estimate A_2^* in (3.45). Note that B_2^* and A_2^* have dimensions of volume per particle. In Table 3.4 we present some selected values of A_2^* (presumably in cm^3/mol). For our purpose the exact units are of no importance since we

Table 3.4

Values of B_2^* and of A_2^* at $t = 0°C$ for Some Solutes in Water[a]

Solute	B_2^* [b]	A_2^* [b]
Methanol	47 ± 12	110
Ethanol	28 ± 4	192
1-Propanol	20 ± 5	261
2-Propanol	37 ± 6	250
1-Butanol	1 ± 7	347
2-Butanol	48 ± 25	295
Formic acid	20	108
Acetic acid	20 ± 2	176
Propionic acid	10 ± 2	249
Butynic acid	7 ± 8	308

[a] From Kozak *et al.* (1968).
[b] Presumably in cm^3/mol.

are interested only in the general trends in the variation of A_2^* within an homologous series of solutes.

From Table 3.4 it is clear that values of A_2^* tend to increase with an increase in the chain length of the alkyl residue. This presumably reflects the pairwise HI between the solute molecules. Note, however, that A_2^* is determined by both the direct and the indirect parts of $\Delta G(R)$, and it is difficult, if not impossible, to separate these two contributions. The mere fact that A_2^* increases with the chain length of the alkyl residue may well be due to the *direct* solute–solute interaction, and not to the solvent-induced HI. Thus, in themselves the data of Table 3.4 do not show any peculiarity of liquid water as a solvent for HI. In order to establish the outstanding behavior of water, one must compare the data of Table 3.4 with similar data in nonaqueous solvents. It is only by such a comparison that the possibly unique nature of water with regard to HI can be established.

Kozak *et al.* (1968) have also processed values of the third virial coefficients B_3^* of these solutes. In the author's opinion, these data cannot be viewed as a fruitful source of information on HI, because of the great difficulties in sorting the various contributions to B_3^*, as we have done for B_2^* in equation (3.45).

Comment

The essential weakness of the method used in this section to obtain A_2^* from B_2^* is the uncertainty involved in estimating the "volume" of the solutes. There exists no simple analog to the split of B_2^* in (3.45) for complex molecules such as alcohols, amines, etc. The use of the partial molar volumes at infinite dilution is probably unjustified, since this quantity bears no simple relation to the "volume" of the solutes. Therefore the significance of the quantities A^* is quite obscure. However, a systematic study of B_2^* of certain solutes in aqueous, as well as nonaqueous, solutions should prove helpful in the study of the problem of the HI.

3.6. SOLUTE–SOLUTE AFFINITY AT FINITE CONCENTRATIONS

In a two-component system one may define three different correlation functions and the corresponding affinities, namely,

$$G_{\alpha\beta} = \int_0^\infty [g_{\alpha\beta}(R) - 1]4\pi R^2 \, dR \tag{3.46}$$

where $g_{\alpha\beta}(R)$ is the pair correlation function between the species α and β. The definition of $G_{\alpha\beta}$ is valid for any mixture of two (or more) components of any composition. In this section we restrict ourselves to a single example of water (W) and ethanol (E) mixtures, which is the only system for which sufficient experimental data are available for evaluating all three $G_{\alpha\beta}$'s.

The procedure of computing the $G_{\alpha\beta}$'s in such a system consists of inverting the Kirkwood–Buff (1951) theory of solutions. Here we shall not elaborate on the derivation of the Kirkwood–Buff theory [a very detailed discussion is found in Ben-Naim (1974), Chapter 4]. But because of its extreme generality and because of its potential usefulness in application to aqueous solutions, we present here a brief survey of the main results of the theory.

In essence the Kirkwood–Buff theory provides a recipe for computing some thermodynamic quantities from knowledge of all the $G_{\alpha\beta}$'s of the system. More specifically, for a two-component system we have the following relations.

Let ϱ_W and ϱ_E be the number densities of W and E. We now introduce the two auxiliary quantities

$$\eta = \varrho_W + \varrho_E + \varrho_W\varrho_E(G_{WW} + G_{EE} - 2G_{WE}) \tag{3.47}$$

$$\xi = 1 + \varrho_W G_{WW} + \varrho_E G_{EE} + \varrho_W\varrho_E(G_{WW}G_{EE} - G_{WE}^2) \tag{3.48}$$

The isothermal compressibility $\varkappa_T$, the partial molar volumes $\bar{V}_E$, $\bar{V}_W$, and the derivatives of the chemical potentials may be expressed in terms of the $G_{\alpha\beta}$'s by the following relations:

$$\varkappa_T = \xi/kT\eta \tag{3.49}$$

$$\bar{V}_E = [1 + \varrho_W(G_{WW} - G_{WE})]/\eta \tag{3.50}$$

$$\bar{V}_W = [1 + \varrho_E(G_{EE} - G_{WE})]/\eta \tag{3.51}$$

$$\mu_{EE} = \varrho_W kT/\varrho_E \eta V, \quad \mu_{WW} = \varrho_E kT/\varrho_W \eta V, \quad \mu_{EW} = \mu_{WE} = -kT/\eta V \tag{3.52}$$

where $\mu_{\alpha\beta} = (\partial\mu_\alpha/\partial N_\beta)_{T,P,N_\alpha}$. Here we have written six equations, although only three of these are independent, since we have the following thermodynamic identities:

$$\varrho_W\mu_{WW} + \varrho_E\mu_{WE} = 0 \tag{3.53}$$

$$\varrho_W\mu_{WE} + \varrho_E\mu_{EE} = 0 \tag{3.54}$$

$$\varrho_E\bar{V}_E + \varrho_W\bar{V}_W = 1 \tag{3.55}$$

Thus in essence we have three independent equations relating thermodynamic quantities to the three quantities G_{WW}, G_{EE}, and $G_{EW} = G_{WE}$ (for any given temperature and composition of the system). An inversion procedure is possible in which the three $G_{\alpha\beta}$'s may be computed from the above-mentioned thermodynamic quantities. We now outline a procedure which we found to be most efficient.

First, we assume that the vapor above the water–ethanol mixtures, at room temperature, may be considered to be an ideal gas mixture (the liquid mixture is far from being an ideal solution in any sense of ideality). Thus, for the chemical potential of the water we write

$$\mu_W{}^l = \mu_W{}^g = \mu_W{}^{og} + kT \ln P_W \tag{3.56}$$

Using the equation for μ_{WW} in (3.52) and the thermodynamic identity

$$\left(\frac{\partial \mu_W}{\partial N_W}\right)_{T,P,N_E} = \left(\frac{\partial \mu_W}{\partial x_W}\right)_{T,P}\left(\frac{\partial x_W}{\partial N_W}\right)_{T,P,N_E}$$

$$= \left(\frac{\partial \mu_W}{\partial x_W}\right)_{T,P}\frac{N_E}{(N_E + N_W)^2} \tag{3.57}$$

we obtain

$$\left(\frac{\partial \mu_W}{\partial x_W}\right)_{T,P} = \frac{kT\varrho^2}{\varrho_W\eta} \tag{3.58}$$

where $\varrho = (N_W + N_E)/V$ is the total number density of molecules in the system. Introducing the assumption of ideality for the gaseous phase we obtain, from (3.56) and (3.58),

$$x_W \frac{\partial \ln P_W}{\partial x_W} = \frac{\varrho}{\eta} \tag{3.59}$$

from which we can compute η if we have the partial vapor pressure of one component as a function of the composition.

Note that we can use information on the vapor pressure of either component, since the corresponding derivatives are related by the identity

$$x_W \frac{\partial \ln P_W}{\partial x_W} = x_E \frac{\partial \ln P_E}{\partial x_E} \tag{3.60}$$

Having computed η from (3.59), we use relation (3.49) to compute ξ from the experimental values of the isothermal compressibilities.

Next we use the following identity, which can be easily obtained from (3.50) and (3.51),

$$\bar{V}_W \bar{V}_E = \frac{\xi - \eta G_{WE}}{\eta^2} \qquad (3.61)$$

and which permits the evaluation of G_{WE} in terms of η, ξ, and the product of the two partial molar volumes of E and W. Finally from (3.50) and (3.51) we may get G_{WW} and G_{EE}. This concludes the inversion procedure, which consists of getting the three $G_{\alpha\beta}$'s from experimental data on isothermal compressibilities, partial molar volumes, and partial vapor pressures. In a recent paper the author (Ben-Naim, 1977) described the detailed procedure and results for the system water–ethanol. Here we shall be interested only in one aspect of this work, namely, the ethanol–ethanol affinity G_{EE} in the entire composition range.

Figure 3.9 shows the values of G_{EE} as a function of the mole fraction of ethanol at 25°C (the units of G_{EE} are cm³/mol). Before discussing the implications of this result, it should be noted that the computations were based on very limited experimental data; in particular, the compressibilities and the vapor pressures are available for only about ten compositions. For example, we have values of $\varkappa_T$ at $x_E = 0$ and at $x_E \approx 0.1$, and not in between —therefore all the computed results should be considered as tentative only, pending the availability of more detailed and accurate experimental data.

One consequence of the limited experimental data is the impossibility of determining the precise limit of G_{EE} at $\varrho_E \to 0$. Disregarding for the present the exact limit of G_{EE} at infinite dilution, we note that the most salient feature of this curve is the very steep increase of G_{EE} on addition of ethanol up to about $x_E \approx 0.1$. Beyond $x_E \sim 0.1$ the values of G_{EE} tend smoothly to the limit of G_{EE} for pure ethanol, which is about G_{EE}(pure

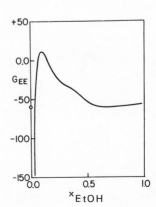

Figure 3.9. Ethanol–ethanol affinity in the water–ethanol system as a function of the mole fraction of ethanol at 25°C [from Ben-Naim (1977)].

ethanol) ≈ -54.5 cm^3/mol. The strong initial behavior of G_{EE} at the low alcohol concentration is undoubtedly a reflection of the enhancement of the strength of the HI brought about by the addition of alcohol. This finding is consistent with our previous results on the dependence of $\delta G^{HI}(\sigma_1)$ on the mole fraction of ethanol, which was discussed in Section 3.4. We stress, however, that this result is unreliable in the region $0 \lesssim x_E \lesssim 0.03$ since the data we have used for our computation are not sufficiently dense in this region. Therefore, one can well expect that if we had more accurate and more detailed data on $\varkappa_T$, P_W, and $\bar{V}_E$, and on $\bar{V}_W$ in the low alcohol concentration region we could have found a completely different behavior of G_{EE}, perhaps an initial minimum followed by a maximum at about $x_E \sim 0.1$.

As in Section 3.5, we can here split G_{EE} into two contributions which we write as

$$G_{EE} = -V_{EE} + I_{EE} \tag{3.62}$$

where V_{EE} is roughly proportional to the molecular volume of ethanol (this is not defined exactly; see also the discussion in Section 3.5) and I_{EE} reflects the average overall attraction between pairs of molecules (again this statement is not precise, as discussed in Section 3.5). It is reasonable to assume that V_{EE} is almost independent of composition (for simple spherical molecules it is four times the volume of the molecule) and hence the major reason for the steep variation in G_{EE} with composition is due to the change of I_{EE}. Thus, we may conclude that on the average the overall attraction between pairs of ethanol molecules increases with the addition of ethanol in the region $0 \lesssim x_E \lesssim 0.1$, with a possible exception at very low concentrations, say $x_E \lesssim 0.03$.

One possible piece of evidence that in the very-low-alcohol concentration region there might be a minimum can be drawn from the following consideration. We use the value of B_2^* for ethanol as estimated by Kozak et al. (1968) (see Table 3.4 of Section 3.5) to estimate the infinite dilution limit of G_{EE}, namely,

$$G_{EE}^\circ = -2B_2^* \approx -56 \pm 8 \text{ cm}^3/\text{mol} \tag{3.63}$$

This point is indicated by a circle in Figure 3.9. If this estimate is sound and if the steep rise of G_{EE} as a function of x_E is real, then it is likely that G_{EE} indeed initially decreases (in say $0 \leq x_E \lesssim 0.03$) and then has a positive slope (up to about $x_E \approx 0.1$). This kind of behavior would be in complete accord with the change of G^{HI} as a function of the mole fraction of ethanol that we have seen in Section 3.4.

We conclude this section by noting that the Kirkwood–Buff theory is an exact theory of solution, it does not involve the assumption of pairwise additivity of the total potential energy, and it is applicable to both spherical and nonspherical molecules. For these reasons this theory is most suitable for treatment of such complex fluids as aqueous solutions.

Comment

The method outlined in this section is very general and does not involve any theoretical assumptions about the system. This is the main reason for its power and potential usefulness in the study of aqueous solutions. Experimentalists are urged to undertake measurements of the appropriate thermodynamic quantities in various water–organic-liquid mixtures. Possible candidates for such a study are mixtures of water with various alcohols, or other homologous series of organic molecules, water–dioxane, water–ethylene–glycol, etc. For comparison, it might be of interest to study also some nonaqueous mixtures. The evaluation of the $G_{\alpha\beta}$'s can be most helpful in understanding the nature of these systems on a molecular level.

3.7. DIMERIZATION EQUILIBRIUM

In Section 3.5 we discussed the first-order deviations from ideal dilute behavior as manifested by the second virial coefficient of the osmotic pressure. Here we present a different approach to such deviations, which can be made formally equivalent to the second virial coefficient but in practice is a different one.

Consider a dilute solution of carboxylic acid, which we denote by R_n—COOH, and for the present consideration we disregard dissociation into ions (this should be taken into account in a proper processing of the experimental data).

In extremely dilute solutions the solute molecules are surrounded only by solvent molecules and no detectable deviations from ideality are possible. If we increase the concentration of the solute, the solutes start to "see each other," and in the very general case this is revealed through deviations from ideal dilute behavior. A special case of such a situation occurs when we have an experimental means of "seeing" pairs of solutes that "see each other," in which case we can measure the concentration of the "dimers" and treat the system as a mixture of monomers (M) and dimers (D) in chemical equilibrium. In the particular example of carboxylic acid we have the dimerization equilibrium

$$2(R_n\text{---COOH}) \rightleftarrows (R_n\text{COOH})_2 \tag{3.64}$$

Let ϱ_M and ϱ_D be the number densities of the monomers and the dimers, respectively. At chemical equilibrium we have

$$\mu_D = 2\mu_M \tag{3.65}$$

If the system is dilute enough such that the dimerization equilibrium (3.64) accounts for most of the deviation from ideal dilute solutions, then we can write

$$\Delta G_D{}^\circ = \mu_D{}^{ol} - 2\mu_M{}^{ol} = -kT \ln K_D \tag{3.66}$$

with

$$K_D = (\varrho_D/\varrho_M{}^2)_{eq} \tag{3.67}$$

It is important to realize that the dimer concentration ϱ_D is the concentration of those "dimers" that are "seen" by our particular instrument. Different experimental techniques may therefore measure different concentrations of dimers. In our particular example, if we have a spectroscopic method of detecting, say, the hydrogen bond formed by two carboxylic acids, then we get a measure of ϱ_D, and from the knowledge of the total solute concentration ϱ_{tot} we get the monomer concentration through the relation

$$\varrho_{tot} = \varrho_M + 2\varrho_D \tag{3.68}$$

It is well known that carboxylic acids may form two kinds of dimers, one involving a *single* and the second involving two hydrogen bonds, as depicted in Figure 3.10. It is known, for instance, that the enthalpy of dimerization of acetic acid in the gas phase is about 14.12 kcal/mol, which is attributed to the formation of two hydrogen bonds, as in the cyclic dimers, each of which involves about 7 kcal/mol. Association equilibria

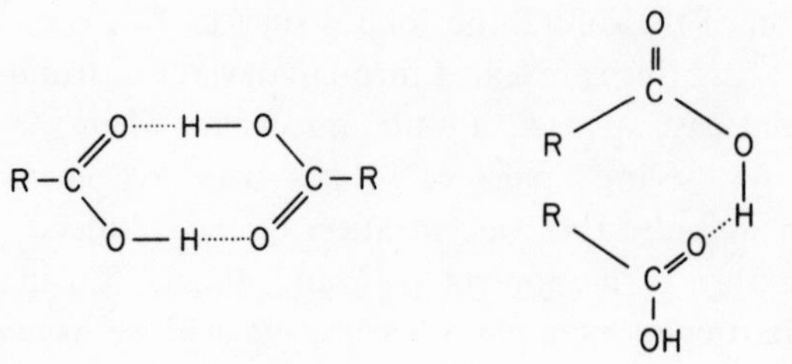

Figure 3.10. Two possible forms of dimers between two carboxylic acids.

Table 3.5

Dimerization Constants of Carboxylic Acids in Benzene at 30°C[a]

	Formic	Acetic	Propionic	Butyric
K_D	130	370	380	430
ΔG_D° (in kcal/mol)	−2.95	−3.58	−3.60	−3.67
$\Delta(\Delta G_D^\circ)$	(—)	−0.63	−0.65	−0.72

[a] From Goodman (1958).

of carboxylic acids in solvents have been measured by various authors; we present only very limited data that are suitable for demonstrating the relation of these data to the problem of hydrophobic interaction.

Table 3.5 presents some values of K_D and ΔG_D°, defined in Equations (3.66) and (3.67), for four carboxylic acids in benzene at 30°C (based on data from Goodman, 1958). The third row gives values of $\Delta(\Delta G_D^\circ)$ defined by

$$\Delta(\Delta G_D^\circ) = \Delta G_D^\circ(\text{carboxylic acid}) - \Delta G_D^\circ(\text{formic acid}) \qquad (3.69)$$

It is instructive to note that the value of K_D for stearic acid [$CH_3(CH_2)_{16}$ COOH] is 520; hence $\Delta G_D^\circ = -3.79$ and $\Delta(\Delta G_D^\circ) = -0.84$. Thus, for all of these acids it seems that values of ΔG_D° are almost independent of the chain length of the alkyl residue (except for formic acid which has a somewhat smaller value of ΔG_D°).

Similar results for association constants and the corresponding free energy of dimerization in CCl_4 are reported in Table 3.6 (data from Weno-

Table 3.6

Dimerization Constants of Carboxylic Acids in CCl_4 at 24°C[a]

	Acetic	Propionic	Valeric	Caproic	Lauric
K_D	2.37×10^3	2.48×10^3	2.72×10^3	1.86×10^3	1.94×10^3
ΔG_D°	−4.61	−4.64	−4.70	−4.47	−4.49

[a] From Wenograd and Spurr (1957).

grad and Spurr, 1957). Here we do not have the value of $\Delta G_D{}^\circ$ for formic acid, but it is clear that the dependence of $\Delta G_D{}^\circ$ on the chain length of the alkyl residue is almost insignificant. It is interesting to note that the value of $\Delta G_D{}^\circ$ for phenylacetic acid is almost the same as for lauric acid, again indicating the independence of $\Delta G_D{}^\circ$ on the type of residue, provided it does not affect the hydrogen-bond energy (i.e., if one takes benzoic acid where the carboxylic group sits directly on the aromatic ring, the values of K_D and $\Delta G_D{}^\circ$ are significantly different from those in Table 3.6, namely, $K_D = 4.4 \times 10^3$ and $\Delta G_D{}^\circ = -4.99$ kcal/mol.)

The above results have been interpreted in terms of a cyclic dimer (Figure 3.10), i.e., in nonaqueous solvents, the main driving force in the dimerization process is the formation of the two hydrogen bonds between the two solutes. In this configuration the two alkyl groups hardly "see" each other, hence the insensitivity to the size of this group.

The results for aqueous solutions seem markedly different from those for the nonaqueous solvents. There are few reports on the dimerization constants for carboxylic acids in aqueous solutions—the agreement between results from different sources is either poor or indeterminable (because of lack of the concentration units or results reported at different temperatures). To demonstrate the difference between water and nonaqueous solutions we present in Table 3.7 values of K_D (presumably in liters/mol; however the precise units are of no importance for our purposes since we shall be interested in *differences* in the standard free energies of dimerization). The data are taken from Martin and Rossotti (1959). It is clearly seen that, in contrast to nonaqueous solvents, in water $\Delta G_D{}^\circ$ depends on the chain length of the alkyl group. The values of $\Delta(\Delta G_D{}^\circ)$ clearly increase (in ab-

Table 3.7

Values of K_D, $\Delta G_D{}^\circ$, and $\Delta(\Delta G_D{}^\circ)$ for Carboxylic Acids in Water at 25°C[a]

	Formic	Acetic	Propionic	Butyric
$K_D{}^b$	0.0575	0.186	0.316	0.549
$\Delta G_D{}^\circ$	1.70	1.00	0.69	0.36
$\Delta(\Delta G_D{}^\circ)$	—	−0.7	−1.01	−1.34

[a] Based on data from Martin and Rossotti (1959).
[b] Presumably in liters/mol.

solute magnitude) almost linearly with the number of methylene groups in the molecules.

Schrier $et\ al.$ (1964), who first used these data in their study of hydrophobic interaction, proposed that the dimers, in water, involve only one hydrogen bond, and the total "driving force" for dimerization consists of two parts: the hydrogen-bond formation, and the hydrophobic interaction between the alkyl groups. The latter are responsible for the chain-length dependence of $\Delta G_D{}^\circ$.

In order to extract information on the hydrophobic interaction from the data of Table 3.7, a method of separating the hydrogen-bond (HB) contribution must be devised. The following procedure, consisting of three steps, was applied to this end:

(1) It is assumed that $\Delta G_D{}^\circ$ for each carboxylic acid may be split into two contributions, one due to hydrogen-bond formation (HB) and one to the hydrophobic interaction (HI):

$$\Delta G_D{}^\circ(R_n\!\!-\!\!COOH) = \Delta G_{HI}^\circ(R_n\!\!-\!\!COOH) + \Delta G_{HB}^\circ(R_n\!\!-\!\!COOH) \quad (3.70)$$

(2) It is assumed that $\Delta G_D{}^\circ(HCOOH)$ is devoid of any contribution from HI, i.e.,

$$\Delta G_D{}^\circ(HCOOH) = \Delta G_{HB}^\circ(HCOOH) \quad (3.71)$$

(3) The contribution of the HB formation is assumed to be independent of the chain length of the alkyl group, namely, for any n we have

$$\Delta G_{HB}^\circ(R_n\!\!-\!\!COOH) = \Delta G_{HB}^\circ(HCOOH) \quad (3.72)$$

Combining the above three assumptions we obtain

$$\begin{aligned} \Delta G_{HI}^\circ(R_n\!\!-\!\!COOH) &= \Delta G_D{}^\circ(R_n\!\!-\!\!COOH) - \Delta G_{HB}^\circ(R_n\!\!-\!\!COOH) \\ &= \Delta G_D{}^\circ(R_n\!\!-\!\!COOH) - \Delta G_{HB}^\circ(HCOOH) \\ &= \Delta G_D{}^\circ(R_n\!\!-\!\!COOH) - \Delta G_D{}^\circ(HCOOH) \quad (3.73) \end{aligned}$$

This is the quantity we have denoted by $\Delta(\Delta G_D{}^\circ)$ in (3.69) and in Tables 3.5 and 3.7.

The above series of assumptions leads to an estimate of the contribution of HI to the free energy of dimerization of carboxylic acid. Note that ΔG_{HI}° [i.e., $\Delta(\Delta G_D{}^\circ)$] for acetic acid seem to have the same magnitude in benzene as in water; the striking difference is in the chain length dependence of $\Delta(\Delta G_D{}^\circ)$ in the two solvents. All the above conclusions clearly depend on the accuracy of the experimental method of determining the association constants, which do not seem to be reliable in view of the inconsistency between results reported by different authors [for more details, see Schrier *et al.* (1964)]. However, a more profound and difficult question is the validity of the assumptions leading to relation (3.73). In particular the split of $\Delta G_D{}^\circ$ in (3.70), although a very appealing one on intuitive grounds, cannot be supported by statistical mechanical arguments. A deeper analysis of the various contributions to $\Delta G_D{}^\circ$ shows that there exists no "obvious" or "natural" split into two terms with the meaning as assigned in (3.70).

We shall devote Appendix A.3 to further elaboration on this problem. It must be noted, however, that the HI as derived from the method of this section is quite different from the HI between two alkyl groups as defined in Section 1.3. There are several reasons for that; one is the fact that in the dimer we do not know the exact relative configuration of the two alkyl groups. Therefore, the quantity we get is a sort of average over many possible configurations. Secondly, the two alkyl residues attached to the carboxylic group see each other in a medium that has been perturbed by the presence of the carboxylic group. In this sense the HI obtained from dimerization equilibria has the character of the *conditional* HI, similar to the conditional free energy of transfer that we discussed in Section 2.3. A more precise treatment of this concept is to be found in Appendix A.3.

Comment

In spite of the shortcomings of the method of extracting information on HI from dimerization equilibria, it is well worth further attention and extension. The data available at present are scarce and data from different sources show large inconsistencies. Therefore there is a need for a systematic measurement of K_D for various solutes that can form dimers (not necessarily carboxylic acids) both in pure water and in nonaqueous solvents. Perhaps it also might be interesting to have this kind of data for mixtures of water and, say, ethanol. There is no doubt that such a systematic study will be rewarding. One may take advantage of the fact that ΔG_{HI} has the character of a conditional HI, and study the dependence of HI on the polar group attached to the alkyl residue. Also the response of the HI to the addition of small amounts of alcohol may be of interest in view of the controversy that exists on this matter. Finally, a more systematic study of the temperature and pressure dependence of K_D will be of value in connection with the problem of the entropy, enthalpy, and volume changes associated with the HI (see also Chapter 5).

3.8. ION-PAIR FORMATION BY DOUBLE LONG-CHAIN ELECTROLYTES

Conceptually this section, like the previous one, is concerned with association equilibria. Whereas in Section 3.7 we considered the process of dimerization between (essentially) neutral molecules to form dimers, here we deal with the association between ionic species forming a neutral ion pair.

The association reaction is

$$A^+ + B^- \rightleftarrows AB \tag{3.74}$$

for which the equilibrium constant is

$$K_a = \frac{a_{AB}}{a_A a_B} \tag{3.75}$$

where a_i is the activity of the species i, and the corresponding standard free energy of association is

$$\Delta G_a^\circ = -kT \ln K_a \tag{3.76}$$

The experimental tool employed for the determination of the association constant, and hence the standard free energy (3.76), is the conductivity of ionic solutions.

The conductivity of a solution σ is defined as the linear coefficient in the relation between the current density, J, and the electrical field (or potential gradient), E,

$$J = \sigma E \tag{3.77}$$

Let c_i be the *molar* concentration of ions of species i, v_i the average (scalar) velocity of the ion, and $|z_i|$ the number of charges it carries. The total charge crossing a unit area perpendicular to the direction of the field per unit time is

$$J = \sum_i \frac{c_i v_i |z_i| e N_0}{1000} \tag{3.78}$$

where e is the electron charge and N_0 is the Avogadro number. Defining the mobility of the ion as $u_i = v_i/E$, we get the general expression for the conductivity of the solution,

$$\sigma = \sum_i \frac{c_i u_i |z_i| e N_0}{1000} = \frac{e N_0}{1000} (c_1 u_1 + c_2 u_2) \tag{3.79}$$

where on the right-hand side of (3.79) we have specified a solution containing a single 1:1 electrolyte.

The equivalent conductance Λ is defined by

$$\Lambda = \frac{\sigma \times 1000}{c} \tag{3.80}$$

where c is the total molar concentration of the salt; hence

$$\Lambda = F\left(\frac{c_1 u_1 + c_2 u_2}{c}\right) \tag{3.81}$$

where F is the Faraday number (or charge per mole of electrons): $F = 96,487\ C\ mol^{-1}$.

At infinite dilution any electrolyte will be completely dissociated into ions (Ostwald's dilution law), in which case we have

$$\Lambda_0 = F(u_1^\circ + u_2^\circ) \tag{3.82}$$

where u_i° is the limiting mobility of the ion of species i at infinite dilution. At a finite concentration c we define the *degree* of *dissociation* α as the fraction of the total concentration of the salt c that has dissociated into ions. In our particular example we have

$$c_1 = c_2 = \alpha c \tag{3.83}$$

and hence for any concentration we write

$$\frac{\Lambda}{\Lambda_0} = \alpha \frac{u_1 + u_2}{u_1^\circ + u_2^\circ} \tag{3.84}$$

In the early history of the theory of conductance it was assumed that the difference between Λ and Λ_0 was due only to the difference in the *number* of conducting ions in the two solutions; hence Λ/Λ_0 was identified with the degree of dissociation α. Today, it is clear that as the total concentration of the solute changes, both α and the ionic mobilities change. The Debye–Hückel–Onsager theory of conductance accounts for the variation of the mobilities with concentration. In the region of very dilute solutions, the general form of these functions is

$$u_i = u_i^\circ + A'(\alpha c)^{1/2} \tag{3.85}$$

After substituting (3.85) into (3.84) we get the general limiting law for the concentration dependence of the conductivity:

$$\Lambda = \alpha[\Lambda_0 + A(\alpha c)^{1/2}] \tag{3.86}$$

A' and A are constants that do not depend on the concentration. A more explicit expression for these constants in terms of the dielectric constant, viscosity, etc., of the solution is available. But for our purposes the general form of the limiting law in (3.86) is sufficient.

It is worth noting that α appears twice in (3.86): once to account for the reduction in the total number of ions due to incomplete dissociation [as in (3.84)], and once [in the term $(\alpha c)^{1/2}$] to account for the dependence of the mobilities on the concentration of the ions. (A is usually negative since the ion–ion interaction causes a net drag for the motion of the ion. Hence the mobility and the conductivity of the solution are initially reduced.)

The infinite dilution limit of the equivalent conductivity Λ_0 is the sum of the equivalent conductivities of the ions: this is the well-known Kohlrausch law of independent ionic migration at infinite dilution.

By taking the limit of (3.86) as $c \rightarrow 0$, one can, in principle, determine Λ_0. A more convenient method is to compute Λ_0 from data on the equivalent conductivities of the single ions.

Plotting Λ as a function of $c^{1/2}$ gives a slope at each point

$$\frac{\partial \Lambda}{\partial c^{1/2}} = \alpha^{3/2} A \tag{3.87}$$

Thus the limiting slope as $c \rightarrow 0$, where $\alpha \rightarrow 1$, gives the constant A, and for any other concentration [within the range of validity of equation (3.86)] one can compute α from (3.87). (In practice other methods of evaluating α that are more efficient are also available.)

The association constant K_a may be expressed in terms of the degree of dissociation α as

$$K_a = \frac{a_{AB}}{a_A a_B} = \frac{(1 - \alpha)c}{\alpha^2 c^2 f_\pm^2} \tag{3.88}$$

Thus from the knowledge of the total salt concentration c, the measurement of α, and the estimation of the mean activity coefficient of the ions (using the Debye–Hückel limiting law), one can compute the association constant K_a.

Note that corrections for nonideality were introduced only for the ionic species—the total concentration is presumed to be small enough so that nonionic solutes form an ideal dilute solution. There is also another important reason for keeping the total solute concentration very low: the ions discussed below form micelles at higher concentrations and all the treatment below is based on the assumption that we are in the region of concentrations below the critical micelle concentration (CMC) (see also Chapter 4).

Oakenfull and Fenwick (1974b) undertook extensive measurements of the conductivity of aqueous solutions of decyltrimethylammonium carboxylate. The cation was kept fixed and the chain length of the anion was varied. Thus

$$A^+ = CH_3(CH_2)_9\overset{+}{-}N(CH_3)_3$$
$$B^- = CH_3-(CH_2)_{n-1}-COO^- \tag{3.89}$$

where $n = 2, 6, 7, 8, 9, 10, 11, 12$.

The values of $\Delta G_a{}^\circ$ computed from (3.76) were processed to extract information on hydrophobic interaction. The assumption made is similar to the one discussed in the previous section, namely, one splits ΔG_a into two contributions

$$\Delta G_a{}^\circ = \Delta G_{EL}^\circ + \Delta G_{HI}^\circ \tag{3.90}$$

where ΔG_{EL}° is due to the interaction between the charged groups, and ΔG_{HI}° is the contribution from HI. As we discussed in Section 3.7 and Appendix A.3, the term ΔG_{HI}° should be interpreted as a sort of *conditional* standard free energy change. This means that the two "hydrophobic groups" see each other not in the pure solvent, but in a solvent perturbed by the presence of the ions. This perturbation is likely to be more pronounced in the presence of charged groups as compared with the association between neutral molecules that we have discussed in Section 3.7. Note that ΔG_{HI}° includes both the direct and indirect interactions between the alkyl groups and therefore it is not possible to compare this value with the values of δG^{HI} discussed in Section 3.3. (One can also interpret ΔG_{EL}° as a conditional free energy—but this is undesirable for our purposes.)

In spite of this limitation, there are two very interesting features of the results obtained by Oakenfull and Fenwick (1974b). The first result is that $\Delta G_a{}^\circ$ seems to be quite linear in n. Figure 3.11 shows several plots of $\Delta G_a{}^\circ$ as a function of n for different solvents. The linearity in n is reported for $n \geq 6$. This means that the methylene groups added to the carboxylate ions are quite far from the electric charge, and perhaps at this distance the methylene groups do not "feel" the electric field of the charge and hence

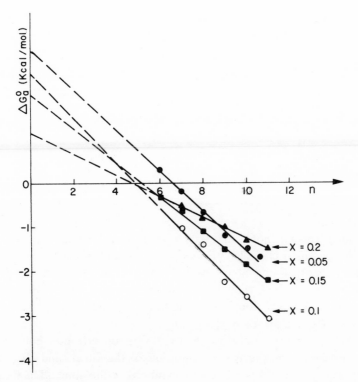

Figure 3.11. Values of ΔG_a° as a function of n for several mixtures of water and ethanol (at 25°C). The mole fraction of ethanol is indicated next to each curve. The dashed lines are extrapolated from the experimental lines for $n \leq 6$ [data from Oakenfull and Fenwick (1974b)].

there is a constant contribution for each methylene group that is given by

$$\frac{\partial \Delta G_a^\circ}{\partial n} \approx \Delta G_{\mathrm{HI}}^\circ \text{ (per CH}_2) \tag{3.91}$$

Values of $\Delta G_{\mathrm{HI}}^\circ$ (per CH$_2$) are given in Table 3.8. Note that the value of $\Delta G_{\mathrm{HI}}^\circ$ (per CH$_2$) for pure water was estimated from a different source, which does not provide exactly the same information as the slope of ΔG_a° against n. We have therefore marked this entry by a question mark (Oakenfull and Fenwick, 1973).

Oakenfull and Fenwick (1974b) have also extrapolated the values of ΔG_a° to $n = 0$ to obtain values of $\Delta G_{\mathrm{EL}}^\circ$. This is a doubtful procedure since we do not know whether or not the linearity of the curves in Figure 3.11 is maintained also for small n ($n \leq 6$). Since these values are of no interest in the present study we do not report them here.

Table 3.8

Values of ΔG_{HI}° (per CH_2) Computed from the Slopes of the Curves of Figure 3.11 for Various Mixtures of Water and Ethanol at 25°C

x_{EtOH}	ΔG_{HI}° (per CH_2) (kcal/mol)
0	−0.33 (?)
0.05	−0.49
0.10	−0.54
0.15	−0.38
0.20	−0.25

It should also be noted that the length of the cation was kept constant, and only the anion chain length was varied. Therefore one cannot expect that the linearity found above would hold for anions which are longer than the cation. Indeed, though some of the measurements of the conductivity were undertaken for n up to 12 (dodecanoate), the plots of ΔG_a° against n extend only for $6 \leq n \leq 11$.

Furthermore, as is always the case with dimerization equilibrium, we do not know the precise configuration of the dimer, and ΔG_a° is in fact an average over all possible configurations of the dimer. The linearity found from these experiments does indicate that ΔG_a° does get a large contribution from those configurations in which there exists a "hydrophobic contact" between the alkyl residues of the two ions. However, we have no information on the relative configuration of the two residues. Thus ΔG_{HI}° (per CH_2) is in fact an average value per CH_2 of the carboxylate interacting with the entire chain of the cation.

The second important observation is the dependence of ΔG_{HI}° (per CH_2) on the mole fraction of ethanol. Although there is some doubt about the value for pure water, it is quite clear that for mixtures of water–ethanol in the range of $0.05 \leq x_{EtOH} \leq 0.1$ the hydrophobic interaction increases with the addition of alcohol. This is consistent with other observations reported in Sections 2.4 and 3.4, but care must be exercised in comparing the values obtained by the different techniques, each of which reports different, though related, information. In particular, in the present case there is some doubt about the meaning of the split of ΔG_a° in (3.90) (see also Appendix A.3), and there is no information on the precise configuration of the dimer and hence on the distance between the methylene groups on the two chains.

Beyond $x_{\text{EtOH}} \sim 0.1$ the hydrophobic interaction becomes weaker as a function of the mole fraction of ethanol, again a result that is consistent with results from other sources.

From these results it is not possible to determine the initial dependence of the HI on the mole fraction of ethanol. For one thing, there is uncertainty in the value of $\Delta G_a{}^\circ$ for pure water. But more important is the fact that we know from other sources that in the region $0 \lesssim x \lesssim 0.05$ the HI becomes first weaker and then stronger relative to pure water. The two possible behaviors are indicated schematically in Figure 3.12.

Of course one could always ask how far one needs to go in dilution in order to obtain the "initial" slope of the HI as a function of x. There is no general answer to this question, and one must examine the situation for each case. In our particular case we already know that a change of the sign of the slope is observed in the dependence of HI on the alcohol concentration (see Section 3.4). Therefore one should be careful about drawing conclusions on the initial slope of this function from data at the two points $x = 0$ and $x = 0.05$.

In another publication Oakenfull and Fenwick (1975) estimated the HI from conductance measurements on micellar solutions. Here the system is far more complicated than in the relatively simple case of ion-pair formation, and extraction of information on HI is difficult, if not impossible. The basic reaction considered here is

$$B^- + \begin{pmatrix} nA^+ \\ mB^- \end{pmatrix} \rightleftarrows \begin{pmatrix} nA^+ \\ (m+1)B^- \end{pmatrix} \tag{3.92}$$

where one anion is added to a micelle built up of n cations and m anions. If the standard free energy of this reaction is plotted as a function of the number of methylene groups on the anion (carboxylate ion) one obtains a straight line, and from its slope one estimates

$$\Delta G^\circ_{\text{mic}} \text{ (per } CH_2) = -0.33 \text{ kcal/mol in } H_2O$$
$$= -0.42 \text{ kcal/mol in } D_2O \tag{3.93}$$

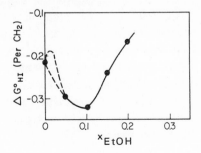

Figure 3.12. Variation of $\Delta G^\circ_{\text{HI}}$ (per CH_2) in kcal/mol with the mole fraction of ethanol at 25°C. Two possible initial slopes at $x = 0$ are indicated which may be consistent with the experimental data [based on data from Oakenfull and Fenwick (1974b)].

From these results they concluded that the HI is stronger in D_2O than in H_2O. Although it is difficult to make an analysis of the various contributions to the free energy of micellization $\Delta G^\circ_{\text{mic}}$ as we have done in Section 3.7 (and in Appendix A.3), we suspect that the main reason for the inconsistency between these results and the ones reported in Sections 2.3, 3.3, and 4.3 is the "conditional" nature of this HI. In fact we have already seen in Section 2.3 that conditional free energy of transfer of an alkyl group between H_2O and D_2O may have a sign different from the free energy of transfer of (almost) the same group. The reason is that the presence of a charge group modifies the properties of the solvent in its immediate surroundings, and therefore the two interacting alkyl groups see each other in a strongly perturbed solvent, which may have properties different from the pure solvent. Of course the situation in the micellar reaction (3.92) is more complicated, and we suspect that in this case it will be impossible to extract information on HI between pairs of alkyl groups in pure H_2O and D_2O.

A similar study was carried out by Packter and Donbrow (1962) on the association between alkyltrimethyl-ammonium ions and arensulfonic anions,

$$\text{C}_6\text{H}_5\text{—SO}_3^- + (CH_3)_3\overset{+}{\text{—N—}}R_n$$

or

$$\text{C}_{10}\text{H}_7\text{—SO}_3^- + (CH_3)_3\overset{+}{\text{—N—}}R_n \qquad (3.94)$$

It was found that for a given anion the values of $\Delta G_a{}^\circ$ increase with the chain length of the cation. And the whole series of values increases with the size of the cation.

Comment

Our comments on the association equilibria discussed in Section 3.7 apply equally well to this section. More systematic measurements are needed in the water–ethanol system, especially in the range of very low concentrations of ethanol. All this applies to the systems where strictly ion pairs, and not higher aggregates, are formed. Even for dimerization processes, extracting information on HI is not easy or straightforward. Nevertheless the general trends in the variation of ΔG_a with changes in the composition of the solvent, and perhaps also changes in temperature and pressure, might help to create a clearer and a more complete picture of the HI phenomenon.

3.9. KINETICS OF AMINOLYSIS BY LONG-CHAIN ALKYLAMINES OF SOME LONG-CHAIN ESTERS OF p-NITROPHENOL

The experimental technique described in this section is quite outstanding in comparison with all the methods we have discussed so far in our study of HI. Here, for the first time we deal with a nonequilibrium phenomenon: the effect of HI on the chemical kinetics of a certain family of reactions. Clearly, the HI is not an exclusively equilibrium phenomenon. We expect that in "real life" it would reveal itself in many diverse processes such as enzyme kinetics, subunit aggregation, transformation of genetic information, etc. Therefore the study of HI in nonequilibrium processes is likely to be extended in various directions in the future. At present, however, there is little we can do to extract direct information on HI from chemical kinetics. The results presented in this section are both interesting and relevant to the problem of the HI, though a direct comparison with results from other sources is difficult to make.

The general idea of the present method is simple. Consider a chemical reaction of the type

$$A + B \xrightarrow{k_1} C \tag{3.95}$$

for which the rate constant k_1 is monitored by some technique. Even without having an explicit theory of reaction rates, it is clear that k_1 should depend on the frequency of the encounters between A and B. More specifically, k_1 depends on the probability that the two reactant molecules will attain a favorable configuration (such as that of an activated complex), before proceeding to the successive steps that lead to the products. (Here we use the term "probability" in the sense of the frequency of occurrence of a certain event, whereas in the equilibrium cases we have used the so-called *a priori* probabilities of these events. These quantities form the basis of the statistical mechanical theory of equilibrium phenomena.)

If the reaction (3.95) occurs in a solvent, we expect the reaction rate to depend on the properties of the solvent. In particular, if A and B contain nonpolar groups, and if the reaction takes place in aqueous media, then the tendency of A and B to approach each other will be partially governed by the HI. Or conversely, a systematic study of reaction rates for a series of reactants A and B can in principle provide information on the HI. This is the basic motivation for the study that has been undertaken by Oakenfull (1973) and Oakenfull and Fenwick (1973, 1974a, 1975, 1977).

Before describing in some detail the work of Oakenfull and Fenwick, it is worth noting that there exists a vast number of studies of the solvent effect on chemical kinetics. We have chosen to discuss only those works that were specifically designed for the study of HI.

In 1967, Knowles and Parsons studied the rate of the two aminolysis reactions

$$\text{para-nitrophenyl decanoate} + \text{1-aminodecane} \xrightarrow{\text{ I }} \text{products}$$
$$\text{para-nitrophenyl acetate} \quad + \text{ethylamine} \quad \xrightarrow{\text{ II }} \text{products} \qquad (3.96)$$

They found that the rate of reaction I was about a hundred times that of reaction II. Later, a more detailed study was also reported by Knowles and Parsons (1969) and by Blyth and Knowles (1971). They found that the rate constant of reaction I was about 300 times that of reaction II. (The kinetics were carried out in carbonate–bicarbonate buffers and 0.99% aqueous acetone solutions at 35°C, and rate constants were obtained by extrapolating to zero buffer concentration.)

A more systematic study of the effect of the chain length on the kinetics of these reactions was undertaken by Oakenfull (1973). We present here a sample of their results, which quite dramatically demonstrates the effect of HI on rate constants.

In Figure 3.13 we present the overall scheme of the reaction under consideration. The three main paths of the reaction are indicated. The first is a hydrolysis of the ester (Es) that does not involve an encounter with the amine. The second involves the combination of the ester with the amine A, to form an intermediate complex I, which then decomposes into the products. The third pathway involves a further encounter between the intermediate complex I with an amine molecule to form a pair of ions, and then proceeds to the final products. The last pathway is referred to as the "amine-catalyzed pathway."

We stress that the three steps depicted in Figure 3.13 cannot be considered the detailed mechanism of the reaction. The actual mechanism is certainly more complicated, but for our purpose we need not be concerned with it.

It is found experimentally that the overall rate of the disappearance of the ester $v = -\partial[\text{Es}]/\partial t$ may be expressed in terms of the concentrations of the ester [Es] and the amine [A] as follows:

$$v = [\text{Es}]\{k_0 + k_1[\text{A}] + k_2[\text{A}]^2\} \qquad (3.97)$$

where k_0 characterizes the rate of the hydrolysis path, and will be of no interest to us here. The two rate constants k_1 and k_2 characterize the two

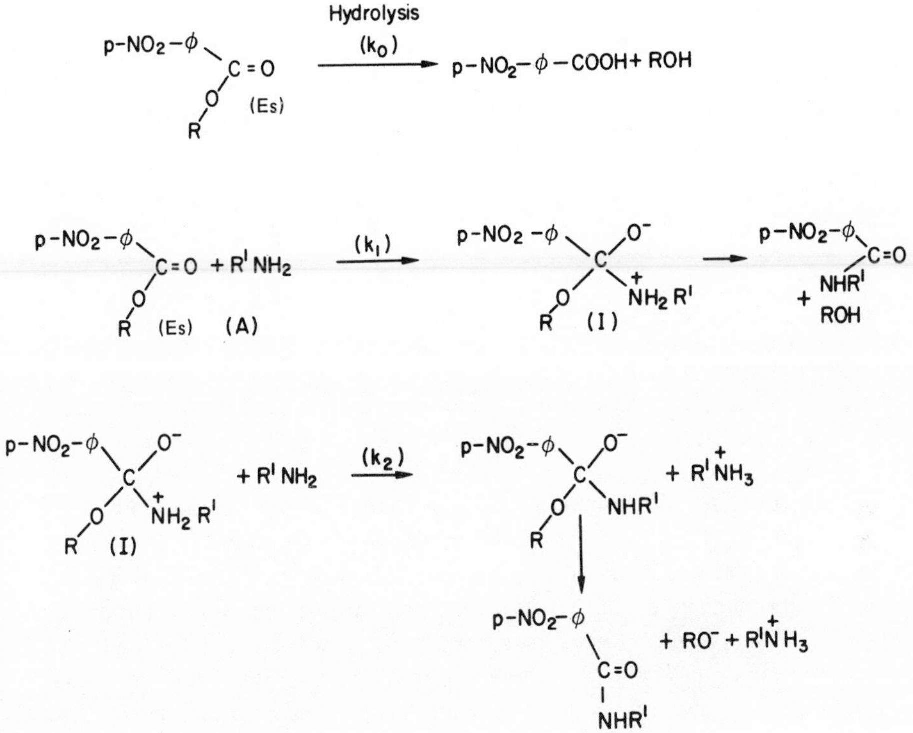

$$p\text{-}NO_2\text{-}\phi \diagdown \underset{R}{\overset{O}{\diagup}}C=O \quad (Es) \quad \xrightarrow[\text{(}k_0\text{)}]{\text{Hydrolysis}} \quad p\text{-}NO_2\text{-}\phi\text{-}COOH + ROH$$

$$p\text{-}NO_2\text{-}\phi \diagdown \underset{R}{\overset{O}{\diagup}}C=O \ (Es) + R'NH_2 \ (A) \quad \xrightarrow{(k_1)} \quad p\text{-}NO_2\text{-}\phi \diagdown \underset{R}{\overset{O}{\diagup}}C \diagup \overset{O^-}{\underset{\overset{+}{NH_2}R'}{}} \ (I) \quad \longrightarrow \quad p\text{-}NO_2\text{-}\phi \diagdown C=O \diagup \overset{NHR'}{+ ROH}$$

$$p\text{-}NO_2\text{-}\phi \diagdown \underset{R}{\overset{O}{\diagup}}C \diagup \overset{O^-}{\underset{\overset{+}{NH_2}R'}{}} \ (I) + R'NH_2 \quad \xrightarrow{(k_2)} \quad p\text{-}NO_2\text{-}\phi \diagdown \underset{R}{\overset{O}{\diagup}}C \diagup \overset{O^-}{NHR'} + R'\overset{+}{N}H_3$$

$$\downarrow$$

$$p\text{-}NO_2\text{-}\phi \diagdown \underset{NHR'}{\overset{|}{C=O}} + RO^- + R'\overset{+}{N}H_3$$

Figure 3.13. Schematic presentation of the three main paths of the aminolysis of esters by alkylamines.

pathways in which one and two encounters with an amine molecule are involved, respectively.

In the following tables the concentrations are expressed as moles per m³ (or 1000 liters), hence k_1 is given in units of $m^3\,mol^{-1}\,sec^{-1}$ and k_2 in units of $m^6\,mol^{-2}\,sec^{-1}$. (All the data referred to are at 25°C, in a borate buffer of 0.1 M, and the pH about 2 units below the pK of the amine.)

Table 3.9 shows values of k_1 and k_2 for the reaction between a fixed ester (p-nitrophenyl acetate) and various amines.

As expected, k_1 is almost unaffected by changing the chain length of the amine. The reason is that k_1 depends on the encounter between a (short) ester and a (varying length) amine. Therefore, even if the HI between the acetate group on the ester and the alkyl group on the amine does play a role, it will not vary markedly with the chain length of the amine.[†]

[†] The basicity of these amines are known to be almost independent of the chain length. It is also assumed that their nucleophilities are very similar.

Table 3.9

Rate Constants for the Reaction of *p*-Nitrophenyl Acetate with Alkyl Amines[a]

Alkyl amine	k_1	k_2
Ethylamine	0.011	6×10^{-5}
n-Hexylamine	0.008	$<6 \times 10^{-3}$
n-Heptylamine	0.008	$\sim 8 \times 10^{-3}$
n-Octylamine	0.007	0.015
n-Nonylamine	0.010	0.0105
n-Decylamine	0.009	0.11

[a] From Oakenfull (1973).

In contrast, we observe a dramatic change in the values of k_2. This is clearly a reflection of the amine–ammonium encounters in the amine-catalyzed pathway. This is a vivid demonstration of the role of the HI in the mechanism of the reaction. Of course, one may suspect that the direct intermolecular forces between the large alkyl groups would also have an effect on the amine–amine encounter. To exclude this possibility, one needs a comparative study in a nonaqueous solvent. As we shall see below, this dramatic trend does not appear if large quantities of ethanol or dioxane are added.

Table 3.10 shows a similar set of data, but now the fixed ester has a long chain (decanoate) and the variation of the amines is as in Table 3.9.

Table 3.10

Rate Constants for the Reaction of *p*-Nitrophenyl Decanoate with Alkyl Amines[a]

Alkyl amine	k_1	k_2
n-Hexylamine	0.006	0.004
n-Heptylamine	~ 0.007	0.0097
n-Octylamine	<0.01	0.272
n-Nonylamine	<0.01	1.93
n-Decylamine	<0.03	30.3
n-Dodecylamine	—	193.0

[a] From Oakenfull (1973).

Table 3.11

Rate Constants for the Reactions between p-Nitrophenyl Acetate and n-Decylamine (S–L for Short and Long) and between p-Nitrophenyl Decanoate and n-Decylamine (L–L for Long and Long)[a]

x_{EtOH}[b]	S–L Reaction		L–L Reaction	
	k_1	k_2	k_1	k_2
0	0.009	0.11	<0.03	30.3
0.05	<0.01	0.76	<0.03	198.0
0.10	<0.01	0.82	<0.03	215.0
0.15	0.082	0.31	<0.03	97.0
0.20	0.086	<0.06	0.075	<0.1
0.25	0.084	—	—	—

[a] From Oakenfull (1973).
[b] x_{EtOH} is the mole fraction of ethanol in the water–ethanol mixture that forms the solvent.

We note again that in spite of the fact that k_1 involves an encounter between a large ester and a large amine, the values of k_1 do not vary in a dramatic fashion[†] as does k_2. This finding might indicate that the formation of the intermediate complex I in Figure 3.13 does not involve an encounter between the alkyl groups of the ester and the amine, at least not in a step that is a rate-determining one. The values of k_2 change by a few orders of magnitude within this series, and as expected all the k_2 values of this table are larger than the corresponding values in Table 3.9. This finding confirms the conjecture that amine–amine interaction, in the amine-catalyzed pathway, is an important factor that determines the value of k_2.

Perhaps the most convincing evidence that what we have observed is a manifestation of HI, i.e., an effect that is due to the presence of the aqueous environment, and not a result of the van der Waals forces between the alkyl groups, is obtained from Table 3.11. Here we have two sets of data: one for the reaction between a short ester and a long amine, the second for a reaction between a long ester and a long amine.

[†] Of course the symbol <0.01 in Table 3.10 does not imply that the k_1 values are constants. Presumably the authors meant to indicate that all values of k_1 in Table 3.10 are of the order of 0.01.

It is clear that k_1 is almost unchanged by the addition of small amounts of ethanol up to $x \approx 0.1$. There seems to be an increase in k_1 for the higher concentrations of ethanol. However, no explanation of this behavior has been provided in the original work (where it is merely said that ethanol has little effect on k_1). We have already established from the data of Tables 3.9 and 3.10 that the important quantity that bears information on HI is k_2 rather than k_1. Therefore we turn to examine the variation of k_2 with x. In both cases we find an "initial" increase in the value of k_2 upon the addition of ethanol, and then the values of k_2 drop sharply at higher concentrations of ethanol.

This behavior is very similar to that found from other measurements, namely, between $x \approx 0.05$ and $x \approx 0.15$, where the strength of the HI increases, but decreases steeply thereafter. This behavior strongly supports the view that the effect of the chain length on the value of k_2 is indeed due to HI. (It is regrettable, however, that values of k_2 are not available for $x = 0.25$ and higher ethanol concentrations.) The "initial" effect of ethanol on the values of k_2 cannot be determined from these results. The reasons are similar to those given in the previous section, namely, one may expect to find a different "initial" behavior within the interval $0 \le x \le 0.05$, as was actually found from other sources. More important, however, is the fact that all these measurements are carried out in a buffer solution of ionic strength 0.1 M. This clearly cannot be considered as a pure water–ethanol system, and therefore comparison with other results on the behavior of the HI in this system is difficult to make. Finally, it should be borne in mind that k_2 is not a straightforward measure of the strength of HI. It is obviously affected by HI, but we do not know exactly how to separate the HI part from the other factors that contribute to the value of k_2. For all these reasons we believe that it is premature to come to a conclusion about the "initial" effect of ethanol on the HI from such data.

It is interesting to note that Blyth and Knowles (1971) compared the rate constant of the reaction between p-nitrophenyl decanoate with ethylamine and with n-decylamine, and found that in water the second is about 300 times faster than the first. But in a 1-to-1 solution (by volume) of water and p-dioxane they found that the rates of both reactions were small and that the rate of the reaction with n-decylamine was marginally smaller than the rate of reaction with the ethylamine. These results further support the view that the effect we are observing is indeed a manifestation of HI.

It is also of interest to cite the values of k_2 reported by Oakenfull and Fenwick (1975) for H_2O and D_2O. Thus, for the reaction between p-nitrophenyl decanoate and n-decylamine (long–long reaction) we have k_2 (in D_2O) $= 306 \ m^6 \ mol^{-2} \ sec^{-1}$, as compared with k_2 (in H_2O) $= 30.3$. On the other hand, for the reaction between p-nitrophenyl acetate and ethylamine (short–short reaction) the corresponding values are k_2 (in D_2O) $< 3 \times 10^{-5}$ and k_2 (in H_2O) $= 6 \times 10^{-6}$ (in the same units). Thus, in the first case, where HI is involved we observe almost a tenfold increase in k_2 upon replacing H_2O by D_2O. It is, however, impossible at present to infer from these results the relative strength of HI in the two solvents, first because many other factors are involved in the determination of k_2, and it is difficult to isolate the contribution from HI. Secondly, the measurements are carried out not in pure solvents, but in buffer solutions, which clearly will have different properties than pure H_2O and D_2O. In spite of these limitations, we suspect that the increase in the value of k_2 found in D_2O is probably due to the *conditional* nature of the HI that is involved here (see Appendix A.3). If this is true, then the trend observed here is similar to that reported in Section 2.3.

We now turn to a slightly different system, which perhaps demonstrates the role of HI in a more dramatic fashion. The reaction is similar to the one described in Figure 3.13, but the alkylamine is now replaced by N-alkylimidazole with varying size of the alkyl group R. The overall reaction is described in Figure 3.14. This kind of reaction was first studied in connection with the problem of HI by Blyth and Knowles (1971), and a more systematic dependence on the chain length of the alkyl group has been examined by Oakenfull and Fenwick (1974a).

Table 3.12 shows the rate constants k for the reaction of p-nitrophenyl decanoate with N-alkylimidazole in pure water at 25°C and in water–ethanol mixtures of composition $x = 0.1$ and $x = 0.32$. Note that the mechanism of this reaction is different from that discussed in connection with the reaction depicted in Figure 3.13. The authors have assumed that the overall reaction may be described by an apparent second-order rate

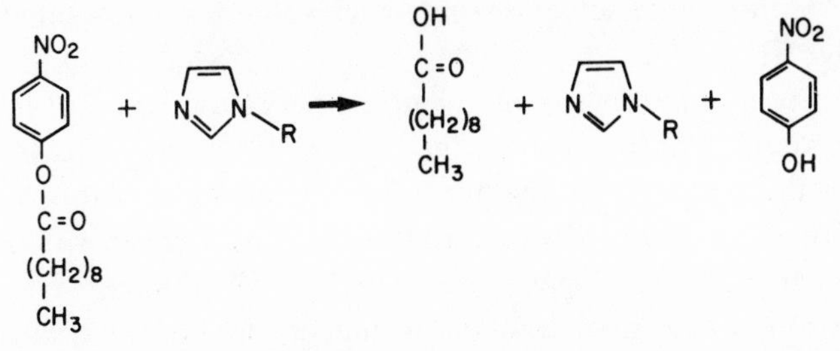

Figure 3.14. The reaction between p-nitrophenyldecanoate and N-alkylimidazole.

Table 3.12

Rate Constants k (in $10^{-3}\ \text{m}^3\ \text{mol}^{-1}\ \text{sec}^{-1}$), for the Reaction of
p-Nitrophenyl Decanoate with N-Alkylimidazoles, in Pure Water and in
Two Mixtures of Water and Ethanol at $25°C^a$

Alkyl group of the imidazole	$x_{\text{EtOH}} = 0$	$x_{\text{EtOH}} = 0.10$	$x_{\text{EtOH}} = 0.32$
H	0.156	0.115	0.0443
Methyl	0.155	0.114	—
Ethyl	0.159	0.112	0.0439
Propyl	0.148	0.115	—
Butyl	0.155	0.113	0.0450
Pentyl	0.381	0.132	—
Hexyl	1.00	0.275	0.0440
Heptyl	3.48	0.467	—
Octyl	12.1	0.835	0.0447
Nonyl	26.3	1.52	—
Decyl	91.7	2.57	0.0458

^{a} All the concentrations are below the CMC.

constant k defined by

$$v = - \frac{d[\text{Es}]}{dt} = k[\text{Es}][\text{A}] \qquad (3.98)$$

where $[\text{Es}]$ and $[\text{A}]$ are the concentrations of the ester and of the amine, respectively.

First we note that the rate constant for the reaction between p-nitrophenyl *acetate* with N-methylimidazole in water is $k = 0.402 \times 10^{-3}\ \text{m}^3\ \text{mol}^{-1}\ \text{sec}^{-1}$; this could serve as a reference value for a reaction that does not involve HI.

Values of k in pure water are relatively small and constant for the first five alkyl groups. For larger ones, the value of k sharply increases with the increase of the chain length. Similar behavior, but somewhat less dramatic, is shown in the mixture of water and ethanol with composition $x = 0.1$. On the other hand, for higher concentrations of ethanol, say, $x = 0.32$, the values of k are all small and almost independent of the chain length of the alkyl group.

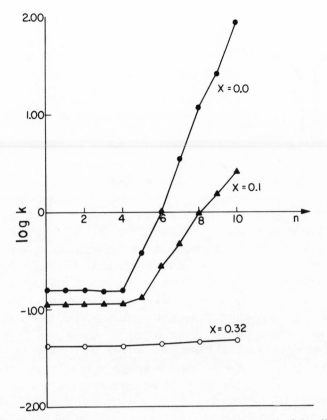

Figure 3.15. The dependence of the rate constant on the chain length of the alkyl group for water and two mixtures of water and ethanol [based on data from Oakenfull and Fenwick (1974a)].

This behavior is also shown graphically in Figure 3.15, where we have plotted log k as a function of n. The quantity $-kT \ln k$ may be interpreted in terms of the free energy of activation for the reaction under consideration.

In fact, one can go one step further and form the ratio of $k(L, L)$ for the long–long reaction and $k(S, S)$ for the short–short reaction, to define

$$\Delta(\Delta G^{\ddagger}) = -RT \ln[k(L, L)/k(S, S)] \tag{3.99}$$

where R is the gas constant. This quantity is interpreted as the hydrophobic contribution to the free energy of activation. However, this procedure is uncertain, since we do not know how to separate HI from other contributions to the rate constant, and in any case this quantity will have the nature

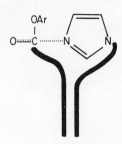

Figure 3.16. A possible configuration for the transition state of the reaction described in Figure 3.14. The hydrocarbon chains (heavy lines) must be long enough to span the imidazole ring before contact between them is possible.

of a *conditional* free energy change similar to the quantities discussed in Section 3.7 and in Appendix A.3. But this further processing of the data is really not necessary. The results of this work are exciting enough even without any further processing. The interpretation given by the authors is interesting and convincing. For short chains $n \leq 4$, because of the bulky imidazole ring the two alkyl groups do not "see each other," and therefore HI does not contribute to the reaction rate, hence k is independent of n. Once the chain length of the alkyl group is long enough to span the imidazole ring (see Figure 3.16), then the two alkyl groups can reach each other, and the HI can play a role in the reaction mechanism. This explains the steep increase of k with n.

At high concentrations of alcohol $x = 0.32$, the HI plays no important role, hence the value of k is almost independent of n for all values of n. This is strong evidence that the effect we have observed in water is indeed a result of the HI, i.e., the result of the properties of the aqueous environment. The values of k (or log k of Figure 3.15) for the mixture of water and ethanol at $x = 0.1$ are somewhat intermediate between pure water and the mixture of $x = 0.32$. Thus, it still has the n dependence similar to pure water, but the slope is now smaller than that for water. This means that the effect of HI still persists at $x = 0.1$. A rough estimate of the slope of the two curves (in the region of $n \geq 4$) gives

$$-\frac{\partial(RT \ln k)}{\partial n} = \Delta G^{\ddagger} \text{ (per } CH_2) \qquad (3.100)$$

and one finds that

$$\Delta G^{\ddagger} \text{ (per } CH_2 \text{ in pure water)} < \Delta G^{\ddagger} \text{ (per } CH_2 \text{ in mixture, } x = 0.1)$$

This seems to indicate that the HI is weakened by alcohol at a concentration of $x = 0.1$, which contradicts our previous findings that at this concentration of alcohol the HI is actually strengthened. We believe that this apparent inconsistency is simply a result of the fact that the quantity defined in (3.100) is not really a measure of the HI. Therefore comparison with other measures of HI is not warranted.

There are two related sets of data relevant to the subject of this section. One is the effect of added urea on k. We report in Table 3.13 values of k from Blyth and Knowles (1971) for water and for solution containing 4 M urea. Note that the values of k reported in Table 3.13 for water are different from those reported in Table 3.12. The reason is that the two sets of experiments were carried out under different conditions. [For details see Oakenfull and Fenwick (1974a).] However, we have here the values of k for pure water and for solution of 4 M urea as reported by Blyth and Knowles, so that there is consistency in the conditions under which the values of k were obtained. Note also that "pure water" means here a solution that does not contain urea. All the reactions were carried out in 0.02 M sodium carbonate–bicarbonate buffers containing 0.99% (by volume) acetone.

The results reported in Table 3.13 on the effect of urea on the reaction rate are inconclusive. In some cases the rate increases, and in others it decreases, upon the addition of urea. The reason for this is probably that Blyth and Knowles did their measurements at high concentrations of the reactants, so that aggregates of molecules could be formed, a fact that should have a marked effect on the kinetics.

One further result from Blyth and Knowles is worth noting. From the temperature dependence of the rate constants, they estimated the free energy, enthalpy, and entropy of activation of the reaction at 25°C. Taking

Table 3.13

Values of k for the Reactions of Figure 3.14 in Pure Water and in Aqueous Solution of 4 M Urea at 25°C[a]

Reactants	Pure water	Solution of 4 M urea
p-Nitrophenyl acetate + N-ethylimidazole	0.397	0.335
p-Nitrophenyl octanoate + N-ethylimidazole	0.045	0.188
p-Nitrophenyl decanoate + N-ethylimidazole	0.014	0.060
p-Nitrophenyl acetate + N-decylimidazole	0.548	0.465
p-Nitrophenyl octanoate + N-decylimidazole	5.53	2.73
p-Nitrophenyl decanoate + N-decylimidazole	11.66	5.66

[a] Units as in Table 3.12.

the differences of these quantities as in (3.99) one obtains

$$\Delta(\Delta G^{\ddagger}) = \Delta G^{\ddagger}(L, L) - \Delta G^{\ddagger}(S, S) = -2.0 \text{ kcal/mol}$$

$$\Delta(\Delta H^{\ddagger}) = +1.1 \text{ kcal/mol} \tag{3.101}$$

$$\Delta(\Delta S^{\ddagger}) = +10 \text{ e.u.}$$

Thus the "hydrophobic free energy of activation" is negative, whereas the corresponding enthalpy and entropy changes are positive. These results are consistent with other results on the thermodynamics of the HI process (see also Chapter 5).

Comment

The results of this section are not directly comparable to the results described in other sections of this chapter. Yet they are very important, and perhaps even more important than measurements of HI in systems at equilibrium. After all the whole subject of HI has evolved from biochemistry, and real biochemical systems are far from equilibrium. Therefore we believe that a more systematic study of the role of HI on chemical kinetics should be undertaken. Of course, at the present level of our understanding of HI it will be more fruitful to study simple model reactions like the ones described in this section, rather than more complex biochemical reactions. The latter, though important in their own right, could give only very qualitative information on the role of the HI, the one factor that we are trying to study in isolation.

3.10. APPLICATION OF NMR AND ESR TECHNIQUES

In this section, we present a brief account of some recent studies of (pairwise) hydrophobic interactions using NMR and ESR methods. In many of the cases considered below, the interpretation of results obtained through these techniques depends on the interpretation of the data pertaining to solute–solvent interactions. To be more specific, suppose we examine the NMR relaxation behavior of the CH_3 protons of t-BuOH in a dilute solution. For a single t-BuOH molecule, the proton relaxation depends on τ_r, the rotational correlation time of the solute molecule in the solvent. As will be noted below, there is experimental evidence that τ_r (t-BuOH) is 2–3 times longer in water than in other isoviscous solvents. Now, if we follow this proton relaxation while increasing the t-BuOH concentration into the region where solute–solute interaction becomes significant, we will find that intermolecular effects also contribute to the proton relaxation. This contribution depends on the relative motion (rotational and diffusional) of the solute molecules in the solvent. Clearly, then, in order to identify solute–solute interactions peculiar to aqueous solutions, we must be able to extract those contributions due to the unique solvation effects of the solutes in water. Since, at present, it is not possible to carry out such an analysis in a satisfactory manner, it seems most appropriate to examine data pertaining to both hydration and solute–solute interactions. As will be seen, most of the results are qualitative but they do conform to the current concepts of hydrophobic hydration and hydrophobic interaction.

It is also important to remark that these methods cannot be applied to very dilute solutions of hydrocarbons in water. Conventional NMR techniques require moderately high concentrations (e.g., $\gtrsim 10^{-2}\,M$), so only solutes containing solubilizing polar functional groups can be studied. ESR methods, although workable at very low concentration (e.g., $\approx 10^{-5}\,M$), require an unpaired electron, which, if we exclude very short-lived species, can be found only in polar molecules or ions. These requirements impose limitations for characterizing "true hydrophobia," but this is not too severe since, in most practical systems where hydrophobic effects contribute, the solutes do contain polar functional groups.

The previous NMR and ESR work relevant to the problem considered here cannot be described in detail, but it seems possible to illustrate the applicability and potential of these methods through a few pertinent studies. A more extensive compilation and discussion of NMR data can be found, for example, in the review by Zeidler (1973). In choosing the examples discussed below, we have restricted ourselves to results of magnetic relaxation studies. Preference of the latter over chemical shift data is based on the fact that the assignment of specific contributions (e.g., due to hydrophobic effects) to the observed chemical shifts is often protracted and remains ambiguous in many cases (Wen and Hertz, 1972; Kingston and Symons, 1973). Before discussing some specific results, it will be helpful, however, to recall some generalities on the phenomenon of magnetic relaxation.

The relaxation of a collection of nuclear (or electron) spins precessing in an external magnetic field will occur because of fluctuations in the local field at the species observed. For some important cases of interest here, fluctuating fields are due to atomic and molecular motions which modulate the magnetic interactions operating on the spin: anisotropic interaction with the external magnetic field, hyperfine or dipolar interactions with other magnetic species, and possibly other effects (zero-field splitting, spin–rotation interactions, quadrupolar, etc.). The efficiency of the various field-modulating mechanisms for inducing NMR or ESR relaxation requires that they occur at a rate comparable to the resonance frequency of the species observed: NMR, 10^6–10^8 sec^{-1}; ESR, 10^9–10^{11} sec^{-1}.

The dynamics of magnetic relaxation processes are usually described by two characteristic times T_1 and T_2 (Abragam, 1961; Wertz and Bolton, 1972; Atherton, 1973). T_1 specifies the exponential decay of the magnetization induced in the sample in the direction of the external magnetic field. Hence, it is referred to as the longitudinal relaxation time; because the process requires exchange of energy between the spins and their environment, T_1 is also termed the spin–lattice relaxation time. It is usually measured by following the return to equilibrium of the spin system after a sudden perturbation, e.g., after a pulse of the radiofrequency field. Likewise, T_2 is the characteristic time for loss of magnetization in the sample, in the plane perpendicular to the external magnetic field; it is therefore denoted the transverse relaxation time. Since the latter is dependent on the loss of the phase coherence between the precessing spins, it is also referred to as the spin–spin relaxation time. The magnitude of T_2 is usually determined from the width of the resonance line, which is inversely related to the lifetime of the spin state (the latter depends on both T_1 and T_2, though from the separation into the longitudinal and transverse components we have $T_1 \geq T_2$).

The theoretical description of magnetic relaxation phenomena in relation to molecular motion is now quite comprehensive and has been applied successfully to a great diversity of chemical systems (Hertz, 1973a,b).

Several NMR investigations on aqueous solutions of organic compounds have indicated that the local solvent viscosity in the neighborhood of alkyl groups is significantly higher than the bulk viscosity of water. This is borne out, for instance, by the work of Howarth (1975) on ^{13}C NMR relaxation of t-BuOH in water and in several other solvents. In this study, the spin–lattice relaxation rates were measured as a function of t-BuOH concentration and the data extrapolated to infinite dilution to eliminate the effects of solute–solute interactions. The zero concentration values $(1/T_1)_0$ plotted against solvent viscosity show a regular viscosity dependence for all solvents except H_2O and D_2O; in the latter, $(1/T_1)_0$ is, strikingly, threefold greater than in other isoviscous liquids.

The ^{13}C relaxation rate in alkyl groups is dominated by dipolar coupling with the C–H protons and has been shown to be proportional to the rotational correlation time τ_c of the C–H dipole–dipole vector. If the rotational motion of the molecule investigated is isotropic, τ_c is a direct measure of the reorientational correlation time of the molecule τ_r, i.e., $\tau_r = 3\tau_c$. From the results examined here, it may thus be inferred that the rotational motion of the t-BuOH molecule is strongly hindered in water compared to other isoviscous solvents (provided the relaxation mechanism and rotational isotropy are comparable in each solvent). This unique ability of water to restrict the rotation of organic molecules has also been shown from NMR studies on other compounds, by comparing the molecular τ_r values in the pure liquid compounds and dilute aqueous solutions.

Investigation of the ESR relaxation of hydrophobic nitroxide radicals provides complementary evidence on the anomalous rotational behavior of the aqueous solutes. Following our earlier remarks, the ESR method should be more sensitive to molecular dynamics, since the resonance frequency involved (10^{10}–10^{11} sec^{-1}) is close to the rates of molecular motions in solutions. However, ESR relaxation times are usually more difficult to measure, so the end result may be of overall lower accuracy. Nonetheless, from ESR spin–spin relaxation, τ_r of the radical 2,2,6,6-tetramethyl-4-piperidone-N-oxide was found to be approximately twofold longer in water than in other isoviscous solvents (Jolicoeur and Friedman, 1971, 1974). Moreover, for a similar (more hydrophobic) radical, 2,2,6,6-tetramethyl-piperidine-N-oxide (TEMPO), a detailed analysis of the relaxation mechanisms suggested that the rotational motion of the radical was less hindered at 5°C than at 25°C. This result is consistent with the picture of a clathrate-like hydration of the hydrophobic solutes, so further studies along these lines would seem warranted.

Based on the evidence presently available, one is strongly compelled to conclude that the dynamic behavior of solutes containing large apolar groups is markedly different in water compared to other media. It remains to be determined, however, whether this is due to the intrinsic properties of liquid water (small molecular size, high particle density, extent of hydrogen bonding) or appears as a result of solvent structural modifications induced by the apolar groups.

By analogy with the ^{13}C results on t-BuOH discussed above, the rotational motion of the water molecules can be monitored through the NMR relaxation rate ($1/T_1$) of the water protons or other isotopic species, namely, ^{2}H and ^{17}O. For the protons, $1/T_1$ is governed by proton–proton dipolar interactions occurring within the water molecule and is dependent on τ_c and τ_r of the water molecule. With ^{2}H or ^{17}O, the nuclear relaxation is due principally to modulation of the interaction between the nuclear quadrupole and the electric field gradient at the observed nucleus. This process also depends on the rotational motion of the water molecules.

The general conclusion derived from studies of NMR relaxation of water is that with most nonelectrolytes and organic ions, $1/T_1$ and τ_c increase with solute concentration and with the size of the alkyl groups. The τ_c values obtained from limiting slopes with various nonelectrolytes (alcohols, tetrahydrofuran, acetone, acetonitrile) are all larger than in pure water [e.g., $\tau_c(H_2O)_{\text{soln } c \to 0}/\tau_c(H_2O)_{\text{liq}}$ is in the range 1–1.65]. Although possible contributions due to the hydration of the polar functional groups (and changes in bulk viscosity) on the concentration dependence of $1/T_1$ still must be sorted out, the results quoted above support the idea that water molecules are structurally more organized in the vicinity of apolar groups than in bulk water. This is fully consistent with evidence from near-infrared spectroscopy (e.g., Philip and Jolicoeur, 1973; Paquette and Jolicoeur, 1977).

To investigate solute–solute interactions of interest in the present context, one can make use of intermolecular magnetic interactions between nuclei on apolar groups. In this way, the intermolecular proton relaxation rates have been measured to evaluate the extent of interaction between carboxylic acid molecules in aqueous solutions (Hertz, 1973a). Using carboxylic acids deuterated at specific positions, the authors advantageously exploited the difference in the NMR relaxation mechanism of ^1H and ^2H. Hence, they were able to monitor selectively the intermolecular relaxation effects for the methyl and each methylene group of the alkyl chain.

The contribution to $1/T_1$ from intermolecular magnetic interactions is related to the translational motions of the solute species, and its first-order formula is

$$(1/T_1)_{\text{inter}} \propto \frac{N}{Db} \qquad (3.102)$$

where N is the concentration of nuclear spins, b the proton–proton distance of closest approach, and D the self-diffusion coefficient of the solute (the *relative* solute–solute diffusion coefficient has been taken as $2D$) (Zeidler, 1973).

The data reported for acetic, propionic, and butyric acid show that the normalized quantity $(D/N)(1/T_1)_{\text{inter}}$ is markedly higher in dilute aqueous solutions than in the pure liquid acids, i.e., up to mole fraction $x \sim 0.2$. However, the $(1/T_1)$ enhancement due to intermolecular effects is greater for the protons α of the —COOH than for the others. For instance, with propionic acid, the enhancement is about twofold greater for the CH_2 protons compared to the CH_3 protons. Thus, the overall intermolecular relaxation rates may be taken as evidence for an increase in the probability of solute–solute proximity in dilute aqueous solutions, but one cannot be certain that the apolar groups have a dominant influence on this behavior.

Similar experiments have been designed to investigate the methyl proton relaxation rates of hydrophobic solutes (e.g., t-BuOH) in the presence of the hydrophobic nitroxide radical, TEMPO (Jolicoeur et al., 1976). Because of the large magnetic moment of the unpaired electron spin, the electron–nuclear-dipolar interaction is extremely effective in inducing proton relaxation. Hence, only low radical concentrations are required ($\sim 10^{-2}$ M), and the results most likely reflect the consequence of pairwise solute–solute interactions. In the cases to be considered here, the proton relaxation rate enhancement is given by (Abragam, 1961)

$$\frac{1}{T_1} - \frac{1}{T_1{}^0} \propto \frac{N\eta}{T} \qquad (3.103)$$

Here, $1/T_1{}^0$ is the proton relaxation rate in the absence of paramagnetic species, and η represents the bulk viscosity.

The t-BuOH–TEMPO experiments performed in water and various other solvents gave the following general result: at fixed concentrations of the solutes (e.g., t-BuOH 0.05 M, TEMPO 0.02 M), the relaxation enhancement in aqueous solution is roughly twofold greater than in other isoviscous media. The same behavior was observed with t-BuNH$_2$, but not with t-BuNH$_3{}^+$ for which the behavior in aqueous and nonaqueous media appeared alike. It may then be inferred that the effects are independent of the nature of the functional group provided the latter is not ionic. These results have been interpreted as evidence that, in water, the large organic solutes experience a greater tendency to attract each other than in other solvents. More specifically, the integrated product of the solute–solute pair correlation function $g(r)$ times a relaxation operator $R(r)$ [i.e., $4\pi \int R(r)g(r)r^2\,dr$] is twofold greater in water than in other solvents. As with the intramolecular relaxation results of Howarth discussed above, it cannot be excluded that the relaxation operator $R(r)$ differs in some specific way in aqueous (H_2O or D_2O) solutions. However, in view of the results obtained with t-BuNH$_3{}^+$, there is strong presumption that the intermolecular relaxation enhancement is due to changes in $g(r)$. Whether such changes occur at the solute–solute contact distance, or at some larger separation, remains to be elucidated. On this aspect, other evidence is afforded by ESR data on the rate of electron spin exchange (Heisenberg spin exchange) in aqueous solution of nitroxide radicals.

In a bimolecular collision between paramagnetic species, the unpaired electrons can undergo exchange of their spin states. By analogy with proton che mical exchange in NMR, this phenomenon will result in line broadening when the two spin states involved correspond to different resonant magnetic fields (or frequencies). The rate constant for this process may be obtained from ESR linewidth measurements using (Miller *et al.*, 1966)

$$\left(\frac{1}{T_2}\right)_{ex} = \frac{3^{1/2}}{2}\,\gamma_e\,\Delta H_{ex} = k_{ex}[x] \qquad (3.104)$$

where γ_e is the gyromagnetic ratio of the electron, ΔH_{ex} is the linewidth due to spin exchange, $[x]$ is the radical concentration, and k_{ex} is a second-order rate constant. Since k_{ex} must be related to the radical–radical collision rate constant k_e, this provides a further method of characterizing the solute–solute interaction. Because the electron-spin exchange process is a result of short-range interaction (orbital overlap), this type of result should complement the information obtained from the longer-range intermolecular dipolar interaction in NMR relaxation.

Following this strategy, Ablett *et al.* (1975) have investigated collision rate constants for the TEMPO radical in water and in several organic solvents. They further correlated their results with k_e values obtained from diffusion rate measurements on diamagnetic analogs of TEMPO using pulsed NMR methods. An important finding of their study is that the spin-exchange rate constants are lower (by approximately 30%) in water than in isoviscous organic solvents. Here again, two aspects can be involved, namely, the frequency of the collisions and their spin-exchange efficiency. The spin exchange occurring at very short radical–radical distances is likely to depend on the relative orientation of the N-oxide molecules and this could be different in water and in other solvents. Alternatively, the results can be explained from the frequency factor, if the hydration shells of the hydrophobic solutes impose an additional barrier to direct radical–radical contact, i.e., clathratelike hydration could tend to prevent solute–solute contact. The latter explanation was preferred by Ablett *et al.* (1975), and it is not inconsistent with a net overall solute–solute attraction evidenced from NMR data quoted above. For instance, the collision rate of a TEMPO molecule with another TEMPO molecule (ESR exchange), or with other hydrophobic solutes such as *t*-BuOH (NMR relaxation), may be lower in water but, at the same time, the lifetime of the pairwise interaction can be much longer.

From the various results quoted above one may conclude that the rotational motion of hydrophobic solutes is markedly different in water (or D_2O) compared to other solvents of similar bulk viscosities. This may, in part, be a consequence of the peculiar properties of bulk liquid water, but there is reliable evidence showing that the water molecules in the vicinity of nonelectrolytes are rotationally hindered, more so than in pure water.

Pertaining to the hydrophobic interaction, NMR relaxation provides evidence towards a net excess attraction between hydrocarbonlike molecules in water, again compared to other solvents. At present these different modes of behavior can only be discussed in terms of changes in a product containing a relaxation operator (mechanistic factor) and solute–solute distances. The true relaxation operator may be very complex (e.g., involving the relative motions and orientations of the solute species) and the unique rotational dynamics of the solute molecules in water may indeed result in specific changes of $R(r)$ in this solvent. Until this aspect can be elucidated the experimental results are interpreted as being due to changes in the solute–solute pair correlation function $g(r)$ at some short solute–solute distance. ESR experiments suggest that these changes in $g(r)$ in water occur mainly at separations greater than contact distance.

Chapter 4

Hydrophobic Interaction among Many Solute Particles

4.1. INTRODUCTION

In the previous chapter we were concerned with the HI between two solute particles. This has been considered to be the first and most important step towards the full characterization and understanding of the phenomenon of HI. It should be remembered, however, that the problem of *pairwise* HI has been isolated as a single factor that contributes to the total "driving force" of very complex biochemical processes.

At present we are still far from having a full, or even a satisfactory, knowledge of the pairwise HI phenomenon. Much is left to be done on both the experimental and the theoretical fronts before we may claim that this goal has been reached. Nevertheless, this fact alone should not hinder our efforts to study more complex processes involving HI. The next step that we have in mind is the study of the HI among many simple solute particles in a solvent. This step serves as a bridge leading from the simplest pairwise HI to the enormously more complex biochemical processes.

This chapter is devoted to surveying the various experimental sources from which we can obtain information on HI among many solute particles. As in the case of pairwise HI we shall find that information on this subject is rather fragmentary and much more should be done before any reasonably clear view of this field emerges.

We shall also devote some space to discussing processes such as micelle formation and conformational changes in biopolymers. All of these certainly involve, in one way or another, the concept of HI. However, care must be exercised to make a distinction between two classes of processes. On one hand we have those processes from which we can *extract* information on HI, and hence contribute to our store of knowledge in this field. On the other hand we have the more complex processes in which many factors besides HI combine to determine their driving force. Here we *use* our knowledge on HI in an attempt to understand the mechanism and the relative importance of the various factors that govern the overall process.

For example, there exists a vast amount of experimental data on the properties of micelles in aqueous solutions. It is clear that HI plays an important role in their formation. It is also clear that there are other factors involved, such as charge–charge interaction and the solvation of polar groups in water. What is not clear is how to extract information on HI from any given piece of information on these systems. This point will be further elaborated upon in Section 4.8.

In our introductory discussion of the pairwise HI, we have pointed out that the solute–solute interaction may be considerably modified when we proceed from the vacuum (or the direct) interaction to the HI. For example we mentioned the temperature and pressure dependence of the HI that may be quite outstanding even when we assume that the direct interaction is temperature and pressure independent. We also noted that these peculiar features of the HI stem from the fact that the HI is an average quantity, the averaging process being carried out over all the possible configurations of the solvent molecules (see also Chapter 5 for further treatment of this subject).

In proceeding to the study of the HI among many particles, there is a new feature that should be recognized from the very outset. This is the non-additivity effect, sometimes also referred to as the cooperative effect of the HI, and its possible dependence on the configuration of the interacting solutes (or nonpolar groups).

It is true, though, that at present we know almost nothing about the extent of the nonadditivity effect of the HI. Nevertheless, it is appropriate at this stage to introduce a precise definition of this concept, hopefully paving the way for future investigation in this field.

Consider for simplicity the case of three simple solute particles at some close configuration, as indicated in Figure 4.1. Let $\mathbf{R}_i$ be the position vector of the center of the ith solute particle, and $R_{ij} = |\mathbf{R}_j - \mathbf{R}_i|$ be the scalar distance between the ith and the jth particles. If these three

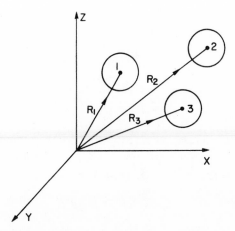

Figure 4.1. Three solutes at a configuration $\mathbf{R}_1$, $\mathbf{R}_2$, $\mathbf{R}_3$.

particles are in vacuum then $U(\mathbf{R}_1, \mathbf{R}_2, \mathbf{R}_3)$ will designate the work required to bring these particles from fixed positions at infinite separation from each other (i.e., $R_{ij} = \infty$ for i, $j = 1$, 2, 3) to the final configuration $\mathbf{R}_1$, $\mathbf{R}_2$, $\mathbf{R}_3$. This work will be referred to as the *direct* interaction among the three solutes particles for the configuration $\mathbf{R}_1$, $\mathbf{R}_2$, $\mathbf{R}_3$.

For simple solutes, such as argon or methane molecules, $U(\mathbf{R}_1, \mathbf{R}_2, \mathbf{R}_3)$ is approximately pairwise additive in the sense that it can be written as a sum of the interactions between each of the pairs i, j, namely,

$$U(\mathbf{R}_1, \mathbf{R}_2, \mathbf{R}_3) = U(\mathbf{R}_1, \mathbf{R}_2) + U(\mathbf{R}_2, \mathbf{R}_3) + U(\mathbf{R}_1, \mathbf{R}_3) \qquad (4.1)$$

For hard-sphere (HS) particles the last property may be considered as part of our definition of the HS. For real molecules we shall assume that this is a good approximation (in fact most of the progress in the theory of simple liquids has been based on the assumption of pairwise additivity of the total potential energy (Hill, 1956; Hansen and McDonald, 1976). There are known cases where large nonadditivity effects do exist— an example might be water molecules. However, in all the forthcoming discussions we shall assume that relation (4.1) strictly holds for all the configurations $\mathbf{R}_1$, $\mathbf{R}_2$, $\mathbf{R}_3$. The reason for doing so is twofold: In the first place we shall be interested in simple solutes for which (4.1) is indeed a good approximation. More important, however, is our desire to stress a new feature of the HI that may appear independently of the assumption of pairwise additivity of the direct interaction.

Next we consider the same process as described above, but now the interparticle space is filled by a solvent. The process is carried out while maintaining a constant temperature and pressure in the system. The total work associated with this process is given by the change of the Gibbs free energy, namely,

$$\Delta G(\mathbf{R}_1, \mathbf{R}_2, \mathbf{R}_3) = G_3(\mathbf{R}_1, \mathbf{R}_2, \mathbf{R}_3) - G_3(\infty) \tag{4.2}$$

where $G_3(\infty)$ stands for the free energy of the system with the three solutes being at fixed positions at infinite separation from each other. Similarly $G_3(\mathbf{R}_1, \mathbf{R}_2, \mathbf{R}_3)$ is the free energy of the system with the solutes being at the configuration $\mathbf{R}_1, \mathbf{R}_2, \mathbf{R}_3$ (the thermodynamic variables T, P are omitted to simplify the notation). The work required to carry out a similar process involving two particles is

$$\Delta G(\mathbf{R}_1, \mathbf{R}_2) = G_2(\mathbf{R}_1, \mathbf{R}_2) - G_2(\infty) \tag{4.3}$$

The quantities defined in (4.2) and in (4.3) are known in the literature as the potentials of average force. We prefer, in the context of this book, to refer to these quantities as free energy changes for the described processes.

The question that may now be asked is to what extent the work $\Delta G(\mathbf{R}_1, \mathbf{R}_2, \mathbf{R}_3)$ may be written as a sum of pairwise terms, in analogy with (4.1). This assumption has indeed been used in the theory of simple fluids and it is known as the Kirkwood superposition approximation (Kirkwood, 1935; Hill, 1956). Nowadays it is generally recognized that this is a poor approximation even for simple fluids, and one can suspect that this is also true for a solution of, say, argon in water. Therefore, we define the extent of nonadditivity in the work $\Delta G(\mathbf{R}_1, \mathbf{R}_2, \mathbf{R}_3)$ by the difference

$$\phi(\mathbf{R}_1, \mathbf{R}_2, \mathbf{R}_3) = \Delta G(\mathbf{R}_1, \mathbf{R}_2, \mathbf{R}_3)$$
$$- [\Delta G(\mathbf{R}_1, \mathbf{R}_2) + \Delta G(\mathbf{R}_2, \mathbf{R}_3) + \Delta G(\mathbf{R}_1, \mathbf{R}_3)] \tag{4.4}$$

In a similar fashion one can extend the definition of nonadditivity for any number of particles. (We discuss here only nonadditivity with respect to *pairs*. One can also define higher-order nonadditivities, but these will not concern us here.)

In classical statistical mechanics one can always split the total work involved in the processes described above into two terms, the *direct* and the *indirect* parts, i.e.,

$$\Delta G(\mathbf{R}_1, \mathbf{R}_2, \mathbf{R}_3) = U(\mathbf{R}_1, \mathbf{R}_2, \mathbf{R}_3) + \delta G^{\mathrm{HI}}(\mathbf{R}_1, \mathbf{R}_2, \mathbf{R}_3) \tag{4.5}$$

$$\Delta G(\mathbf{R}_1, \mathbf{R}_2) = U(\mathbf{R}_1, \mathbf{R}_2) + \delta G^{\mathrm{HI}}(\mathbf{R}_1, \mathbf{R}_2) \tag{4.6}$$

where the indirect part is being referred to as the HI part, since we shall be mainly interested in water as a solvent.

Using the assumption of the additivity of the direct work (4.1) and relations (4.4), (4.5), and (4.6) we find that the nonadditivity defined in (4.4) may be expressed in terms of the indirect parts of the work, namely,

$$\phi(\mathbf{R}_1, \mathbf{R}_2, \mathbf{R}_3) = \delta G^{\mathrm{HI}}(\mathbf{R}_1, \mathbf{R}_2, \mathbf{R}_3) - \delta G^{\mathrm{HI}}(\mathbf{R}_1, \mathbf{R}_2) - \delta G^{\mathrm{HI}}(\mathbf{R}_2, \mathbf{R}_3)$$
$$- \delta G^{\mathrm{HI}}(\mathbf{R}_1, \mathbf{R}_3) \qquad (4.7)$$

The source of the nonadditivity in (4.7) is exactly the same as the one responsible for the peculiar features of the pairwise HI that we have discussed in Chapter 1, namely, the averaging over all possible configurations of the solvent molecules. Since our solvent of interest is water, an anomalous liquid in many respects, we may expect to find that the nonadditivity of the HI is somewhat outstanding in water as compared with other solvents. It is difficult to see exactly how the nonadditivity effect arises, or how it may be related to any known property of the solvent. We shall demonstrate, however, one very simple case of the nonadditivity effect in Section 4.2.

For any real solution we have, at present, no experimental information on the extent of the nonadditivity effect. We shall mention in Sections 4.4 and 4.6 two possible ways of studying this problem by experimental means. Other new methods are urgently needed. One potential source of "experimental" information might be the computer experiments, which may be used to study the nonadditivity effect. This has indeed been applied to the case of simple fluids, and its extension to aqueous solutions should be encouraged.

4.2. A SPECIFIC EXAMPLE FOR DEMONSTRATING THE NONADDITIVITY EFFECT

We demonstrate in this section, by a simple example, how the nonadditivity of the HI may affect the probability of occurrence of certain configurations of a group of solute particles. In a similar fashion it may also affect the preference of one particular conformation of a single complex molecule in the solvent. We shall also see that such a preference effect cannot always be ascribed to an *intramolecular* hydrophobic interaction between nonpolar groups of a single molecule such as in proteins.

Consider four simple particles at two specific configurations as depicted in Figure 4.2. In order to emphasize effects that do not originate from the *direct* interaction between these particles we assume that the total potential energy of interaction is pairwise additive and that the pair potential $U(R_{ij})$

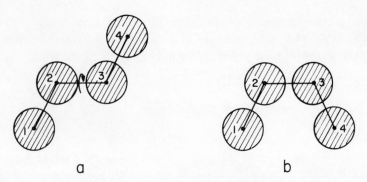

Figure 4.2. Two possible configurations of four solute particles (or groups in one molecule). The configuration b is obtained from a by rotation about the line connecting the centers of particles 2 and 3.

is of very short range. Hence for each of these configurations we write

$$U(\mathbf{R}_1, \mathbf{R}_2, \mathbf{R}_3, \mathbf{R}_4) = U(R_{12}) + U(R_{13}) + U(R_{23}) + U(R_{24}) + U(R_{34}) \quad (4.8)$$

Note that we have neglected $U(R_{14})$, presuming that in both configurations R_{14} is larger than the range of the direct interaction between the particles.

The two configurations a and b in Figure 4.2 are obtained from each other by rotation about the line connecting the centers of particles 2 and 3. Therefore all the distances R_{ij} except R_{14} are the same in the two configurations, hence

$$U(a) = U(b) \quad (4.9)$$

where $U(a)$ and $U(b)$ are the total interaction energies of the four particles at the configurations a and b, respectively.

A straightforward result from (4.9) is that the probability of occurrence of the configuration a is the same as that of b, provided that these are the only particles in the system, i.e.,

$$\frac{P(a)}{P(b)} = \frac{\exp[-U(a)/kT]}{\exp[-U(b)/kT]} = 1 \quad (4.10)$$

The situation becomes markedly different if the same four particles are surrounded by solvent molecules. Referring again to the same configurations a and b of Figure 4.2, and using the notation of Section 4.1, we write the HI among these solute particles as

$$\delta G^{\mathrm{HI}}(a) = \sum_{i<j} \delta G^{\mathrm{HI}}(R_{ij}^a) + \phi(a) \quad (4.11)$$

$$\delta G^{\mathrm{HI}}(b) = \sum_{i<j} \delta G^{\mathrm{HI}}(R_{ij}^b) + \phi(b) \quad (4.12)$$

where R_{ij}^a and R_{ij}^b are the interparticle distances in the configurations a and b, respectively, *including* R_{14} (i.e., $i, j = 1, 2, 3, 4$). $\phi(a)$ and $\phi(b)$ designate the nonadditivity of the HI for the two configurations.

In this system the ratio of the probabilities of finding the two configurations a and b is given by

$$\frac{P(a)}{P(b)} = \frac{\exp[-\Delta G(a)/kT]}{\exp[-\Delta G(b)/kT]} = \frac{\exp[-\delta G^{\mathrm{HI}}(a)/kT]}{\exp[-\delta G^{\mathrm{HI}}(b)/kT]} \qquad (4.13)$$

The second equality on the right-hand side of (4.13) is a result of the equality of the direct potential energies in (4.9).

We shall now examine two particular cases. First we assume that the nonadditivity effects are negligible, i.e., we put $\phi(a) = \phi(b) = 0$ in (4.11) and (4.12). Since we have the equalities $R_{ij}^a = R_{ij}^b$ for all i, j except for R_{14} we obtain from (4.13)

$$\frac{P(a)}{P(b)} = \exp\left[-\frac{\delta G^{\mathrm{HI}}(a) - \delta G^{\mathrm{HI}}(b)}{kT}\right]$$
$$= \exp\left[-\frac{\delta G^{\mathrm{HI}}(R_{14}^a) - \delta G^{\mathrm{HI}}(R_{14}^b)}{kT}\right] \qquad (4.14)$$

This result means that in spite of the equalities of the direct potential energies of the two configurations (4.9), the ratio of the probabilities of the two configurations a and b in the *solvent* may be different from unity. In this particular case we may ascribe this new feature to the difference in the HI between particles 1 and 4, which is the only pair of particles the distance between which is different in the two configurations a and b. Clearly these two configurations may be viewed as representing two possible conformations of a single butane molecule. In the latter case we can ascribe the above result to an *intramolecular* hydrophobic interaction. However, as we shall soon demonstrate, this kind of assignment is not always possible. We recall that relation (4.14) has been based on the assumption of pairwise additivity of the HI. Clearly the same result would have been obtained if nonadditivity effects exist, but they are of equal magnitude for the two configurations a and b, i.e., $\phi(a) = \phi(b) \neq 0$.

Before turning to the more general case it should be noted that our initial assumption was that the *direct* interaction has a very short range, so that $U(R_{14})$ is practically zero in the two configurations. The possibility that the exponent in (4.14) might be nonzero is equivalent to the statement that the HI might have a larger range compared to the direct interaction. This is again a new feature of the HI that may arise from the averaging over all the configurations of the solvent molecules.

Next we turn to the more general case where additivity of the total HI is not granted. In this case we have

$$\frac{P(a)}{P(b)} = \exp\left[-\frac{\delta G^{\mathrm{HI}}(R_{14}^{a}) - \delta G^{\mathrm{HI}}(R_{14}^{b})}{kT}\right] \exp\left[-\frac{\phi(a) - \phi(b)}{kT}\right] \quad (4.15)$$

Again in order to emphasize a new feature we assume that the HI is of a short range so that the first factor in (4.15) is unity. In this case the difference in the probabilities of the occurrence of the two configurations is ascribed to the difference in the extent of the nonadditivity effect of the two configurations a and b. Since $\phi(a)$ and $\phi(b)$ are properties of the configuration of the *entire* set of solute particles, it is impossible to claim that the deviation from unity of the ratio $P(a)/P(b)$ is due to a HI between any specific pair of solute particles. This is an important observation, which should be borne in mind in any discussion of the concept of intramolecular HI between nonpolar groups hanging on a polymer. We have, for simplicity, discussed the case of four solute particles at two configurations, a and b. Clearly the same considerations would have been relevant to the case where our solutes are replaced by methylene groups and chemical bonds exist along lines 1–2, 2–3, and 3–4. In this case all our conclusions apply to the two possible conformations of a single butane molecule. Here again we may find that one of the conformations may have a higher probability than the other. To analyze the reasons for such a preference we must first examine the intramolecular potential function of the molecule, i.e., whether the same effect is also observed in the gaseous phase. If this is not the case, then we turn to the indirect free energy change for the transformation from one conformation to the other. This, in general, would involve nonadditivity effects. Therefore, the reason for observing such a preference effect could not be ascribed simply to an intramolecular HI, say between the two end methyl groups of the molecule. Extending the same type of argument to biopolymers would lead to the following conclusion. Consider two conformations of a biopolymer, say the helix–coil pair. We find that one conformation is much more probable than the other. If nonadditivity effects are large, then it would be impossible to explain such a phenomenon by invoking the concept of *intramolecular* HI between pairs of nonpolar groups.

For this reason we believe that a systematic study of the nonadditivity effect, along either theoretical or experimental routes, should be given a high priority. The importance of this effect might turn out to be decisive to our understanding of the driving forces behind biochemical processes.

At this point it is instructive to present a specific example in which the ratio of the probabilities in (4.15) may be computed exactly. To do that we consider four hard-sphere (HS) solutes, in a "solvent" which consists of only *one* additional HS of a different diameter. Denoting the position vector of the "solvent" molecules by R_5, we can write the ratio of the probabilities of the two configurations as

$$\frac{P(a)}{P(b)} = \frac{\int d\mathbf{R}_5 \exp\{-[U(a) + U(5 \mid a)]/kT\}}{\int d\mathbf{R}_5 \exp\{-[U(b) + U(5 \mid b)]/kT\}} \tag{4.16}$$

where $U(a)$ and $U(b)$ have the same meaning as before and are presumed to be equal to each other. $U(5 \mid a)$ is the total interaction energy between the "solvent" molecule at R_5 and the four solute molecules at the configuration a. A similar meaning applies to $U(5 \mid b)$. The integrations in (4.16) extend over the entire volume V of our system. This integration is the simplest example of an average over all the configurations of the "solvent" molecules.

Let σ_1 and σ_2 be the diameter of the solute and the solvent molecules, respectively. The integrand in (4.16) has the obvious property

$$\exp\left[-\frac{U(5 \mid a)}{kT}\right] = \begin{cases} 0 & \text{if at least one distance} \\ & |\mathbf{R}_5 - \mathbf{R}_i| < (\sigma_1 + \sigma_2)/2 \\ 1 & \text{if } all \text{ distances} \\ & |\mathbf{R}_5 - \mathbf{R}_i| > (\sigma_1 + \sigma_2)/2 \end{cases} \quad \text{for } i = 1, 2, 3, 4 \tag{4.17}$$

In words, the integrand is zero whenever the "solvent" particle penetrates into the excluded volume $V_{ex}(a)$ produced by the solute molecules, and is unity everywhere else. The two excluded volumes are depicted in Figure 4.3.

The property (4.17) renders possible the immediate integration of the two integrals in (4.16). The result is

$$\frac{P(a)}{P(b)} = \frac{V - V_{ex}(a)}{V - V_{ex}(b)} \tag{4.18}$$

Thus the configuration that produces the larger excluded volume will have the smaller probability of occurrence. In our particular example it is evident from Figure 4.3 that

$$V_{ex}(a) > V_{ex}(b) \tag{4.19}$$

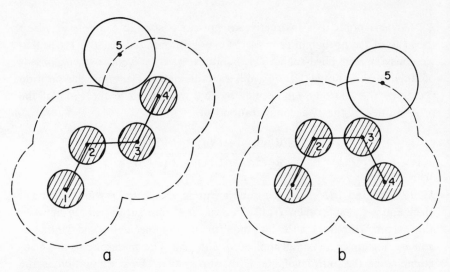

Figure 4.3. The excluded volume of four hard spheres for the two configurations *a* and *b* in Figure 4.2. A hard sphere "solvent" cannot penetrate into the region bounded by the dashed line.

Hence

$$P(a) < P(b) \tag{4.20}$$

Clearly the excluded volume of a given configuration is a property of the configuration of the entire set of the solutes (or the set of groups in a single molecule such as butane). For this reason an inequality of the form (4.20) may not be interpreted in terms of a pairwise HI between any specific pair of solutes (or groups in a single molecule).

In the above discussion we have made a distinction between the effect of the long range of the HI and its nonadditivity. Such a distinction was made only to stress two aspects of the same phenomenon. The important quantity is always the *total* HI among the set of solute molecules. In general, it is not clear under which conditions one can assume complete pairwise additivity of this quantity. Hence the distinction between the two effects may not be possible in practical examples.

The above illustration was made for a very simple "solvent." It is obvious that the situation becomes more complicated if the "solvent" consists of many molecules and far more so when the solvent is liquid water. For such cases it is impossible to predict, even in a qualitative fashion, which configuration might be favored by any real solvent. This problem forms an interesting, though difficult, challenge for future work.

Comments

The reader may wonder, at this point, on the usefulness of the general strategy of studying HI as outlined in Section 1.1. There we started with a very complex process of a conformational change of a biopolymer. This process involves several factors that combine to determine its overall driving force. As a first step in the study of such complex processes we have decided to separate the various factors and study each of them in isolation, by using simple model systems. One of these factors is the pairwise HI, to which we have devoted an entire chapter in this book.

The present section raises some doubts about the way the various factors cooperate in the complex process. It is, in principle, possible that, if nonadditivity effects are large, all our information on pairwise HI might become totally irrelevant to the understanding of the complex process. We have stressed here the nonadditivity of the HI only. Clearly, different factors such as HI and charge–charge interaction might also combine in a nonadditive manner.

In spite of the above somewhat discouraging comment, we believe that the study of each of the factors separately is unavoidably the best strategy we can adopt before we can even hope to understand the mechanism of real biochemical processes. Besides, whatever the relevance of the pairwise HI to real processes is, we can always view this topic as one aspect of the properties of aqueous solutions. This in itself is sufficient reason for pursuing further the study of pairwise HI, as well as any other single factor that contributes to the more complicated processes.

4.3. THE RELATION BETWEEN HI AND THE STANDARD FREE ENERGY OF AGGREGATION

In anticipation of the discussion of micellization processes in Section 4.8, we present here a general relation between the HI among m solute particles and what is conventionally referred to as the standard free energy of aggregation.

Consider a system consisting of m simple solutes in a solvent at a given T and P. As before, we are interested in the process of bringing these solute particles from some fixed positions at infinite separation from each other (a configuration that will be symbolically denoted by $\mathbf{R}^m = \infty$) to some close configuration $\mathbf{R}^m = \mathbf{R}_1, \ldots, \mathbf{R}_m$. The process is carried out within the solvent, keeping the temperature and pressure fixed.

The free energy change associated with this process is

$$\Delta G(\mathbf{R}^m) = G_m(\mathbf{R}^m) - G_m(\infty) \tag{4.21}$$

Clearly, this process cannot be carried out as an actual experiment in the laboratory. Its importance stems from the fact that the free energy change associated with this process is directly related to the probability density of observing the configuration $\mathbf{R}^m$, namely,

$$P(\mathbf{R}^m) = C \exp[-\Delta G(\mathbf{R}^m)/kT] \tag{4.22}$$

where C is a normalization constant. This fact establishes the relevance of $\Delta G(\mathbf{R}^m)$ to the general problem of HI.

As we have done in Sections 1.3 and 4.1 we can split $\Delta G(\mathbf{R}^m)$ into two contributions; the direct and the indirect parts, namely,

$$\Delta G(\mathbf{R}^m) = U(\mathbf{R}^m) + \delta G^{\mathrm{HI}}(\mathbf{R}^m) \tag{4.23}$$

Generalizing the idea of the thermodynamic cycle of Section 1.3 we can first transfer all the m solute particles to the gaseous phase. Then we bring the solutes to the configuration $\mathbf{R}^m$, in the gaseous phase. Finally we transfer the aggregate, as a single entity, from the gas into the liquid. From the equality of the free energy change of the process along the two routes (see Figure 4.4), we obtain the generalized relation for $\delta G^{\mathrm{HI}}(\mathbf{R}^m)$

$$\delta G^{\mathrm{HI}}(\mathbf{R}^m) = \Delta\mu_A{}^\circ - m\,\Delta\mu_M{}^\circ \tag{4.24}$$

where $\Delta\mu_M{}^\circ$ is the standard free energy of transferring a single solute M from a fixed position in the gas to a fixed position in the liquid. Similarly, $\Delta\mu_A{}^\circ$ refers to the process of transferring the aggregate A as a single entity from the gas to the liquid. The last relation may be obtained either by the use of a generalized thermodynamic cycle as we have done in Section 1.3 or directly by using statistical mechanical arguments (Ben-Naim, 1974).

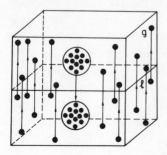

Figure 4.4. A thermodynamic cycle involving m solute particles. Instead of bringing the m solute particles from infinite separation from each other to a close configuration, we first transfer all the particles to the gaseous phase, then we bring them to the close-packed configuration, and finally we transfer the entire aggregate from the gas into the liquid.

Next we consider an experimental system consisting of the same solvent as before, at the same T and P, but now there is a chemical equilibrium between the monomers and the aggregates of m monomers (for simplicity we assume here the existence of one kind of aggregate, with the interparticle distances between its monomers the same as in $\mathbf{R}^m$). We assume that we have an experimental means by the use of which we can determine the concentration of the monomers ϱ_A. If the solute is very dilute in the solvent we may write the equilibrium condition

$$\mu_A = m\mu_M \qquad (4.25)$$

in the form

$$\Delta G_A{}^\circ \equiv \mu_A{}^\circ - m\mu_M{}^\circ = -kT \ln [\varrho_A/\varrho_M{}^m]_{\text{eq}} \qquad (4.26)$$

where $\mu_A{}^\circ$ and $\mu_M{}^\circ$ are the standard chemical potentials of A and M, respectively. Thus from the knowledge of the concentrations of A and M at equilibrium, we can determine the corresponding standard free energy of aggregation $\Delta G_A{}^\circ$.

The question we pose now is how the experimental quantity $\Delta G_A{}^\circ$ is related to the quantities of interest $\Delta G(\mathbf{R}^m)$ or $\delta G^{\text{HI}}(\mathbf{R}^m)$ in our study of HI.

To do this we must use the statistical mechanical expressions for the chemical potentials of A and M. These are (for details, see Appendixes A.1 and A.7)

$$\mu_A = W(A \mid \text{sol}) + kT \ln \Lambda_A{}^3 q_A{}^{-1}\varrho_A = \mu_A{}^\circ + kT \ln \varrho_A \qquad (4.27)$$

$$\mu_M = W(M \mid \text{sol}) + kT \ln \Lambda_M{}^3 q_M{}^{-1}\varrho_M = \mu_M{}^\circ + kT \ln \varrho_M \qquad (4.28)$$

where $W(A \mid \text{sol})$ is the coupling work of A against the solvent and a similar meaning is assigned to $W(M \mid \text{sol})$. $\Lambda_A{}^3$ and $\Lambda_M{}^3$ are the momentum partition functions of A and M, respectively, and q_A and q_M are the remaining parts of the internal partition functions of A and M. For simplicity we assume that the solutes are structureless particles, and that the aggregate as an entity has a rigid structure (no internal motions of the monomers in A), hence we take

$$q_M = 1, \qquad q_A = q_{\text{rot}} \exp[-U(\mathbf{R}^m)/kT] \qquad (4.29)$$

where q_{rot} is the rotational partition function of A.

In an ideal gas system containing A and M, the coupling work in (4.27) and in (4.28) are evidently zero, hence we have

$$\mu_A{}^{\circ g} = kT \ln \Lambda_A{}^3 q_A{}^{-1} \tag{4.30}$$

$$\mu_M{}^{\circ g} = kT \ln \Lambda_M{}^3 q_M{}^{-1} \tag{4.31}$$

Combining the expressions for the standard chemical potentials in (4.27), (4.28), (4.30), and (4.31) we obtain

$$
\begin{aligned}
\Delta G_A{}^\circ &= \mu_A{}^\circ - m\mu_M{}^\circ = \Delta\mu_A{}^\circ - m\,\Delta\mu_M{}^\circ + \mu_A{}^{\circ g} - m\mu_M{}^{\circ g} \\
&= \delta G^{\mathrm{HI}}(\mathbf{R}^m) + kT \ln(\Lambda_A{}^3 q_A{}^{-1}/\Lambda_M{}^{3m}) \\
&= U(\mathbf{R}^m) + \delta G^{\mathrm{HI}}(\mathbf{R}^m) + kT \ln(\Lambda_A{}^3 q_{\mathrm{rot}}^{-1}/\Lambda_M{}^{3m}) \\
&= \Delta G(\mathbf{R}^m) + kT \ln(\Lambda_A{}^3 q_{\mathrm{rot}}^{-1}/\Lambda_M{}^{3m})
\end{aligned}
\tag{4.32}
$$

This is the required relation between the standard free energy of aggregation $\Delta G_A{}^\circ$ and the free energy of the process described at the beginning of this section. The physical meaning of this relation is quite simple. The standard free energy of aggregation consists of the free energy of bringing m solutes from fixed positions at infinite separation to some close configuration $\mathbf{R}^m$. In addition, in the real process of aggregation we "lose" m times the momentum partition function of the monomers, and we "gain" one momentum partition function and one rotational partition function of the aggregate. This is essentially the meaning of the second term on the right-hand side of Equation (4.32).

In other words, in order to form a *free* aggregate from m *free* solutes, we may first "freeze in" the translational degrees of freedom of the solutes, form the aggregate at a *fixed* configuration, and then release translational and rotational freedom of the aggregate.

From relation (4.32) it is evident that more information is required in order to extract information on HI from experimental data. One simple way of supplementing this information is to measure, if possible, the standard free energy of aggregation of the same aggregate in the gaseous phase. If this is possible, then the corresponding standard free energy of aggregation is

$$\Delta G_A{}^{\circ g} = U(\mathbf{R}^m) + kT \ln(\Lambda_A{}^3 q_{\mathrm{rot}}^{-1}/\Lambda_M{}^{3m}) \tag{4.33}$$

Hence from (4.32) and (4.33) we may obtain

$$\Delta G_A{}^\circ - \Delta G_A{}^{\circ g} = \Delta G(\mathbf{R}^m) - U(\mathbf{R}^m) = \delta G^{\mathrm{HI}}(\mathbf{R}^m) \tag{4.34}$$

which is the required measure of the HI.

In practice, the above procedure is not useful. The reason is that in real cases even if aggregates are formed in two phases, they would usually have different configurations in each phase. We have seen in Section 3.7 that even the structure of the dimers of carboxylic acids in two phases are different. The situation is, of course, far more complicated for larger aggregates.

Finally, we note that in our treatment of the aggregates in this section we have ignored, for simplicity, internal motions in the aggregate (such as vibrations and internal rotations). In real examples the corresponding partition functions should also be taken into account in relations such as (4.32) and (4.33).

This complication renders the whole procedure outlined above useless from the practical point of view. For this reason we shall devote the next four sections to other routes by way of which we may obtain information on the HI among m solute particles. We shall then return to micellization processes for which $\Delta G_A{}^\circ$ may be determined experimentally. It is true that many attempts have been made to extract information on HI from such experimental data. Unfortunately, these procedures are not well founded. We shall elaborate further on these difficulties in Section 4.8.

4.4. APPROXIMATE MEASURE OF THE HI AMONG m SOLUTE PARTICLES

In this section we present a straightforward generalization of the method of Section 3.3 to devise a measure of the HI among a large number of solute particles. As in Section 3.3 our treatment here is based essentially on thermodynamics. A more detailed statistical mechanical treatment may be found in Ben-Naim (1971b, 1974). The nature of the approximation that we shall use in this section is the same as the one we have used in Section 3.3. However, in the following section we shall show how to improve upon this approximation to obtain more reliable measurements of the HI at more realistic configurations.

Consider again the process of bringing m solute particles from fixed positions, at infinite separation from each other, to some close configuration that we denote by $\mathbf{R}^m = \mathbf{R}_1, \ldots, \mathbf{R}_m$. As in the previous sections we write the free energy change associated with this process as

$$\Delta G(\mathbf{R}^m) = U(\mathbf{R}^m) + \delta G^{\mathrm{HI}}(\mathbf{R}^m) \qquad (4.35)$$

where the two terms on the right-hand side of (4.35) are referred to as the *direct* and the *indirect* parts of the free energy change. The *indirect* part is also referred to as the HI among the *m* solute particles at the configuration $\mathbf{R}^m$.

Using the same thermodynamic cycle as in Section 4.3 (see Figure 4.4), we may express $\delta G^{\mathrm{HI}}(\mathbf{R}^m)$ as

$$\delta G^{\mathrm{HI}}(\mathbf{R}^m) = \Delta\mu_A{}^\circ - m\,\Delta\mu_M{}^\circ \tag{4.36}$$

which is a generalization of relation (3.13) of Section 3.3. Here, $\Delta\mu_M{}^\circ$ is the experimental standard free energy of solution of the monomers M, and $\Delta\mu_A{}^\circ$ is the standard free energy of transferring the aggregate A from the gas to the liquid. The latter is not a measurable quantity, since A is not a molecular entity. Therefore, we seek an approximate version of relation (4.36) which leads to an experimentally determinable quantity. To this end, we exploit the fact that the quantity $\delta G^{\mathrm{HI}}(\mathbf{R}^m)$ is a smooth function of the configuration $\mathbf{R}^m$ even for configurations that are experimentally inaccessible. Specifically, suppose we start with three methane molecules as the monomers, and we bring these solutes to a configuration in which the centers of the monomers occupy the same relative positions as the centers of the three carbon atoms in propane.

We recall that $\Delta\mu_A{}^\circ$ depends essentially on the binding energy of A with the solvent. More specifically, the binding energy of A is defined by

$$B_A = \sum_{j=1}^{3} \sum_{i=1}^{N} U(\mathbf{R}_j, \mathbf{X}_i) \tag{4.37}$$

where $U(\mathbf{R}_j, \mathbf{X}_i)$ is the interaction energy between the jth solute at $\mathbf{R}_j$ and the ith solvent molecule at a given configuration (location and orientation) $\mathbf{X}_i$. The standard free energy of solution $\Delta\mu_A{}^\circ$ is given by

$$\Delta\mu_A{}^\circ = -kT \ln\langle\exp[-B_A/kT]\rangle_0 \tag{4.38}$$

where the symbol $\langle\ \rangle_0$ stands for an average over all the configurations of the solvent molecules.

Our approximation involves the replacement of B_A, the binding energy of the three methane molecules, by the binding energy of *one* propane molecule, which we denote by B_{Pr}, i.e., we assume

$$B_A \approx B_{\mathrm{Pr}} \tag{4.39}$$

If (4.39) is a good approximation for all the configurations of the solvent molecules that contribute to the average in (4.38), then we have the following approximate replacement:

$$\Delta\mu_A{}^\circ \approx \Delta\mu_{Pr}^\circ \qquad (4.40)$$

and hence, instead of (4.36) we write the approximate relation

$$\delta G_3{}^{HI} = \Delta\mu_{Pr}^\circ - 3\Delta\mu_{Me}^\circ \qquad (4.41)$$

where $\Delta\mu_{Pr}^\circ$ and $\Delta\mu_{Me}^\circ$ are the standard free energies of solution of propane and methane, respectively. What we have achieved by this approximation is a relation between HI at some particular configuration of three methane molecules and experimentally determinable quantities. As in Section 3.3 we note again that our generalized measure of HI among three or more solute particles is not related to any realizable configuration of a real system. Therefore, the main use of these quantities is not to estimate the absolute magnitude of the HI but to compare the relative strength of the HI in different solvents. We also note here that all of the $\delta G_m{}^{HI}$ values for molecules that have internal rotations should be understood as averages over all possible conformations of the molecules (see also Appendix A.7 for more details).

In Figures 4.5 and 4.6 we present some values of $\delta G_m{}^{HI}$ for m methane particles in water and in methanol. The final configuration at which the HI is measured is indicated next to each of the curves. Two important features should be noted. In the first place the absolute magnitude of the HI, at any specific configuration, in water is larger than in methanol. Second, the temperature dependence of δG^{HI} is distinctly more pronounced and negative in water as compared with the corresponding curves in methanol. These two features have already been observed in the behavior of the pairwise HI, reported in the previous chapter.

In Tables 4.1 and 4.2 we present some further values of $\delta G_m{}^{HI}$ for various m, at one temperature, $t = 25°C$. Perhaps the most interesting aspect of the results of these tables is the following. Let m_B be the number of the nearest-neighbor carbon atoms in the final configuration (i.e., the number m_B is equal to the number of chemical bonds in the molecule which is used to replaced the m methane molecules in the final configuration). If we divide $\delta G_m{}^{HI}$ by m_B, we obtain a measure of the HI per pair of nearest neighbors. This quantity seems to tend to an almost constant value of about -1.88 kcal/mol for all the reported data in Tables 4.1 and 4.2. Does this indicate some kind of additivity of the HI? It is difficult to answer this question affirmatively. One reason for this is that the data on which the results of these tables are based are not sufficiently accurate to draw

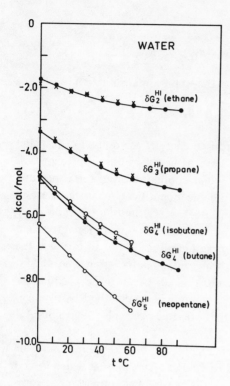

Figure 4.5. Values of $\delta G_m{}^{HI}$ in water for various numbers of methane molecules brought to the final configuration as indicated next to each curve. Based on data from Morrison and Billet (1952) and from Wetlaufer *et al.* (1964).

such a conclusion. More important, however, is the observation that the constant value of $\delta G_m{}^{HI}/m_B \approx -1.88$ kcal/mol is considerably different from the pairwise HI, i.e., $\delta G_2{}^{HI} = -2.16$ kcal/mol. Therefore, even if we trust the data on which these results are based, we cannot conclude that $\delta G_m{}^{HI}$ is pairwise additive in the sense that it is a sum of m_B times the pairwise HI, $\delta G_2{}^{HI}$. It is also difficult to imagine that in such close-packed

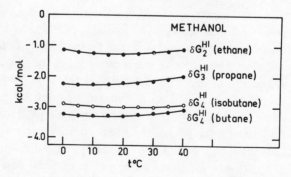

Figure 4.6. Values of $\delta G_m{}^{HI}$ for the same configurations as in Figure 4.5, but in methanol.

Table 4.1

Values of δG_m^{HI} (in kcal/mol at 25°C) for the Interaction of m Methane Molecules Brought to a Final Configuration Similar to an Existing Molecule Containing m Carbons[a]

Hydrocarbon	m	$-\delta G_m^{HI}$	$-\delta G_m^{HI}/m_B$
Ethane	2	2.16	2.16
Propane	3	4.01	2.00
n-Butane	4	5.87	1.96
Isobutane	4	5.70	1.90
n-Pentane	5	7.53	1.88
Isopentane	5	7.59	1.89
2,2-Dimethylpropane	5	7.34	1.83
n-Hexane	6	9.43	1.88
2-Methylpentane	6	9.48	1.89
3-Methylpentane	6	9.41	1.88
2,2-Dimethylbutane	6	9.39	1.88
n-Heptane	7	11.34	1.89
2,4-Dimethylpentane	7	11.07	1.85
n-Octane	8	13.08	1.87
2,2,4-Trimethylpentane	8	13.10	1.87

[a] In the last column, m_B is the number of chemical bonds (or nearest neighbors) in the hydrocarbon. [Computations based on data from Wen and Hung (1970) and McAuliffe (1966).]

Table 4.2

Values of δG_m^{HI} (in kcal/mol, at 25°C) for Cyclic Molecules[a]

Cycloparaffin	m	$-\delta G_m^{HI}$	$-\delta G_m^{HI}/m_B$
Cyclopropane	3	3.49	1.16
Cyclopentane	5	8.80	1.76
Cyclohexane	6	10.72	1.78
Cycloheptane	7	13.17	1.88
Cyclooctane	8	15.13	1.89
Methylcyclopentane	6	10.37	1.73
Methylcyclohexane	7	12.25	1.75
1-cis-2-Dimethylcyclohexane	8	14.40	1.80

[a] Here $m = m_B$. (From Ben-Naim, 1972a.)

configurations only the "nearest-neighbor" pairs contribute to the total HI. We note that the second nearest neighbors are at a distance of about 2.51 Å and this cannot be regarded as large compared to the possible range of the HI.

Perhaps the best example that may be used to study the extent of additivity (or nonadditivity) of the HI is provided by the example of cyclopropane, for which we find

$$\delta G_3{}^{HI} = -3.49 \text{ kcal/mol} \tag{4.42}$$

If we assume that each C–C bond in cyclopropane is of the same length as the C–C bond of ethane, then we may estimate the extent of nonadditivity of the HI by

$$\phi(\text{cyclopropane}) = \delta G_3{}^{HI}(\text{cyclopropane}) - 3 \times \delta G_2{}^{HI}(\text{ethane})$$
$$= -3.49 + 3 \times 2.16 = 2.99 \text{ kcal/mol} \tag{4.43}$$

Thus, if our figures for $\delta G_3{}^{HI}$ and $\delta G_2{}^{HI}$ are reliable, we can conclude that the nonadditivity of the HI is positive and quite large (having the same order of magnitude of the HI itself).

The above example was presented mainly to illustrate the possibility of studying the nonadditivity effect of the HI by experimental means. Of course, more accurate and detailed data are needed in order to reach any significant conclusions from such measurements. Clearly one can extend the method for studying nonadditivity effects in higher cycloparaffin molecules. We believe, however, that this should be postponed until better experimental data become available.

Comment

We believe that more extensive and accurate data on the solubilities of various simple hydrocarbons in water and in other solvents should be sought. Such data could well be used to study the extent of nonadditivity of the HI and its dependence on configuration (e.g., the comparison between the butane and the isobutane configurations).

Similar data may be used to study the effect of a polar group on the HI. For example, the following expression

$$\delta G^{HI} = \Delta\mu°[CH_3-(CH_2)_n-P] - \Delta\mu°[H-(CH_2)_n-P] - \Delta\mu°(CH_4) \tag{4.44}$$

is a measure of the indirect work required to bring a methane molecule to one end of a hydrocarbon, which has on its other end a polar group P. Thus, with a given group P (say, halogen or carboxylic groups) one can study the range of the HI by varying n. The question is how large n must be so that the above work becomes independent of P. A second study could be to fix n and examine the relative effects of different groups P on the HI.

4.5. AN IMPROVED APPROXIMATE MEASURE OF THE HI

The measures of the strength of the HI that were introduced in Sections 3.3 and 4.4 contain two flaws. One is the approximation, on which we have not elaborated in any detail [see Ben-Naim (1974) and Appendix A.6]. In essence, this approximation involves the neglect of the field of force produced by the two inner hydrogens of the pair of methane molecules brought to the separation $R = 1.53$ Å. The second is concerned with the final configuration of the solutes, which in practice have almost zero probability of occurrence.

In this section we present a modified measure of the HI which is based on essentially the same type of arguments as before but can, in principle, provide improved information on the HI. The improvement is achieved in both the nature of the approximation and the realizability of the final configuration of the solutes. The extent of the improvement depends on the availability of relevant experimental data.

As a prototype of our new measure we consider the neopentane molecule. According to the procedure of constructing a measure of the HI as discussed in Section 4.4, we can write the HI among *five* methane molecules, brought to the final configuration of a neopentane molecule (see Figure 4.7) as

$$\delta G_5^{\mathrm{HI}}(\text{neopentane}) = \Delta\mu^\circ(\text{neopentane}) - 5\Delta\mu^\circ(\text{methane}) \qquad (4.45)$$

This measure involves the same kind of approximation as indicated above.

Next we consider a different process. We start with *four* methane molecules at fixed positions at infinite separation from each other. These solutes are brought (within the solvent, keeping T and P constant) to the final configuration of the four peripheral methyl groups in neopentane. These are indicated in Figure 4.8. At this configuration we have the exact

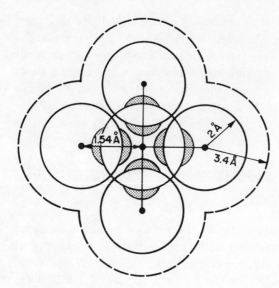

Figure 4.7. A schematic, two-dimensional description of five methane molecules in the configuration of neopentane. All the inner hydrogens are indicated by the dark areas. The boundaries of the excluded volume are indicated by the dashed curve (assuming a radius of 2 Å for the methane and 1.4 Å for the water molecule).

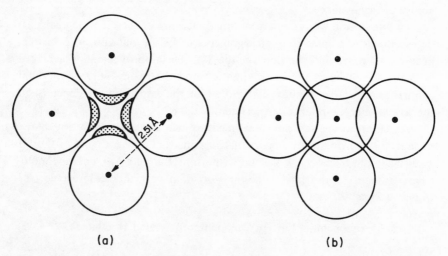

(a)   (b)

Figure 4.8. A schematic, two-dimensional description of the replacement procedure corresponding to Equations (4.50) and (4.51). (a) Four methane molecules are brought to the position of the four peripheral methyl groups of a neopentane molecule. The four hydrogens pointing towards the center are indicated by the dark areas. The distance of closest approach between any two of the methane molecules is about 2.51 Å. (b) The four methane molecules are replaced by a single neopentane molecule. A new carbon nucleus is added, which partially compensates for the loss of the four inner hydrogens.

relation

$$\delta G^{\mathrm{HI}}(\text{agg.}) = \Delta\mu^{\circ}(\text{agg.}) - 4\Delta\mu^{\circ}(\text{methane}) \tag{4.46}$$

where $\Delta\mu^{\circ}(\text{agg.})$ is the standard free energy of transferring the aggregate (agg.) of Figure 4.8(a) from the gas to the liquid. Clearly, this is not a measurable quantity. The statistical mechanical expression for $\Delta\mu^{\circ}(\text{agg.})$ is

$$\Delta\mu^{\circ}(\text{agg.}) = -kT\ln\langle\exp(-B_{\mathrm{agg.}}/kT)\rangle_0 \tag{4.47}$$

where $B_{\mathrm{agg.}}$ is the total binding energy of the aggregate to the solvent molecules; more explicitly,

$$B_{\mathrm{agg.}} = \sum_{j=1}^{4}\sum_{i=1}^{N} U(\mathbf{R}_j, \mathbf{X}_i) \tag{4.48}$$

where $U(\mathbf{R}_j, \mathbf{X}_i)$ is the solute–solvent pair potential between the jth solute at $\mathbf{R}_j$ and the ith solvent molecule at the configuration $\mathbf{X}_i$.

We notice that since the aggregate has a compact structure, no solvent molecule can penetrate into the interior of this aggregate. [Note, however, that the average $\langle\ \rangle_0$ is over *all* possible configurations of the solvent molecules. This also includes configurations for which some solvent molecules do penetrate into the region that is occupied by the solute molecules. However, for each of the configurations in which a solvent molecule penetrates into this region, $B_{\mathrm{agg.}}$ becomes very large and positive and hence $\exp[-B_{\mathrm{agg.}}/kT]$ becomes practically zero]. In Figure 4.7 we indicated by the dashed line the boundary of the so-called excluded volume (assuming that the solvent is water, with a molecular diameter of 2.8 Å).

Clearly, the average in (4.47) gets nonzero contributions only from those configurations for which no solvent molecules penetrate into the excluded region produced by the four solute molecules. Therefore, if we insert in the center of this aggregate any particle or a group that produces a short-range field of force—such that it is not felt outside the excluded volume, the value of the average in (4.47) will not be affected. We exploit this fact to introduce into the center of the aggregate an "agent" that binds the four solute molecules in such a way that a *real* molecule is formed and for which the standard free energy of solution is measurable.

In our particular example we replace the aggregate of *four* methane molecules by *one* neopentane molecule. This replacement is shown schematically in Figure 4.8. The approximation that is employed is

$$B(\text{agg.}) \approx B(\text{neopentane}) \tag{4.49}$$

If this is valid for all the configurations of the solvent molecules that have nonzero contribution to the average in (4.47), then we have the approximation

$$\Delta\mu^\circ(\text{agg.}) \approx \Delta\mu^\circ(\text{neopentane}) \qquad (4.50)$$

Hence the exact relation (4.46) is transformed into the approximate, but more useful, relation

$$\delta G^{\text{HI}}(\text{agg.}) = \Delta\mu^\circ(\text{neopentane}) - 4\Delta\mu^\circ(\text{methane}) \qquad (4.51)$$

This should be compared with (4.45), which is a measure of the HI among *five* solute particles. Here we have a measure of the HI among *four* solutes at a configuration that is more realizable than the ones we have treated in Section 4.4. Furthermore, the nature of the approximation involved in (4.51) is different from the one used in Sections 3.3 and 4.4. Here we have replaced the four inner hydrogens by one carbon center. In a sense we have partially compensated for the loss of the field of force produced by these hydrogens on the solvent. It is clear that had we started with four bulkier molecules, say four benzene molecules, and used the same procedure as above, we would have reached the relation

$$\delta G_4^{\text{HI}} = \Delta\mu^\circ(\text{TPM}) - 4\Delta\mu^\circ(\text{benzene}) \qquad (4.52)$$

which measures the HI among *four* benzene molecules holding the positions of the four benzyl radicals in tetraphenylmethane (TPM). In this case the boundaries of the excluded volume are quite far from the center of the aggregate. The effect of any replacement made at the center of this aggregate on the solvent becomes negligible. Clearly, the bulkier the four molecules the better is the replacement approximation that is used in (4.49) or (4.50).

Figure 4.9 presents some values of $\delta G_4^{\text{HI}}(\text{agg.})$ defined in (4.51) as a function of the temperature. These values are compared with two other measures of the HI among *four* methane molecules at the configuration of butane and isobutane. Note that the latter are systematically more negative than the corresponding values of $\delta G_4^{\text{HI}}(\text{agg.})$. This is probably a result of the fact that the HI becomes larger as the particles come closer together in the final configuration. [For further discussion on this aspect of the HI, the reader is referred to Ben-Naim (1974).]

The quantity $\delta G^{\text{HI}}(\text{agg.})$ in (4.51) measures the HI interaction among four methane molecules: the closest distance between any pair of molecules

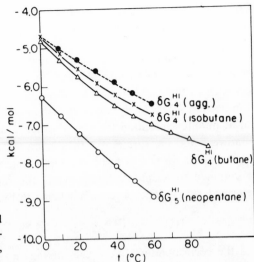

Figure 4.9. Values of δG_4^{HI} and δG_5^{HI} as a function of temperature for various configurations, as indicated next to each curve.

is about 2.51 Å, as compared to 1.5 Å between some of the pairs of methane molecules in Figure 4.7. This is an improvement towards a more realistic configuration of solute molecules in real systems. One can make further improvement in this direction by taking four bulkier molecules, such as benzene or long-chain paraffin molecules, to form tetraalkyl or tetraphenyl methane. In such cases the final configuration is very similar to an actually realizable configuration.

Also, in Figure 4.9 we have plotted values of δG_5^{HI}(neopentane) as defined in (4.45). These values are distinctly larger than the corresponding δG_4^{HI} values. The reason is that in the former case we are concerned with the HI among *five* molecules, whereas in the latter case only *four* solute molecules are involved.

Comment

We recommend the extension of the method outlined in this section to bulkier molecules. This is the only available experimental method to obtain information on HI among several solute particles at some close-packed configuration that is very close to a realizable configuration. Such information may indicate to what extent the HI is an important ingredient in the driving force for the formation of micelles in aqueous solutions.

4.6. APPLICATION OF THE SCALED-PARTICLE THEORY (SPT)

In the previous sections we presented two measures of the HI in which we made use of experimental data. In this section a partial theoretical approach to the problem of HI is described. The basic process is the same as in Section 4.3. Namely, we start with m solute particles at fixed positions but at infinite separation from each other in a solvent at some given temperature T and pressure P. We then bring these particles to a close-packed configuration. More specifically we require that the centers of all the particles be confined to a spherical region S_A, the radius of which is chosen as described below. The process is schematically written as

$$(\mathbf{R}^m = \infty) \to (\mathbf{R}^m \in S_A) \tag{4.53}$$

and the corresponding free energy change is

$$\Delta G_m = U_m + \delta G_m{}^{\text{HI}} \tag{4.54}$$

where U_m and $\delta G_m{}^{\text{HI}}$ are the *direct* and the *indirect* parts of the work required to carry out the process indicated in (4.53). Using the same argument as in Section 1.3 (see also Figure 4.4), we write for the indirect, or the hydrophobic interaction, part the exact relation

$$\delta G_m{}^{\text{HI}} = \Delta \mu_A{}^\circ - m \, \Delta \mu_M{}^\circ \tag{4.55}$$

$\Delta \mu_M{}^\circ$ is the experimental standard free energy of solution of the monomers M, and $\Delta \mu_A{}^\circ$ is the standard free energy of transferring the close-packed aggregate A, viewed as a single entity, from a fixed position in the gas into a fixed position in the liquid.

In the previous sections we endeavored to find approximations for $\Delta \mu_A{}^\circ$ using experimental sources. Here, however, we appeal to theory to find an estimate for $\Delta \mu_A{}^\circ$.

In Section 2.5 we mentioned one possible application of the scaled-particle theory (SPT) to the problem of HI. Here the same theory is used in a different way to estimate the quantity $\Delta \mu_A{}^\circ$. This application is based on a recent work by Ben-Naim and Tenne (1977). In Appendix A.4 we present some details on the elements of the SPT. We feel that applicability of the SPT to liquid water is somewhat dubious. Therefore we shall be using this theory only to compute $\Delta \mu_A{}^\circ$, whereas $\Delta \mu_M{}^\circ$ is taken from experimental sources. In this way we base our calculations of the HI only partially on this theory.

The procedure of estimating $\Delta\mu_A^\circ$ by the SPT is the following. First we split $\Delta\mu_A^\circ$ into two terms

$$\Delta\mu_A^\circ = \Delta\mu_A^\circ(\text{cav}) + \Delta\mu_A^\circ(\text{soft}) \qquad (4.56)$$

where the first term on the right-hand side of (4.56) is the work required to create a cavity of a suitable size (see below) in the solvent. The second term is due to the "turning on" of the "soft" (or the attractive) part of the interaction between the aggregate A and the solvent. Such a split of $\Delta\mu_A^\circ$ into two terms may be carried out in a rigorous fashion by using a consecutive double-charging process for introducing A into the solvent [for details see Ben-Naim (1974)].

We further assume that m is a large number, that the solute monomers are simple (e.g., argon, methane), and that the aggregate A has a spherical shape and consists of closely packed monomers. Following these assumptions we expect that the soft part of the field of force of A will originate essentially from those molecules that are in direct contact with the solvent, i.e., the molecules that form the surface of the aggregate A. If the number of monomers m is large, the contribution of $\Delta\mu_A^\circ(\text{soft})$ to $\Delta\mu_A^\circ$ becomes small compared to $\Delta\mu_A^\circ(\text{cav})$. Thus for sufficiently large m we use the approximation

$$\Delta\mu_A^\circ \approx \Delta\mu_A^\circ(\text{cav}) \qquad (4.57)$$

where $\Delta\mu_A^\circ(\text{cav})$ may be computed from the SPT. Note that for hard-sphere solutes $\Delta\mu_A^\circ(\text{soft})$ is zero and (4.57) is an equality. We therefore expect that for a simple solute such as methane, (4.57) is a good approximation.

To proceed we must now estimate the size of the appropriate cavity in which the aggregate is to be accommodated. Let σ_M be the effective hard-core diameter of methane, which we take as equal to the Lennard-Jones diameter of methane $\sigma_M = 3.82$ Å. If m solutes of diameter σ_M are packed compactly in such a way that they form a sphere of diameter σ_A, it is well known that the ratio of the volume of the m particles to the volume of the sphere S_A is

$$\frac{m\pi\sigma_M^3/6}{\pi\sigma_A^3/6} = 0.7405 \qquad (4.58)$$

From which we may eliminate σ_A:

$$\sigma_A = (m\sigma_M^3/0.7405)^{1/3} \qquad (4.59)$$

Let the diameter of the solvent molecules be σ_S, then the radius of the cavity produced by the aggregate A is given by

$$R_{\text{cav}} = (\sigma_A + \sigma_S)/2 \qquad (4.60)$$

The situation is schematically depicted in Figure 4.10. Once we have the radius R_{cav}, the molecular diameter, and the number density of the solvent, we can use the SPT to estimate $\Delta\mu_A{}^\circ(\text{cav})$. Here we present only some results of these calculations. For more details on the SPT see Appendix A.4, and for the calculation procedure see Ben-Naim and Tenne (1977).

Thus, in essence we have replaced the exact result in (4.55) by the approximate relation

$$\delta G_m{}^{\text{HI}} \approx \Delta\mu_A{}^\circ(\text{cav}) - m\,\Delta\mu_M{}^\circ \qquad (4.61)$$

where $\Delta\mu_M{}^\circ$ is taken from experimental sources, and $\Delta\mu_A{}^\circ(\text{cav.})$ is computed from the SPT.

As we have noted in Section 4.3, the process described in the beginning of this section is not a *real* process, i.e., one cannot carry out such a process of aggregation in the laboratory. The relevance of the quantity $\delta G_m{}^{\text{HI}}$ to real processes is through its relationship to the probability of finding such an aggregate made up of m *free* solute particles in a solvent (see Section 4.3). It is only in the latter sense that the quantity $\delta G_m{}^{\text{HI}}$ might be of relevance to the process of micelle formation, which will be discussed in Section 4.8.

In the following numerical examples we always use methane, with a molecular diameter of $\sigma_M = 3.82$ Å, as our monomer. The solvents that

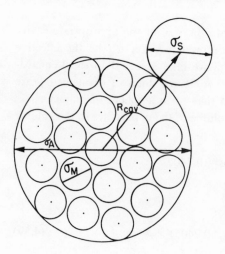

Figure 4.10. A cavity of radius R_{cav} is formed by an aggregate of diameter σ_A in a solvent; the diameter of the solvent molecules is σ_S.

are used in these illustrations are water (and heavy water) with $\sigma_S = 2.90$ Å, methanol with $\sigma_S = 3.69$ Å, ethanol with $\sigma_S = 4.34$ Å, and cyclohexane with $\sigma_S = 5.63$ Å. All the "effective" hard-core diameters are taken from the literature [see, for example, Reiss (1966) and Wilhelm and Battino (1971)]. It should be borne in mind, however, that for nonspherical molecules σ_S has no clear-cut physical meaning, as in the case of simple spherical molecules.

In addition to the effective molecular diameter, the solvents are characterized by their number densities at the given temperature and pressure. This information is sufficient for the computation of the free energy change associated with the formation of a cavity $\Delta\mu_A^\circ(\text{cav})$, and hence the computation of the HI through relation (4.61).

In order to compute the entropy and the enthalpy changes that correspond to δG_m^{HI} we use the following relations:

$$\delta S_m^{\text{HI}} = -\frac{\partial \delta G_m^{\text{HI}}}{\partial T} = \Delta S_A^\circ(\text{cav}) - m\,\Delta S_M^\circ \tag{4.62}$$

and

$$\delta H_m^{\text{HI}} = \delta G_m^{\text{HI}} + T\delta S_m^{\text{HI}} = \Delta H_A^\circ(\text{cav}) - m\,\Delta H_M^\circ \tag{4.63}$$

where again we use the SPT to compute $\Delta S_A^\circ(\text{cav})$ and $\Delta H_A^\circ(\text{cav})$ but use experimental sources for ΔS_M° and ΔH_M°. The latter are the standard entropy and enthalpy changes corresponding to the process of transferring a monomer from a fixed position in the gas to a fixed position in the liquid. These are different from the conventional standard quantities as used in the literature. [For more details see Section 5.2, and Ben-Naim (1974) and (1978a)].

In using the SPT to compute $\Delta S_A^\circ(\text{cav})$ and $\Delta H_A^\circ(\text{cav})$ we must also consider the temperature dependence of the effective hard-core diameter of the solvent molecules. (Only for hard-sphere particles is the molecular diameter, by definition, strictly temperature independent.) As we have noted above, the effective diameter, especially for nonspherical molecules, is not a uniquely defined quantity, and clearly the same is true for its temperature dependence. There are several procedures that have been suggested in the literature to obtain this temperature dependence, but none is satisfactory from the theoretical point of view. This fact is another quite serious flaw of the SPT when applied to complex solvents such as water, methanol, ethanol, etc.

Finally we note that δS_m^{HI} as defined in (4.62) is the same as the *total* entropy change for the process indicated in (4.53). On the other hand,

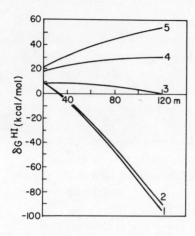

Figure 4.11. Values of δG_m^{HI} as a function of the number of monomers m for different solvents. The solute is methane and the solvents are (1) H_2O; (2) D_2O; (3) methanol; (4) ethanol; and (5) cyclohexane. All values are for atmospheric pressure at $t = 30°C$.

δH_m^{HI} is only the indirect part of the enthalpy change that corresponds to this process. The relation between the total and the indirect enthalpy changes is

$$\Delta H_m[(\mathbf{R}^m = \infty) \to (\mathbf{R}^m \in S_A)] = U_m + \delta H_m^{HI} \qquad (4.64)$$

which should be compared with relation (4.54).

In Figures 4.11 and 4.12 we present some computed values of δG_m^{HI} as a function of m, the number of monomers, for two temperatures, 30°C and 60°C. It is quite clear that as m becomes large enough (say $m \gtrsim 100$) the values of δG_m^{HI} in water (and heavy water) become large and negative. In methanol, ethanol, and cyclohexane the corresponding values are either positive or slightly negative. As we have noted above, we do not particularly

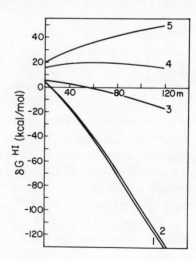

Figure 4.12. Same as Figure 4.11 but for $t = 60°C$.

trust the absolute results of $\delta G_m{}^{HI}$ for each solvent, but we believe that the difference between such values between two solvents is a more reliable quantity. Such differences may be easily transformed into ratios of probabilities as we have discussed in Section 4.3. To be more precise, suppose we take two solvents, say water and methanol, both at the same temperature T and pressure P. Also we assume that the solute M forms a very dilute solution in these two solvents, in such a way that the number density ϱ_M is the same in the two solvents, i.e., ϱ_M(in water) $= \varrho_M$(in methanol). (ϱ_M is the number of solute molecules per unit volume of the solvent.) In such a solution we may ask what the probability is of finding a close-packed aggregate containing m solute molecules. Clearly, if we have given a precise configuration $\mathbf{R}^m = \mathbf{R}_1, \ldots, \mathbf{R}_m$ to these particles, then the probability of its occurrence is zero (since one point in a continuous space of events has a zero measure). However, the ratio of such probabilities in two solvents is a *finite* quantity, which provides information on the difference in the solvation properties of the two solvents.

This ratio is given by

$$\xi = \frac{P_m(\text{water})}{P_m(\text{methanol})} = \exp\left[-\frac{\delta G_m{}^{HI}(\text{water}) - \delta G_m{}^{HI}(\text{methanol})}{kT} \right] \quad (4.65)$$

Note that this ratio should be understood as a limit of a ratio of two *finite* quantities, namely,

$$\lim_{d\mathbf{R}_1 \cdots d\mathbf{R}_m \to 0} \left\{ \frac{\Pr[1 \text{ in } d\mathbf{R}_1 \text{ at } \mathbf{R}_1 \cdots m \text{ in } d\mathbf{R}_m \text{ at } \mathbf{R}_m(\text{in water})]}{\Pr[1 \text{ in } d\mathbf{R}_1 \text{ at } \mathbf{R}_1 \cdots m \text{ in } d\mathbf{R}_m \text{ at } \mathbf{R}_m(\text{in methanol})]} \right\} \quad (4.66)$$

where $\Pr[\ ]$ means the probability of finding the event specified in the square brackets.

As an example we choose $m = 100$ and compute the ratio ξ at three temperatures. The results are

$$\xi(t = 10°C) \approx 4 \times 10^{32}, \quad \xi(t = 30°C) \approx 5 \times 10^{53}, \quad \xi(t = 60°C) \approx 2 \times 10^{57} \quad (4.67)$$

These results indicate that, in a dilute solution of the monomers M, as described above, the probability of finding a close-packed aggregate in water is far larger than the corresponding probability in methanol. Furthermore, this ratio becomes *larger* as the temperature increases in the range of temperatures of say, $0 \lesssim t \lesssim 80°C$. (One would have expected to find that, as the temperature increases, water would tend to become more "normal" and hence $\xi \to 1$. This may be true at higher temperatures.

There is evidence that indicates that the opposite effect is true at around room temperature. This aspect will be further discussed in Chapter 5, where we shall also present a qualitative molecular reason for this kind of behavior.)

Another interesting aspect of the results reported in Figures 4.11 and 4.12 is the difference between light and heavy water. In spite of our reservation about the applicability of the SPT to liquid water, we believe that whatever the nature of the approximation introduced in these calculations, they are likely to be of the same order of magnitude in light and heavy water. If this reasoning is sound, we should trust the difference in δG_m^{HI} between the two solvents, rather than the magnitude of each of these in a single solvent.

We have seen in Section 3.4 that the pairwise HI between methane molecules is stronger in H_2O as compared to D_2O. We also noted that conclusions to the contrary have been reached from other sources (see Sections 3.8 and 3.9). From Figures 4.11 and 4.12 we see that the HI among m solute particles in D_2O is weaker than in H_2O, in conformity with the behavior of pairwise HI. At present there is no molecular interpretation to this finding. However, in Section 5.8 we shall present a qualitative rationalization of this result, which will depend on a particular definition of the concept of the "structure of water." We shall also see that in some sense the replacement of H_2O by D_2O has an effect similar to that of decreasing the temperature of H_2O.

Next we turn to assessing the relative extent of the contribution of the two terms in (4.61). The question that we pose is the following: We have used a theoretical source for the computation of $\Delta\mu_A{}^\circ(\text{cav})$ and an experimental source for $\Delta\mu_M{}^\circ$. Which of the two terms is the dominating one?

Figure 4.13 shows the variation with m of the three quantities involved in (4.61). We see that for small m the values of $\Delta\mu_A{}^\circ(\text{cav})$ and $m\,\Delta\mu_M{}^\circ$ are of comparable magnitude. As m increases it is clear that the term $m\,\Delta\mu_M{}^\circ$ becomes the dominating one. This means that for large values of m our computed results rest more heavily on the experimental rather than on the theoretical source. This conclusion may also be understood on intuitive grounds. To see this, let us make a distinction between two kinds of monomers that build up our aggregate; let m_S be the number of solute monomers that form the surface of the aggregate (i.e., those that are in contact with the solvent) and m_I be the number of solute monomers that are in the interior of the aggregate (i.e., those that are surrounded by other solute monomers only).

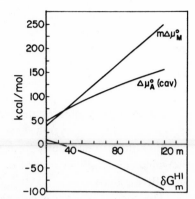

Figure 4.13. Values of δG_m^{HI}, $\Delta\mu_A^\circ$(cav) and $m\Delta\mu_M^\circ$ as a function of the number of monomers m in water at $P = 1$ atm and $t = 30°\mathrm{C}$.

Thus the overall process of aggregation

$$m \text{ monomers} \rightarrow \text{aggregate} \qquad (4.68)$$

may be viewed as being split into two parts:

$$m_S \text{ monomers} \rightarrow \text{surface of the aggregate} \qquad (4.69)$$

$$m_I \text{ monomers} \rightarrow \text{interior of the aggregate} \qquad (4.70)$$

Clearly, for very large aggregates, the number of surface particles may be neglected with respect to the number of interior particles. This means that the "reaction" (4.70) will dominate the overall process (4.68). Hence the free energy change of the overall process will be determined by the free energy of transferring m_I monomers from the solvent into the interior of the aggregate. Furthermore, since we have eliminated the *direct* solute-solute interaction in the definition of δG_m^{HI} (see 4.54), the process (4.70) is the same as transferring m_I solutes from the liquid to the gas. Hence, this process is approximately represented by $-m\,\Delta\mu_M^\circ$.

The above considerations are valid for very large m's, in which case δG_m^{HI} becomes essentially equal to m times $-\Delta\mu_M^\circ$. As we see from Figure 4.13, for m of the order of 100, both terms in (4.61) contribute to δG_m^{HI}, hence for such a size of aggregate we are still far from the limiting behavior that we mentioned above. This means that δG_m^{HI}, with m of the order of 100, is more relevant to an aggregation process rather than to a mere reversal of the solubility of the monomer.

In the above examples we have used a mixture of experimental and theoretical sources to compute δG_m^{HI}. It is worth noting that similar quantities may, in principle, be obtained by either purely theoretical or purely experimental sources.

First, consider m hard-sphere solutes of diameter σ_M brought to a compact configuration to form a sphere of diameter σ_A. We can apply the SPT to compute *both* $\Delta\mu_A{}^\circ(\text{cav})$ and $\Delta\mu_M{}^\circ(\text{cav})$ and define

$$\delta G_m{}^{\text{HI}} = \Delta\mu_A{}^\circ(\text{cav}) - m\,\Delta\mu_M{}^\circ(\text{cav}) \qquad (4.71)$$

Figure 4.14 compares the results obtained from (4.61) with those of (4.71) for water and cyclohexane. It is clear that though the magnitude of $\delta G_m{}^{\text{HI}}$ changes significantly from one method of computation to the other, the values of $\delta G_m{}^{\text{HI}}$ in water are systematically lower than in cyclohexane in both methods.

The second, purely experimental, way of computing $\delta G_m{}^{\text{HI}}$ is noted here, though no relevant computations have been carried out. Suppose that we could find a *real* compact polymer which has a roughly spherical shape, and for which the interaction energy with the solvent is similar to the interaction between our aggregate of m solutes and the solvent. In such a case we could write the approximate relation

$$\delta G_m{}^{\text{HI}} = \Delta\mu_P{}^\circ - m\,\Delta\mu_M{}^\circ \qquad (4.72)$$

where $\Delta\mu_P{}^\circ$ is the experimental standard free energy of solution of the polymer P. This relation relies only on experimental results for estimating $\delta G_m{}^{\text{HI}}$. It might be interesting to explore the practicability of this method in the future.

In Figure 4.15 we present further results computed by the mixed method of relation (4.61) for the system of water and ethanol. These

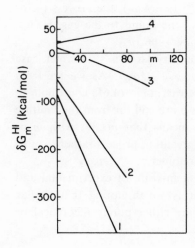

Figure 4.14. Comparison between the results obtained from equations (4.71) and (4.61). (1) Water, using only SPT. (2) Cyclohexane, using only SPT. (3) Water, using partial experimental data. (4) Cyclohexane, using partial experimental data. All values are for $P = 1$ atm at $t = 30°C$.

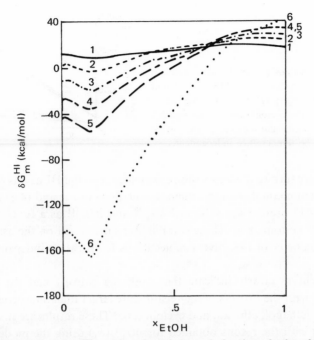

Figure 4.15. Values of δG_m^{HI} as a function of the mole fraction of ethanol in mixtures of water and ethanol. The various curves correspond to different values of m: (1) $m = 20$; (2) $m = 40$; (3) $m = 60$; (4) $m = 80$; (5) $m = 100$; (6) $m = 200$. All values are for $P = 1$ atm and $t = 10°C$.

calculations are based on an extension of the SPT to a mixture of solvents (Lebowitz *et al.*, 1965; Tenne and Ben-Naim, 1977). The interesting trend that we observe in Figure 4.15 is that as m increases to the order of 100 particles the HI, as measured by δG_m^{HI}, has a behavior very similar to the one we have found in Section 3.4. Namely, when we add ethanol, the HI becomes initially *weaker* than in water; thereafter there is a pronounced increase in the strength of the HI, and finally (above, say, $x_{EtOH} \approx 0.2$) the HI gradually decreases to its limiting value in pure ethanol. We believe that this behavior is due mainly to our use of experimental data through $\Delta\mu_M°$. It has been demonstrated that the SPT alone, as used in a relation of the form (4.71) *does not* show the characteristic dependence of the HI on the composition of the solvent that we have mentioned above. [Details are to be found in Tenne and Ben-Naim (1977).] We therefore believe that the SPT, as devised to deal with a mixture of simple solvents (say two kinds of hard spheres), is not applicable, as it stands, to a mixture of complex fluids such as water and ethanol.

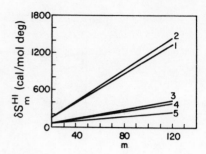

Figure 4.16. Values of $\delta S_m{}^{HI}$ as a function of m, at $P = 1$ atm and $t = 30°C$, in different solvents: (1) H_2O; (2) D_2O; (3) methanol; (4) ethanol; (5) cyclohexane. In all the calculations the molecular diameter of the solvent is taken to be temperature independent.

Next we turn to the temperature dependence of the HI as it is computed by the mixed method (4.61) through the relations (4.62) and (4.63). Figures 4.16 and 4.17 present the values of $\delta S_m{}^{HI}$ and $\delta H_m{}^{HI}$ as a function of the number of monomers m. These computations are based on the assumption that the diameter of the solvent molecules is temperature independent (see below).

The curves clearly indicate that both the entropy and the enthalpy associated with the process of aggregation are larger than the corresponding values in methanol, ethanol, and cyclohexane. These results are in complete agreement with the results obtained for $\delta G_2{}^{HI}(\sigma_1)$ using the model of Section 3.3.

It must be noted, however, that the question of which temperature dependence for the molecular diameter one must employ in these computations is not yet settled. It is obvious that only for hard spheres is the diameter of the particles a well-defined quantity and temperature independent (by definition!). For simple fluids, say argon, methane, etc., one may reasonably argue that the effective hard-core diameter should be a decreasing function

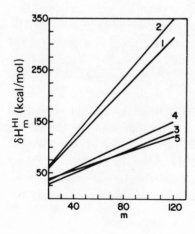

Figure 4.17. Values of $\delta H_m{}^{HI}$ as a function of m, at $P = 1$ atm and $t = 30°C$, in different solvents as in Figure 4.16.

of the temperature. The physical idea is that as one increases the temperature, the kinetic energy of the particles increases. Hence, on the average, interparticle collisions would lead to more extensive penetration into the repulsive region of the pair potential for the two particles.

In fact, it has been demonstrated that if such a negative temperature dependence of σ_S is adopted, then one can obtain a good agreement between the prediction from the SPT and experimental results.

The situation is far more complicated for nonspherical, or more complex, solvent molecules. In the first place the very concept of a hard-core diameter is not a well-defined quantity. For water, for instance, one may conveniently *choose* the effective diameter of the water molecule as the location of the first peak in the radial distribution function $g(R)$ for pure water. If we adopt this definition, we find that there exists a small *positive* temperature dependence of the molecular diameter of water. The rationalization of this behavior is quite simple. It is known that in liquid water at room temperature most of the water molecules are engaged in hydrogen bonds. The optimal distance for a hydrogen bond is about 2.76 Å, which is well within the effective hard-core diameter assigned to a water molecule, about 2.9 Å. Now as we increase the temperature we should consider at least two competing effects. On the one hand, we have the kinetic effect that was described above, which tends to *decrease* the effective hard-core diameters of *free* water molecules. On the other hand, hydrogen-bonded pairs are broken as we increase the temperature: hence fewer pairs of molecules will be found at the relatively short distance of 2.76 Å. This tends to *increase* the effective hard-core diameters of the bonded molecules. We believe that in water at room temperature the second effect mentioned above is the dominant one. This belief rests on two findings. In the first place the location of the first peak of the radial distribution function increases by about 0.005 Å as the temperature is raised by 10°C. Second, we have recently found (Ben-Naim and Tenne, 1977) that if one takes a positive temperature dependence for σ_M the computed results from the SPT are more consistent with experimental findings.

In any event the question of the sign or the extent of the temperature dependence of the diameter of a water molecule is far from being resolved. What is usually done is to choose a $\partial\sigma_S/\partial T$ that leads to the best fit between the computed and the experimental result. This kind of approach clearly is equivalent to feeding the theory with quantities that are characteristic of the solvent (in addition to the solvent density and its temperature dependence, which are to be supplied from experimental sources). In this sense the results of the computation of the entropies and the enthalpies based on the SPT should not be considered as emerging from a pure molecular theory of liquids.

Similarly when extending the applicability of the SPT to a mixture of solvents, such as water and ethanol, one must consider the possibility that the effective hard-core diameter of both molecules be composition dependent (at a given temperature). The results exhibited in Figure 4.15 were computed on the assumption of a *fixed* molecular diameter for both water and ethanol molecules. This choice is inevitable since we have no information to guide us in the choice of the composition dependence for these diameters. We believe that this is the major reason for the failure of the theory to predict the "correct" trend of the dependence of δG_m^{HI} on the composition of the solvent at small values of m. However, for large m, the term $-m\,\Delta\mu_M^\circ$ becomes dominant and the results of the computations rely more on the experimental than the theoretical sources. This is the reason for the more plausible results obtained for the larger m's in Figure 4.15.

Finally, we turn briefly to the pressure dependence of δG_m^{HI}. We use again Equation (4.61), where $\Delta\mu_A^\circ(\text{cav})$ is estimated from the SPT and $\Delta\mu_M^\circ$ is estimated as follows. We use the relation

$$\frac{\partial\Delta\mu_M^\circ}{\partial P} = \bar{V}_M^\circ \tag{4.73}$$

where $\bar{V}_M^\circ$ is the *local* standard partial molar volume of the solute at infinite dilution. [This is different from the conventional standard partial molar volume. For more details see Section 5.2 and Ben-Naim (1978a).]

Assuming that $\bar{V}_M^\circ$ is approximately constant over a certain range of pressures, we write

$$\Delta\mu_M^\circ(P) \approx \Delta\mu_M^\circ(P = 1 \text{ atm}) + (P - 1)\bar{V}_M^\circ \tag{4.74}$$

This quantity is used in Equation (4.61) to compute the pressure dependence of δG_m^{HI}. Two sets of results are shown in Figure 4.18. One is based on the method described above. The second is based solely on the SPT, i.e., both $\Delta\mu_A^\circ$ and $\Delta\mu_M^\circ$ are computed by the SPT. We note again that some kind of pressure dependence of σ_S should be included in such calculations, but because of lack of any reliable information on this we have simply taken a constant value for σ_S.

The two sets of results presented in Figure 4.18 are considerably different in their magnitude. In both cases, however, the general result is that the strength of the HI increases with pressure. We shall further discuss the pressure dependence of the HI in Chapter 5.

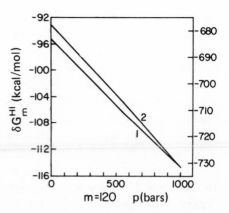

Figure 4.18. The dependence of δG_m^{HI}, with $m = 120$, on the pressure in water at $t = 30°C$. Curve (1) was obtained by using only the SPT (4.71). Curve (2) was obtained by using the combination of experimental data and the SPT as in (4.61). The right and left scales refer to curves (1) and (2), respectively.

Comments

In spite of our general reservation on the applicability of the SPT to complex solvents such as water, methanol, and the like, we believe that the procedure outlined in this section provides reasonable information on the HI among a large number of solute particles. One should also be aware of the fact that, at present, we have no other source that provides information of this kind.

Of course, if m is very large, then the results obtained from these calculations are equivalent to the standard free energy of solution of the monomers. In this respect our process of aggregation provides the same information as the reversal of the dissolution process. This is not the case, however, for m of the order of 100, in which both $\Delta\mu_A^\circ$ and $m\,\Delta\mu_M^\circ$ are of comparable magnitude. In the future, when δG_2^{HI} at contact distance $R \approx \sigma_M$ between two solutes may be available, one could use the computed values of δG_m^{HI} to estimate the extent of nonadditivity of the HI. A further possible application of the SPT is to compute δG_m^{HI} for configurations other than spherical, and therefore to gain some idea of the dependence of the HI on the configuration of the aggregate.

4.7. A DIRECT MEASURE OF INTRAMOLECULAR HI

In Section 3.3 we introduced a quantity, $\delta G^{HI}(\sigma_1)$, that measures the HI between two simple solutes at a very small distance $R = \sigma_1$. It was stressed there that this quantity should be useful for comparing the HI in different solvents, and not for estimating the strength of the HI in any specific solvent. The derivation of the relation between $\delta G^{HI}(\sigma_1)$ and experimental quantities also involved an approximation. It is, therefore, desirable to construct a new measure of HI that does not involve such an approximation.

We now present a new, and exact, relation between a quantity that conveys information on HI and experimental quantities. Furthermore, this relation provides information on the HI between two alkyl groups at realistic configurations. Perhaps these configurations are not the ones that nonpolar groups are actually found in in real biopolymers, but we certainly are making a step forward towards that end.

The method described below starts from the recognition of the fact that pairwise HI in a single biopolymer is an *intramolecular* phenomenon. Referring to Figure 1.1 we realize that in the process of a conformational change two alkyl groups are brought from a large to a small separation. In both of these conformations the two alkyl groups are anchored on the same molecular backbone. Having this prototype process in mind, we now replace the biopolymer, as a carrier for the alkyl groups, by a relatively simple molecule. By doing that, we have freed ourselves from the formidable complexities of the real biopolymer. We are thus left with a simple process of transferring an alkyl group between two states of a small carrier. Hence, the term *intramolecular* HI seems to be appropriate for this process.

We now turn to describe the theoretical background of the method. The idea is very similar to the one described in Section 3.3. The reader should realize that for a full appreciation of the method described below, it is essential to appeal to some statistical mechanical arguments. The final result is indeed a relation between thermodynamic quantities, but its derivation relies on statistical mechanics.

Consider a molecule such as 1,4 dialkylbenzene, which we shall denote by ϕ_{14}. Suppose that we "cut" this molecule into four groups as indicated schematically in Figure 4.19. The exact location of the "cut" is of no importance since in our final expression we shall deal with the whole molecule. For convenience, however, we assume that the "cut" is done at

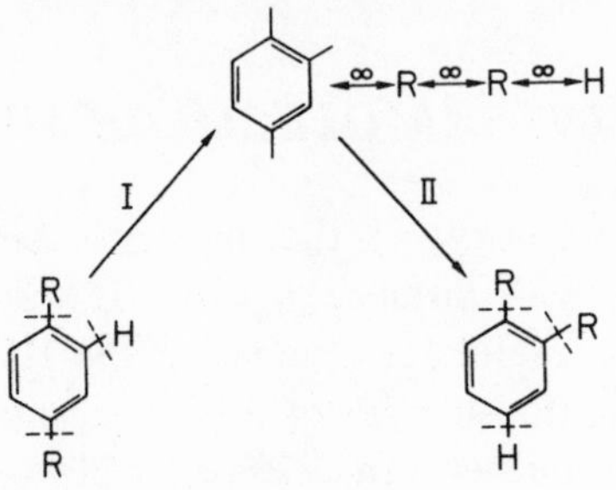

Figure 4.19. I is a schematic process of breaking 1,4-dialkylbenzene into four radicals: benzyl ring ϕ, two alkyl groups R, and a hydrogen atom H. These radicals are recombined in process II to form the new molecule 1,2-dialkylbenzene. In the intermediate stage all the radicals are at fixed configurations and at infinite separation from each other.

the center of the C–C and of the C–H bond. Also for simplicity we assume that the molecule as a whole is rigid.

We now consider the following process: We start with a single solute ϕ_{14} at some fixed configuration (location and orientation) in a solvent at a given temperature T, pressure P, and total number of molecules N. We "cut" the molecule into four radicals and remove them to fixed configurations but at infinite separation from each other. This process is indicated by I in Figure 4.19.

The Gibbs free energy change for this process is written, in the T, P, N ensemble, as

$$\Delta G(\mathrm{I}) = kT \ln\left[\frac{\Delta(T, P, N; \phi_{14})}{\Delta(T, P, N; \phi, H, R, R)}\right] \tag{4.75}$$

where in the numerator and in the denominator we wrote the partition functions of the system at the initial and the final states considered above. Dividing the numerator and the denominator by the partition function of the pure solvent $\Delta(T, P, N)$ we obtain

$$\Delta G(\mathrm{I}) = kT \ln \frac{\langle\exp[-\beta B(\phi_{14})]\rangle_0 \exp[-\beta U(\phi_{14})]}{\langle\exp[-\beta B(\phi) - \beta B(H) - \beta B(R) - \beta B(R)]\rangle_0} \tag{4.76}$$

where $B(A)$ represents the "binding energy" of the solute (or radical) A to the rest of the system which is at a fixed configuration $\mathbf{X}^N$. $\beta = (kT)^{-1}$ with k the Boltzmann constant, and the average $\langle \ \rangle_0$ is over all the configurations and volumes of the pure solvent. More specifically the binding energy of A is defined by

$$B(A) = \sum_{i=1}^{N} U(\mathbf{X}_A, \mathbf{X}_i) \tag{4.77}$$

where $(\mathbf{X}_A, \mathbf{X}_i)$ represents the configuration of the solute A and of the ith solvent molecule. By $U(\phi_{14})$ we denote the total *direct* interactions between the four radicals at the final configuration of the molecule ϕ_{14}. Alternatively, $-U(\phi_{14})$ is the total work required to perform the same process, as described above, but in the absence of a solvent. The two alkyl groups R are identical (i.e., methyl, ethyl, etc.). However, in (4.76) they are understood to be at infinite separation from each other.

Since we have assumed that the radicals ϕ, H, R, R are at infinite separation from each other we may factor the average in the denominator on the right-hand side of (4.76) into a product of four average quantities, namely,

$$\langle\exp[-\beta B(\phi) - \beta B(H) - \beta B(R) - \beta B(R)]\rangle_0$$
$$= \langle\exp[-\beta B(\phi)]\rangle_0 \langle\exp[-\beta B(H)]\rangle_0 \langle\exp[\beta B(R)]\rangle_0^2 \tag{4.78}$$

Using this factorization we rewrite (4.76) as (see also Appendix A.1)

$$\Delta G(\text{I}) = -U(\phi_{14}) - \Delta\mu^\circ(\phi_{14}) + \Delta\mu^\circ(\text{H}) + \Delta\mu^\circ(\phi) + 2\Delta\mu^\circ(\text{R}) \qquad (4.79)$$

where $\Delta\mu^\circ(A)$ is the free energy change for transferring the solute (or radical) A from a fixed configuration in the gaseous phase to a fixed configuration in the liquid. Of course in (4.79) the only directly measurable quantity is $\Delta\mu^\circ(\phi_{14})$. All the other terms will now be eliminated by the following considerations. First we define the indirect part of the total work $\Delta G(\text{I})$ by

$$\delta G^{\text{HI}}(\text{I}) = \Delta G(\text{I}) + U(\phi_{14}) \qquad (4.80)$$

i.e., we define the excess work for process I in the liquid relative to the gaseous phase. This quantity has been referred to as the *indirect*, or the *hydrophobic interaction* (HI), part of the total work. From (4.80) and (4.79) we obtain

$$\delta G^{\text{HI}}(\text{I}) = -\Delta\mu^\circ(\phi_{14}) + \Delta\mu^\circ(\text{H}) + \Delta\mu^\circ(\phi) + 2\Delta\mu^\circ(\text{R}) \qquad (4.81)$$

Similarly for the process indicated as II in Figure 4.19 we write the indirect part of the total work as

$$\delta G^{\text{HI}}(\text{II}) = \Delta\mu^\circ(\phi_{12}) - \Delta\mu^\circ(\text{H}) - \Delta\mu^\circ(\phi) - 2\Delta\mu^\circ(\text{R}) \qquad (4.82)$$

Adding (4.81) to (4.82) we obtain

$$\delta G^{\text{HI}}[(1,4) \to (1,2)] = \delta G^{\text{HI}}(\text{I}) + \delta G^{\text{HI}}(\text{II})$$
$$= \Delta\mu^\circ(\phi_{12}) - \Delta\mu^\circ(\phi_{14}) \qquad (4.83)$$

Thus, on the right-hand side of (4.83) we have two measurable quantities: the standard free energies of solution of the two dialkylbenzene molecules. All the "standard free energies" of the solution of the radicals in (4.81) and (4.82) have been canceled out. On the left-hand side of (4.83) we have the indirect part of the work required to transfer an alkyl group from position 4, where it does not "see" the alkyl group at position 1, to position 2, where it is close to the group at position 1.

The quantity defined by (4.83) also has an important probability interpretation. Suppose we have a single dialkylbenzene in a solvent in such a way that one alkyl group at position 1 is fixed, whereas the second alkyl group may attain one of the two positions: 2 or 4. We assume that the alkyl group is free to move between these two states. The ratio of the probabilities of finding the second alkyl group in the two states is given by

$$\frac{Pr(2)}{Pr(4)} = \exp[-\beta U(\phi_{12}) + \beta U(\phi_{14})] \exp\{-\beta \delta G^{HI}[(1,4) \rightarrow (1,2)]\} \quad (4.84)$$

The first factor on the right-hand side of (4.84) is the probability ratio in vacuum, i.e., in the absence of the solvent. The second factor is due to the presence of the solvent. This factor determines the probability ratio in a system in which the direct interactions $U(\phi_{12})$ and $U(\phi_{14})$ have been "switched off." Alternatively, if we take two phases a and b in which the direct interactions do exist but are not affected by the solvent, we obtain

$$\left[\frac{Pr(2)}{Pr(4)}\right]_a \bigg/ \left[\frac{Pr(2)}{Pr(4)}\right]_b = \frac{\exp\{-\beta \delta G^{HI}[(1,4) \rightarrow (1,2)]\}_a}{\exp\{-\beta \delta G^{HI}[(1,4) \rightarrow (1,2)]\}_b} \quad (4.85)$$

and in particular if a is a liquid and b is an ideal gas we have

$$\frac{y(2)}{y(4)} \equiv \left[\frac{Pr(2)}{Pr(4)}\right]_l \bigg/ \left[\frac{Pr(2)}{Pr(4)}\right]_g = \exp\{-\beta \delta G^{HI}[(1,4) \rightarrow (1,2)]\}_l \quad (4.86)$$

which is the probability interpretation of the quantity defined in (4.83). We have denoted the ratio of the two probabilities of the same state in the two phases by $y(2)$ and $y(4)$. This quantity has a significance similar to $y(R)$ introduced in Section 1.4. Before turning to some numerical illustrations we note that relation (4.83) is an exact relation and does not involve an approximation similar to the one used in Section 3.3. Furthermore, the quantity $\delta G^{HI}[(1,4) \rightarrow (1,2)]$ means essentially the difference between the HI in the 1,4 relative to the 1,2 configuration. Both of these configurations are "realistic" ones and do not involve extensive penetration of one group into the other, as was the case in $\delta G^{HI}(\sigma_1)$.

We will now describe some illustrative examples of the application of relations (4.83) and (4.86).

The solubilities and partition coefficients of dialkylbenzene solutes between water and n-hexane were measured spectroscopically (Ben-Naim and Wilf, 1979). From these measurements one may easily calculate the various standard free energies of transfer of the solutes from the gas to water, $\Delta\mu°(G \rightarrow W)$, from n-hexane to water, $\Delta\mu°(H \rightarrow W)$, and from the gas to n-hexane, $\Delta\mu°(G \rightarrow H)$. These values are reported in Table 4.3.

From the standard free energies of solution we compute the quantities $\delta G^{HI}[(1,4) \rightarrow (1,2)]$ as defined in (4.83). These are shown in Table 4.4. It is clearly seen from this table that in n-hexane the values of δG^{HI} are either positive or very small and probably within the limits of the ex-

Table 4.3

Standard Free Energies of Transfer between the Gas and Water, between
n-Hexane and Water, and between the Gas and n-Hexane for Different
Solutes at Two Temperatures and 1 atm[a]

Solute	t (°C)	$\Delta\mu^\circ(G \to W)$ (kcal/mol)	$\Delta\mu^\circ(H \to W)$ (kcal/mol)	$\Delta\mu^\circ(G \to H)$ (kcal/mol)
Benzene	10	−1.175	2.862	−4.038
	20	−0.981	2.868	−3.849
Methylbenzene	10	−1.224	3.632	−4.856
	20	−0.974	3.741	−4.715
Ethylbenzene	10	−1.232	3.678	−4.910
	20	−0.956	3.754	−4.710
1,2-Dimethylbenzene	10	−1.478	4.189	−5.667
	20	−1.243	4.321	−5.564
1,4-Dimethylbenzene	10	−1.240	4.496	−5.736
	20	−0.943	4.593	−5.536
1,2-Diethylbenzene	10	−1.908	3.926	−5.834
	20	−1.540	3.727	−5.267
1,4-Diethylbenzene	10	−1.305	4.888	−6.193
	20	−0.924	4.788	−5.712

[a] Data from Ben-Naim and Wilf (1979).

Table 4.4

Values of the Indirect Part of the Work Required to Transfer an Alkyl
Group from Position 4 to Position 2 at Two Temperatures

Alkyl group	t (°C)	$\delta G^{HI}[(1,4) \to (1,2)]$ (cal/mol) in water	$\delta G^{HI}[(1,4) \to (1,2)]$ (cal/mol) in n-hexane
Methyl	10	−238	+69
	20	−300	−28
Ethyl	10	−603	+359
	20	−616	+445

perimental error. However, in water we find negative values of δG^{HI} which seem to increase with the chain length of the alkyl group.

It should be noted that the HI reported in Table 4.4 are not *pairwise* HI in the sense of Chapter 3 (this is why this method belongs to this chapter). The reason is that we have started our considerations with the HI among *four* radicals, and by taking differences we have ended up with a quantity that measures the indirect part of the work of transferring an alkyl group from position 4 to position 2; the process is carried out in the *presence of the benzene ring*. This is the reason for referring to δG^{HI} as a measure of the *intramolecular* HI. This process should therefore be clearly distinguished from the process of bringing two alkyl groups from infinite separation to some close configuration in the solvent. Here, the proximity of the benzene ring must have some effect on the structure or properties of the medium in which the two alkyl groups "see" each other. We believe that this is one reason for finding small values of δG^{HI} between two methyl groups in water. In this case the two groups are very close to the benzene ring and hence their surroundings greatly differ from that of pure water. Once we take a longer alkyl group, such as ethyl (or longer chains), the medium for the HI between the two groups is farther away from the benzene rings and hence closer to that of pure liquid water.

We also note that the values of $|\delta G^{HI}|$ in water are slightly larger at 20°C compared to 10°C. This is in agreement with our previous conclusions on the temperature dependence of the HI. However, we believe that in this particular example the difference is well within the experimental error involved in the estimation of δG^{HI}.

It is now instructive to translate the same data reported in Table 4.4 into the language of probabilities. This reinterpretation is contained in Equation (4.86). The relevant situation is the following.

Suppose we have a dialkylbenzene molecule at some fixed configuration in the solvent. Let us fix one alkyl group at position 1 and assume that the second group can attain either position 2 or position 4. We may now ask: what is the ratio of the probabilities of finding this group in the two positions? The answer to this question requires a knowledge of the direct interaction between the alkyl groups as well as the HI part. Since we are interested only in the latter part we may assume that the direct interactions are being "switched off," hence the quantity $y(2)/y(4)$ defined in (4.86) gives the ratio of the two probabilities in such a system.

We see from Table 4.5 that the entries for *n*-hexane are either of the order of unity or smaller than unity, indicating a preference for the 1,4 configuration. On the other hand, in water there is a clear-cut preference

Table 4.5

Ratios of the (Solvent-Induced) Probabilities of Finding an Alkyl Group in Positions 2 and 4 in a Molecule with a Fixed Alkyl Group at Position 1

Alkyl group	t (°C)	$y(2)/y(4)$ in water	$y(2)/y(4)$ in n-hexane
Methyl	10	1.527	0.885
	20	1.675	1.049
Ethyl	10	2.924	0.528
	20	2.878	0.465

for the 1,2 configuration. This is another way of describing the phenomenon of HI.

Another way of processing the data presented in Table 4.3 follows. Consider the disproportionation "reaction" depicted in Figure 4.20. This is not a "real" reaction, but one in which the solute molecules are devoid of their translational and rotational degrees of freedom. We start with two monoalkylbenzenes ϕ—R at fixed configurations and at infinite separation from each other in the solvent. Next we exchange the alkyl group R of one molecule with the hydrogen atom at position 2 of the second molecule. As a result of this exchange we obtain a 1,2-dialkylbenzene molecule ϕ_{12} and a benzene molecule B. This process is indicated by a in Figure 4.20.

The corresponding indirect part of the free energy change is

$$\delta G^{\mathrm{HI}}(a) = \Delta\mu^\circ(\phi_{12}) + \Delta\mu^\circ(B) - 2\Delta\mu^\circ(\phi - \mathrm{R}) \tag{4.87}$$

and similarly for the "reaction" b in Figure 4.20 we obtain

$$\delta G^{\mathrm{HI}}(b) = \Delta\mu^\circ(\phi_{14}) + \Delta\mu^\circ(B) - 2\Delta\mu^\circ(\phi - \mathrm{R}) \tag{4.88}$$

Values of $\delta G^{\mathrm{HI}}(a)$ and $\delta G^{\mathrm{HI}}(b)$ were computed from the data in Table 4.3 and are presented in Table 4.6.

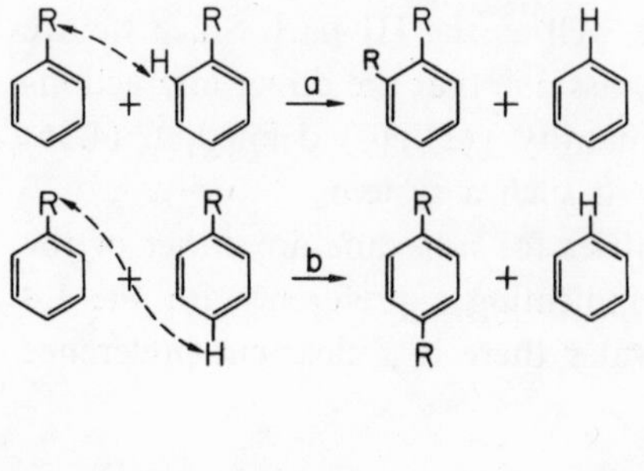

Figure 4.20. Two disproportionation reactions: In (a) two monoalkylbenzenes are used to form a 1,2-dialkylbenzene and a benzene molecule. In (b) a 1,4-dialkylbenzene and benzene are formed from the same initial molecules.

Table 4.6

Values of $\delta G^{HI}(a)$ and $\delta G^{HI}(b)$ (in kcal/mol) as Defined in Equations (4.87) and (4.88), Respectively

Alkyl group	t (°C)	In water		In n-hexane	
		$\delta G^{HI}(a)$	$\delta G^{HI}(b)$	$\delta G^{HI}(a)$	$\delta G^{HI}(b)$
Methyl	10	−0.205	+0.033	+0.007	−0.062
	20	−0.276	+0.024	+0.040	+0.068
Ethyl	10	−0.619	−0.016	−0.160	−0.411
	20	−0.609	+0.007	+0.304	−0.141

The values of $\delta G^{HI}(a)$ and $\delta G^{HI}(b)$ may be assigned a probability meaning similar to the one given by relation (4.86). Briefly, we look at the two sides of the "chemical reactions" in Figure 4.20 as being two states of the system (i.e., the solvent with the two solutes at infinite separation from each other). We may ask about the relative probabilities of these two states. Excluding, as before, the direct interaction energies, we can focus only on the solvent effect on the relative probabilities of finding these two states of the system, which is given by $\exp[-\beta \delta G^{HI}]$. Thus, from Table 4.6 we see that for the two "ethyls in water," the *right*-hand side of "reaction" a will be about 3 times more probable (at 20°C) than the left-hand side. On the other hand, the opposite is true for "reaction" b, namely, the *left*-hand side is about 1.3 times more probable than the right-hand side. This is clearly another manifestation of the HI between the two ethyl groups at the positions 1,2—an aspect that bears important relevance to the role of HI in chemical reactions.

Finally, we note that the HI, as measured by the quantity defined in (4.83), seem to be larger in H_2O as compared with D_2O (Wilf and Ben-Naim, 1979), a result that is consistent with the conclusion arrived at in Section 3.4.

Comment

We believe that the extension of the method described in this section will provide important information on the intramolecular HI and on its dependence on temperature, on pressure, and on the addition of solutes. Furthermore, by changing the "carrier molecule," one can gain some ideas about the effect of the carrier on the HI. Hopefully, one could eventually extrapolate to real carriers, such as proteins or nucleic acids, in order to understand the role of HI in biopolymers.

4.8. HI *IN AQUEOUS MICELLAR SOLUTIONS*

In this and in the following section we present a very brief discussion of some properties of more complex aqueous systems. In all of these systems the concept of HI has been involved, in one way or another, in order to explain some of their outstanding properties. The point we shall emphasize here is that though HI certainly plays a role in the determination of the properties of these systems, it is impossible, at present, to extract information on HI from the study of such systems. In this sense we deviate here from the prevailing attitude maintained in this book. The reader should realize that each of the topics touched upon in these two sections consists of a large field of research in its own right. We mention here only a few points that bear some relevance to the problem of HI.

Aqueous micellar solutions may be viewed as intermediate systems bridging the gap between the simple aqueous solutions that we have discussed before, on the one hand, and the more complex biological systems, on the other. In this sense, the study of micelles provides an excellent model through which one can infer, and perhaps understand, the more complex biological solutions. The literature on micelles is quite voluminous; some general reviews are Winsor (1954), Mukerjee and Mysels (1971), Kresheck (1975), and Mittal (1977).

The basic experimental observation is quite simple. A surface-active molecule usually contains a polar "head" group and a nonpolar "tail." These molecules are known to reduce the surface tension when they are added to water—hence the term "surface-active" or "surfactants." The main reason for their doing so is ascribed to the tendency of these molecules (more precisely, the nonpolar part of them) to avoid contact with water and to seek, as far as possible, a nonaqueous environment. It is here that the concept of "hydrophobicity," in the sense of Chapter 2, enters into this field.

When the concentration of the surfactant is gradually increased, one observes systematic deviations from the behavior of ideal dilute solutions. This phenomenon may be ascribed to the formation of small aggregates of solute molecules. This is a common phenomenon shared by many concentrated solutions. What makes aqueous surfactant solutions so remarkable is that at some small concentration range one finds an abrupt change in the properties of the solution. The concentration (or better the range of concentration) at which this "turning point" occurs is referred to as the "critical micelle concentration" (CMC).

There are many physical properties that may be followed in order to determine the CMC. The most common ones are surface tension and conductivity of the solution.

As an illustrative example, consider the equivalent conductance of aqueous solutions of sodium dodecylsulfate. At very small concentrations of the surfactant there is almost no change in Λ as a function of $c^{1/2}$ (where c is the concentration of the surfactant). Beyond a certain concentration one finds a sharp decrease of the equivalent conductance as a function of $c^{1/2}$. This behavior is demonstrated in Figure 4.21. By drawing the two asymptotes to this curve one determines the CMC from their intersection. As is clear from Figure 4.21, the experimental points do not indicate a well-defined point but a range of concentrations, which is referred to as the CMC.

It is not uncommon to find in the literature statements referring to the "discontinuity" at the CMC. However, though in some cases there is a remarkably sharp transition at the CMC, the function and its derivatives are quite continuous. [See, also, Mukerjee and Mysels (1971), who stressed this point.]

Instead of the conductivity of the solution, one may follow other physical properties of the solution. In all of these the determination of the CMC gives almost the same value of the CMC within 3–5%. This observation indicates that the solution undergoes some fundamental changes in this concentration range. It is now well established that at the CMC large aggregates of surfactant molecules having compact shapes are formed. These are called micelles. A typical structure of a spherical micelle is depicted in Figure 4.22. The main feature of the mode of packing of the

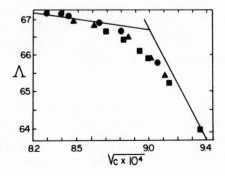

Figure 4.21. Equivalent conductance of aqueous solutions of sodium dodecyl-sulfate as a function of the square root of the surfactant concentration [reproduced with changes from Mukerjee and Mysels (1971)].

solute molecules in the micelle is that the nonpolar "tails" occupy the interior of the micelle, whereas the "head" groups are exposed to the aqueous environment.

The qualitative rationalization for the formation of this particular structure of the micelle is based on the idea that the nonpolar groups tend to avoid the aqueous environment. By clustering together they obviously achieve that end. Here, again, we have a phenomenon akin to the HI.

Experimental evidence indicates that below the CMC micelles are not formed (or at least are undetectable by all experimental means). Above the CMC it has been established that most of the added surfactant is used to build up micelles. The concentration of the monomeric solute remains fairly constant. If it was exactly constant, then, as was suspected by several authors, we would have a phenomenon similar to a phase transition at the CMC. However, to the best of the author's knowledge, there exists no experimental evidence to support the phase-separation contention. The very fact that the CMC is not a singular point but rather a small range of concentration, at which the properties of the solution change smoothly precludes the possibility of the existence of a phase transition. In fact, Mukerjee and Mysels (1971) have stressed that the very term "CMC" might be misleading, because of its implication that there exists such a singular point—which is not the case, however. Further discussion of the various theories of micelle formation may be found in a review by Hall and Pethica (1967). We shall demonstrate below that by using a quite simple "mass-action" model for micelle formation one can simulate the characteristic change in the monomer concentration as a function of the total concentration of the surfactant.

Regarding the nature of the interior of the micelles, there is experimental evidence showing that it has the character of both a nonpolar and a water mixture. We present here some evidence that is based on NMR chemical shifts and nuclear spin–lattice relaxation times.

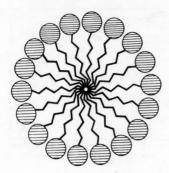

Figure 4.22. Schematic structure of a micelle in aqueous solution.

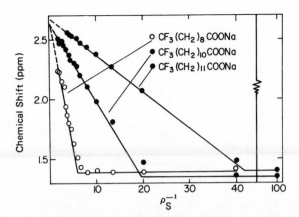

Figure 4.23. Fluorine chemical shifts as a function of ϱ_S^{-1}, where ϱ_S is the surfactant concentration (in mol/liter). The three surfactant molecules are indicated. [Reproduced with changes from Muller and Birkhahn (1967).]

Muller and Birkhahn (1967) measured the chemical shift of fluorine in solutes of the form $CF_3(CH_2)_n COONa$. We reproduce some of their results in Figure 4.23. From this figure one observes several characteristic features of micellar solutions. In the first place the sharp transition at the CMC is demonstrated. Secondly, we find, as is commonly found by other experimental means, that the larger the alkyl chain the smaller the CMC. In fact, there exists a general correlation between the ability of the surfactant to reduce the surface tension of the solution, and the tendency to form micelles at lower concentration. For a given homologous series, the larger the alkyl group, the more "anxious" the molecules are to form micelles, hence the lower their CMC.

Finally, it is seen from Figure 4.23 that the fluorine chemical shift in very dilute aqueous solutions is the same and about 1.38 for the three solutes and it is almost unchanged as a function of concentration. At the CMC the chemical shift changes abruptly and reaches the value of about 2.66 for the three solutes. The interesting finding is that the latter value of the chemical shift is about midway between that of water and that of pure hydrocarbon. This suggests that the interior of the micelles has the character of a water and hydrocarbon mixture.[†]

Similar data on shorter alkanoates have been reported by Ödberg *et al.* (1972) and by Henriksson and Ödberg (1976). We present one of

---

[†] Of course one should note that the C–F bond is quite polar and therefore it is likely to drag water molecules into the micelle. Hence, from such measurements spurious conclusions may be drawn about water penetration in simple hydrocarbon micelles.

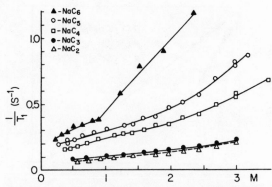

Figure 4.24. Proton spin–lattice relaxation rate for sodium alkanoate as a function of the solute concentration [reproduced with changes from Ödberg *et al.* (1972)].

their results in Figure 4.24. Here, the proton spin–lattice relaxation rate of sodium alkanoate in D_2O solutions was followed as a function of the solute concentration. For the first two alkanoates there is a very slight concentration dependence of T_1^{-1}. For butyrate and valerate, the dependence is more pronounced, indicating the formation of small aggregates. It is only for the caproate solution that a clear-cut CMC is observed. The steep increase of the relaxation rate beyond the CMC is clearly a result of the proximity and high concentration of the protons of the alkyl chains that form the micelles.

We now turn briefly to the thermodynamic description of surfactant solutions, to see where HI might be involved and why it is not a simple matter to extract information on HI from the study of these systems.

First we note that if the micelles are viewed as a separate phase, then the chemical potential of the surfactant S in the two phases is (assuming ideality of the aqueous solution)

$$\mu_S(\text{in micelle}) = \mu_S(\text{in water}) = \mu_S^{\circ\varrho} + kT \ln \varrho_S \qquad (4.89)$$

If $\mu_S(\text{in micelle})$ is treated as the chemical potential of a "pure" phase, then one would have predicted that ϱ_S is constant and equal to the CMC. Hence, one often writes an equation in the form

$$\Delta G^\circ = \mu_S(\text{in micelle}) - \mu_S^{\circ\varrho} = kT \ln(\text{CMC}) \qquad (4.90)$$

where ΔG° is interpreted as the "free energy" of transferring S from water into the micelle. This interpretation is unsound, however, for the same reasons given in Appendix A.1 (see also Section 2.6). For our present purposes, the fact that the CMC is not a singular point and the fact that the solute concentration does not remain strictly constant above the CMC are sufficient reasons to abandon the phase-separation model of micellar solutions.

The next, more realistic approach, is to assume a sequential series of association reactions of the form

$$nM \rightleftarrows A_n, \qquad n = 2, 3, \ldots \tag{4.91}$$

where A_n is an aggregate consisting of n monomers M. Of course one could, in principle, make a finer distinction between aggregates with the same size n but having different shapes (or packing structures). However, since there are no experimental means to make such a distinction we can lump in A_n all aggregates containing n monomers.

A further assumption that is customarily made for these solutions is that they form an *associated ideal dilute* solution. This means that the solution, when viewed as a two-component system of water W and surfactant S, is *not* an ideal dilute solution. The deviation from the ideal behavior results from the solute–solute interactions. However, if these interactions are of short range, then any n-tuplet of simultaneously interacting solutes may be identified as an aggregate and is assigned the symbol A_n. Thus, by definition, all the solute–solute interactions will be contained within the various aggregates A_n. Now, if we view the system as a multicomponent mixture, $W, M, A_2, A_3, \ldots$, we can, to a good approximation, ignore the interactions between these species. This is the basic argument that leads to the idea of an associated ideal dilute solution. Within this model one may write for each solute species the chemical potential in the form

$$\mu_M = \mu_M{}^{o\varrho} + kT \ln \varrho_M \tag{4.92}$$

$$\mu_{A_n} = \mu_{A_n}^{o\varrho} + kT \ln \varrho_{A_n}, \qquad n = 2, 3, \ldots \tag{4.93}$$

where the standard chemical potential $\mu_i{}^{o\varrho}$ of the ith solute species contains essentially the internal partition function of that species and the coupling work of the ith species to an essentially pure water environment (for more details see Appendix A.1). Now, from the condition of chemical equilibrium

$$n\mu_M = \mu_{A_n}, \qquad n = 2, 3, \ldots \tag{4.94}$$

we obtain the well-known result

$$K_n = \frac{\varrho_{A_n}}{\varrho_M{}^n} = \exp\left(-\frac{\mu_{A_n}^{o\varrho} - n\mu_M{}^{o\varrho}}{kT}\right) \tag{4.95}$$

where K_n is the equilibrium constant for the nth chemical reaction in (4.91). $\mu_{A_n}^{Oe} - n\mu_M^{Oe}$ is the standard free energy of formation of the aggregate A_n. This quantity is related to the free energy of transferring a monomer M from water into the micelle. However, in order to be more precise, one should specify under what conditions this process is carried out. This point is of crucial importance if we want to attempt to extract information relevant to HI from the quantity $\mu_{A_n}^{Oe} - n\mu_M^{Oe}$.

Before elaborating on this, however, it is appropriate at this stage to comment on the practice of using a relation of the form (4.95) with "mole fractions" as concentration units. This practice is quite commonplace and some authors claim that it has some advantages. We shall now show that the use of mole fractions is quite ambiguous, for the following reason: Writing an equation of the form

$$K_n' = x_{A_n}/x_M^n \tag{4.96}$$

and referring to x_M and x_{A_n} as the "mole fractions" of M and A_n leaves a certain ambiguity in the meaning of these "mole fractions." One can think of at least two possible definitions of the "mole fractions" in this system, namely,

$$x_M = \frac{N_M}{N_W + N_M + \sum_{i=2} N_{A_i}} \tag{4.97}$$

and

$$x_M' = \frac{N_M}{N_W + N_M + \sum_{i=2} i N_{A_i}} \tag{4.98}$$

In the first definition we count each aggregate as a different molecular species, whereas in the second we count only the total number of solvent and surfactant molecules in the system. To avoid this kind of ambiguity, one has to specify which definition of the "mole fraction" has been chosen. However, now we face another difficulty, since there is neither a theoretical nor a practical argument that may be used to guide us in making the "proper" choice of the mole fraction. Thus, the ambiguity already exists at the stage of making a choice of the "best" definition for the mole fractions in this system. These difficulties are avoided by using number or molar densities for all the species involved.

Following Mukerjee (1974) we demonstrate in Figure 4.25 the variation of the monomer concentration as a function of the total concentration of the surfactant. In these calculations we have solved Equation (4.95) for different values of n as indicated next to each curve. (The equilibrium constant was arbitrarily chosen to be unity.)

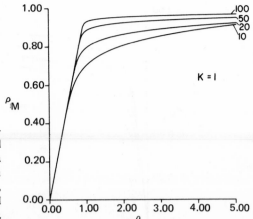

Figure 4.25. Variation of the monomer concentration ϱ_M with the total surfactant concentration ϱ_T for a single aggregation equilibrium with a fixed n. K_n was taken to be unity, and the different curves correspond to different aggregation numbers n.

The remarkable finding is that, with the use of a *single* equilibrium condition (4.95), one finds that a fairly sharp transition in the dependence of ϱ_M on ϱ_T is manifested, provided that n is large enough. We see that in this particular case for $n = 100$ the monomer concentration ϱ_M is almost (but not exactly) constant beyond $\varrho_T \geq 1.0$.

Of course, in a real system, there is no reason to exclude all the intermediate-size aggregates. If one takes a sequential series of aggregates and the corresponding equilibrium conditions (4.95), one can easily show that the change of ϱ_M as a function of the total concentration $\varrho_T = \varrho_M + \sum_{i=2} i\varrho_{A_i}$ will be much smoother. This is demonstrated in Figure 4.26. Clearly, one can proceed from the latter to the former case by taking K_n to be small for all n's except for one, say $n = 50$, for which K_n is large. This is probably what really occurs in micellar solutions. Namely, some aggregation number n^* has a particularly large equilibrium constant, or equivalently, a particularly large (and negative) standard free energy of aggregation $\mu_{A_n}^{Oo} - n\mu_M^{Oo}$. The exact reason for singling out such a specific n, or a small range of n's, is not known. However, we shall indicate below what might be its origin and how this origin is connected to the problem of HI.

We next turn to analyzing the content of the standard free energy of micellization $\Delta G^\circ(A_n)$. In most theoretical treatments of this problem one starts from the assumption that $\Delta G^\circ(A_n)$ may be split into two additive contributions. The particular notation and the meaning assigned to the two terms differs from one theory to another [see, for example, Mukerjee (1977), Tanford (1974), Birdi (1977)]; but the common idea is very similar. We write here a general form of such a split of $\Delta G^\circ(A_n)$ into two terms

$$\Delta G^\circ(A_n) = \Delta G_{RR}^\circ + \Delta G_{HH}^\circ \qquad (4.99)$$

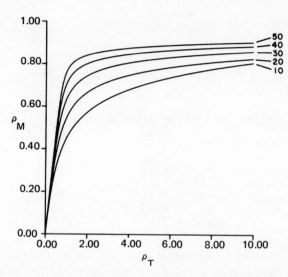

Figure 4.26. Variation of the monomer concentration ϱ_M with the total concentration of the surfactant ϱ_T for sequential aggregation processes. A series of equations (4.95) were solved for $n = 2, 3, 4, \ldots, n_{\max}$. The equilibrium constants were chosen to increase with n as $K_n = \exp[(n - n_{\max})/10]$. This was repeated for $n_{\max} = 10, 20, 30, 40, 50$ as indicated next to each curve.

where ΔG°_{RR} is the contribution due to the alkyl–alkyl free energy of interaction. This term is often described as the negative contribution to $\Delta G^\circ(A_n)$ and is related to the HI. The second, ΔG°_{HH} (often assumed to be positive) is associated with the free energy of interaction between the head groups. (There might be some variations in the meaning of this term according to whether the heads are ionic or nonionic.) The motivation for making the assumption (4.99) clearly stems from the desire to identify the contribution of the HI to $\Delta G^\circ(A_n)$. The argument of "additivity" used in (4.99) is similar to the one used for the dimerization of carboxylic acids, discussed in Section 3.7. The same criticism that we have raised there applies here also: namely, that there exist no theoretical grounds to support such a split, with the specific interpretation assigned to each of the terms ΔG°_{RR} and ΔG°_{HH}. It is at this point that a serious difficulty arises which precludes the extraction of information on HI from the study of micellar formation. We shall demonstrate now the origin of this difficulty and leave some further details to Appendix A.8.

To simplify the argument we consider a rigid monomer surfactant and a rigid spherical micelle built up of n molecules. With this assumption, both the monomers and the micelles have translational and rotational degrees of freedom, but we ignore any specific reference to internal motions of each species (otherwise we should take proper averages over all possible internal conformations of each species—this complicates the presentation without adding to the argument). Using classical statistical mechanics, one can write $\Delta G^\circ(A_n)$ as

$$\Delta G^\circ(A_n) \equiv \mu_{A_n}^{\circ\varrho} - n\mu_M^{\circ\varrho} = [U(A_n) + kT \ln(q_M{}^n/q_{A_n})] + \delta G(A_n) \quad (4.100)$$

with

$$\delta G(A_n) = W(A_n \mid W) - nW(M \mid W) \quad (4.101)$$

Here $U(A_n)$ is the direct interaction energy of the n monomers at the configuration of A_n. The terms q_M and q_{A_n} include the translational and the rotational partition functions of the monomer and of the micelle, respectively. Thus the term in the square brackets is the standard free energy of formation of the micelle in the absence of the solvent. The contribution of the solvent is included in the term $\delta G(A_n)$. This is essentially the difference between the coupling work of one micelle and n monomers to the solvent, W.

In previous sections when we have dealt with nonpolar solutes we have used the notation δG^{HI} for the solvent contribution to $\Delta G^\circ(A_n)$, and referred to that term as the HI. Here we refrain from doing so since the monomers consist of at least two parts: head and tail. Hence the term $\delta G(A_n)$ may not be properly referred to as the HI term. We can still refer to this term as the solvent contribution to $\Delta G^\circ(A_n)$, whatever it may include.

We now focus on the direct interaction $U(A_n)$. This is the work required to bring n monomers from infinite separation to the final configuration of A_n in vacuum. If one assumes that the heads and tails interact in an additive manner, then we may write

$$U(A_n) = U(HH) + U(RR) + U(HR) \quad (4.102)$$

where $U(HH)$ is the total contribution to the energy of interaction due to the head–head interactions. A similar meaning applies to the other two terms. In particular, we note that we also have the cross term $U(HR)$.

Now, even if to a good approximation the direct interaction is additive (with respect to heads and tails), and even if we can ignore the cross term $U(HR)$, so that

$$U(A_n) \approx U(HH) + U(RR) \quad (4.103)$$

this cannot then lead to a similar split of $\Delta G^\circ(A_n)$ as proposed in (4.99). The reason is that in the first place the term that includes the translational and rotational partition functions in (4.100) may not be interpreted as "belonging" to either ΔG°_{RR} or to ΔG°_{HH}. Second, the quantity $\delta G(A_n)$ is a property of the interaction free energy of the aggregate A_n and the monomers M with the solvent. This again may not be viewed as composed of separate contributions due to head–head and tail–tail interaction as suggested in (4.99). We shall present some more detailed arguments on this point in Appendix A.8. Here, we can conclude that theory does not provide any sound argument to suggest a split of the form (4.99). This split, though admittedly appealing on intuitive grounds, probably stems from the desire to imitate the corresponding additive behavior of the interaction energy, such as in (4.103). The fact that "additivity" may not be transferred from the energy into the free energy of a process has already been demonstrated in Section 4.2.

Thus, although HI certainly plays an important role in the process of micellization, the above analysis shows that there exists no simple way of extracting information on HI from the study of $\Delta G^\circ(A_n)$, as might be suggested from a relation of the form (4.99).

The question of why micelles are formed and what molecular mechanism is responsible for singling out only a small range of aggregate sizes is still unanswered. As we have demonstrated above, one may simulate a sharp transition in the properties of the solution if one assumes a single equilibrium equation of the form (4.95) provided that n is large enough. Such an argument is valid for any phase in which the aggregation occurs, including even an ideal gas. However, we know that singular aggregates of one size do not occur even in water. The more realistic picture is that there exists a sequential series of equilibrium reactions, but one of these n's, or a small range of them, has a particularly large equilibrium constant (or a particularly large negative standard free energy of aggregation). The question may then be asked, which term in (4.100) might be responsible for such a singular behavior of $\Delta G^\circ(A_n)$ as a function of n? We know that micellization is a phenomenon that almost exclusively occurs in water. [There are claims that inverted micelles exist in organic liquids. These are usually much smaller aggregates and some doubts regarding their existence have been raised; see, e.g., Kertes (1977).] Therefore, we suspect that it is $\delta G(A_n)$ in (4.101) that is responsible for this phenomenon. In particular, it is probably the nonadditivity effect of the indirect part of the free energy change that causes the singular preference for some specific values of the aggregation numbers n. However, since we know nothing of the nonad-

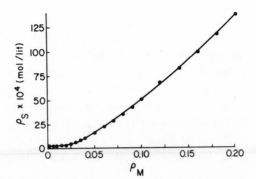

Figure 4.27. Solubility of naphthalene as a function of the concentration of sodium cholate in aqueous solution at 25°C. [Redrawn from Mukerjee and Cardinal (1976).]

ditivity of the HI, even for simpler systems (see Section 4.2), we cannot, at present, go beyond this speculation to find out how this nonadditivity arises from the peculiarities of aqueous systems.

We now briefly turn to an important aspect of micellar solutions which is also relevant to the problem of HI in the sense of Chapter 2. This is the phenomenon of solubilization. The experimental observation is the following. We take a nonpolar solute, such as naphthalene, which is sparingly soluble in water (about 2.55×10^{-4} mol/liter), and dissolve it in aqueous surfactant solution. One finds that in the premicellar region the solubility of the naphthalene changes very slowly with the concentration of the surfactant. At the CMC we find an abrupt change in the solubility of the solute as we add more surfactant. This behavior is illustrated in Figures 4.27 and 4.28. The common and well-justified interpretation of this phenomenon is quite simple. Once micelles are formed, they provide some "pockets" of nonpolar environments in which the nonpolar solute might enter. Since

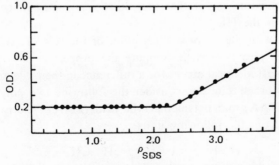

Figure 4.28. Dependence of the optical density (O.D.) of naphthalene in aqueous solutions of sodium dodecylsulfate (SDS) at 25°C. ϱ_{SDS} is the total concentration, in grams per liter of SDS. These measurements were carried out in the presence of excess naphthalene. [Redrawn with changes from Birdi (1976).]

these solutes prefer, or feel more comfortable in, the nonpolar environment as we have demonstrated in Chapter 2, they will preferentially dissolve into the micelles. This explains the sharp increase of the solubility of these solutes in such solutions. From a study of the solubility of simple solutes in aqueous micellar solutions, one can estimate the free energy of transfer of the solute from water into the micelles. This process is similar to the one discussed in Chapter 2. However, since the micelles are not really well-defined phases, some difficulties in defining the free energy of transfer arise. We defer to Appendix A.9 a more detailed discussion of this topic.

4.9. HI *IN SYNTHETIC AND BIOLOGICAL POLYMERS*

Open any modern textbook of biochemistry, look for the interpretation of molecular processes such as conformational changes of biopolymers, or association of subunits to form multisubunit enzymes, and you are likely to find the concept of HI invoked to explain the driving forces that govern these processes. There is certainly an element of truth in such interpretations, and as a matter of fact the whole field of study of HI has evolved from the need to explain such processes in biological solutions that could not be explained by the conventional and well-established interactions between molecules, or between groups within a molecule. We shall survey in this section some typical processes in which HI might be involved. As in Section 4.8, we use here the concept of the HI as "input" to, rather than "output" from, the study of these processes.

The examples given below were selected randomly from an immense literature that exists on each of the topics. They do not form any coherent pattern of behavior, except that they have one factor in common—the involvement of the HI.

We start with the simplest illustration of the role of HI in synthetic polymers. The idea here is to compare the properties of two polymers, which are almost identical except for a difference in their "degree of hydrophobicity." More specifically, consider the following two polymers: polyacrylic acid (PAA) and polymethacrylic acid (PMA). The two polymers are built up from the following units:

$$\left(\begin{array}{c} \text{H} \quad \text{H} \\ | \quad | \\ -\text{C}-\text{C}- \\ | \quad | \\ \text{H} \quad \text{COOH} \end{array} \right)_n \qquad \left(\begin{array}{c} \text{H} \quad \text{CH}_3 \\ | \quad | \\ -\text{C}-\text{C}- \\ | \quad | \\ \text{H} \quad \text{COOH} \end{array} \right)_n \qquad (4.104)$$

(PAA)   (PMA)

Thus the two polymers are essentially polyelectrolytes; in PMA we have an additional methyl group that is missing in PAA. Therefore, the difference in their behavior in aqueous solutions is largely attributed to the effect of the additional methyl group—hence the relevance to the problem of HI.

Experimental evidence indicates that the PMA in an acidified aqueous medium (i.e., when all the carboxylic groups are not charged) attains a compact structure. Upon increasing the pH of the solution, the compact structure breaks down and we get a random coil polyelectrolyte. The difference in the state of packing of the polymer may be followed, for example, through measurements of the intrinsic viscosity of the solution [η]. The interpretation of this phenomenon is quite simple. The intrinsic viscosity is related to the radius of gyration of the molecules, and hence gives a rough measure of the compactness of the polymer. At high pH values, the carboxylic groups are ionized and the charge–charge interaction between them tends to open up the conformation of the polymer. At low pH values, when the carboxylic groups are not ionized, the intramolecular hydrophobic interaction drives the conformational equilibrium towards the compact structure. The situation is schematically depicted in Figure 4.29. We have here a simple demonstration of the two competitive effects (the charge–charge vs. the hydrophobic interaction effects) on the conformational equilibrium of the polymer.

A similar effect is observed upon the addition of methanol, ethanol, or higher alcohols to aqueous solutions of PMA. Figure 4.30 shows the reduced intrinsic viscosity of aqueous solutions of PMA as a function of the mole fraction of alcohols. The overall effect of alcohol is to increase the intrinsic viscosity of the solution—a rough indication that the compact structure of the polymer breaks down at high concentration of the alcohol.

At very low alcohol concentration, however, one finds an opposite effect of the alcohol, namely, a decrease of the reduced intrinsic viscosity, indicating a stabilization of the compact form of the polymer. (All of

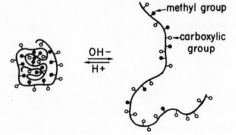

Figure 4.29. The two conformations of PMA. Increasing the pH shifts the equilibrium to the right, towards an open conformation. Decreasing the pH shifts the equilibrium to the left, towards a more compact conformation.

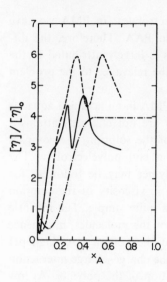

Figure 4.30. Reduced intrinsic viscosity of PMA in 0.002 N HCl, as a function of the mole fraction of methanol (—·—); ethanol (— —); and n-propanol (——). $[\eta]_0$ is the intrinsic viscosity in 0.002 N HCl. All measurements were carried out at 30°C. [Redrawn with changes from Priel and Silberberg (1970).]

these effects are observed in acidic solutions in which the charge–charge repulsion is not significant.)

The interpretation of the effect of the alcohol on the conformation of the PMA is based on the effect of alcohol on the HI. Thus, small amounts of alcohol ($x \lesssim 0.02$) seem to *increase* the strength of the HI, hence the conformational equilibrium is shifted towards the more compact form, as revealed by the lowered intrinsic viscosity. At high concentrations of alcohols the HI are weakened and the polymer opens up. This interpretation is consistent with what we have learned already on the effect of alcohol on the HI (except for the very small region of alcohol concentration where we have found, in Section 3.4, an initial weakening of the HI; this region is not revealed in these experiments, probably because of the presence of the HCl in the solution; thus the initial effect of the alcohol measured here is not relative to pure water, but relative to water and HCl).

It should be noted that similar effects are not observed in PAA. This lends further support to the interpretation that the observed effect of alcohol is largely due to the presence of the methyl groups in PMA. Priel and Silberberg (1970) have studied, in great detail, the properties of these solutions. As we have noted before, the concept of HI was found very useful in the interpretation of the observed trends. Unfortunately, one cannot extract any quantitative information on the strength of the HI from such studies.

A closely related phenomenon, which is reminiscent of the solubilization phenomenon, is also observed in aqueous solutions of PMA. Barone *et al.* (1966) have compared the solubility of low paraffins in aqueous solutions of PMA and PAA. The most striking difference between the two polymers is the following: An unneutralized solution of PMA causes a net solubilization effect on the alkenes. This effect increases with the molecular weight of the PMA. No similar effect has been observed for solutions of PAA.

We report in Table 4.7 some of their results for both alkenes and for large aromatic molecules. Note that the effect of addition of PMA is rather small for the low alkenes. It is quite pronounced for *n*-hexane, the solubility of which increases about fourfold compared with pure water. The solubilization effect is far more dramatic for the aromatic molecules, where an increase in solubility of more than an order of magnitude is observed. (Of course, that depends on the concentration and the molecular weight of the polymer.)

The interpretation of this effect is similar to the solubilization effect of micellar solutions. There, we encountered a solution of surfactant molecules which above some concentration form nonpolar regions that can absorb nonpolar solutes. Here, on the other hand, we have a polymer, which under certain conditions (say low pH) attains a compact form, in

Table 4.7

Solubility of Alkenes (mol/liter at 25°C) in Water and in Unneutralized Solution of $0.4\,N$ PMA (mol wt $= 1.4 \times 10^4$)[a] and the Solubilities of Aromatic Hydrocarbons in Water and in Unneutralized Solution of $0.1\,N$ PMA (mol wt $= 85 \times 10^4$)[b]

Solute	Solubility in water	Solubility in PMA solution
Cyclopropane	11.10×10^{-3}	11.24×10^{-3}
Propane	1.50×10^{-3}	1.83×10^{-3}
n-Butane	1.09×10^{-3}	1.37×10^{-3}
n-Pentane	1.03×10^{-3}	2.27×10^{-3}
n-Hexane	0.96×10^{-3}	3.77×10^{-3}
Phenanthrene	9.0×10^{-6}	2.7×10^{-4}
Anthracene	4.47×10^{-7}	0.31×10^{-4}
Pyrene	7.7×10^{-6}	2.08×10^{-4}
1,2-Benzpyrene	2.9×10^{-8}	0.67×10^{-4}
3,4-Benzpyrene	1.6×10^{-8}	1.25×10^{-4}

[a] From Barone *et al.* (1966).
[b] From Barone *et al.* (1967).

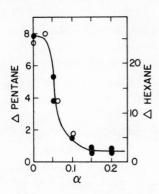

Figure 4.31. Dependence of the enhancement of the solubility Δ of n-pentane and of n-hexane as a function of the degree of neutralization of PMA at 25°C; open circles are for n-pentane, filled circles for n-hexane. [Redrawn with changes from Barone *et al.* (1966).]

the interior of which nonpolar molecules may be accommodated. The capability to form such a compact conformation is largely ascribed to the HI among the methyl groups of the PMA. No such effect has been observed in PAA solutions.

It is interesting to follow the extent of the solubilization effect upon changing the pH of the solution.

We define the relative enchancement of the solubility of a nonpolar solute S per unit concentration of the polymer by

$$\Delta = \frac{\varrho_S(\text{in PMA solution}) - \varrho_S(\text{in water})}{\varrho_S(\text{in water})\varrho_{\text{PMA}}} \tag{4.105}$$

where ϱ_S is the solubility (in mol/liter) of the solute S, and ϱ_{PMS} is the polymer concentration.

Figure 4.31 shows the values of Δ for n-pentane and n-hexane as a function of the degree of neutralization α. The case $\alpha = 0$ corresponds to the unneutralized PMA, i.e., the case when the polymer has a compact structure and hence is capable of absorbing nonpolar solutes. As we increase the value of α, which corresponds to increasing the pH of the solution, the relative enhancement of the solubility, Δ, drops sharply to almost zero value. This is clearly a manifestation of the conformational change that takes place—from the globular to the random coil structure. Similar behavior has also been reported for aromatic solutes (Barone *et al.*, 1967). Again, it was reported by the same authors that no such effects were observed in the case of PAA.

Similar effects of solubilization of simple solutes, such as butane and pentane, by proteins have been observed [see for example Wishnia (1962), Wishnia and Pinder (1964, 1966)]. We have chosen to illustrate the effect of solubilization in a relatively simple polymer where many of the complexities of real biopolymers are eliminated.

We now turn to a slightly more complicated system. Dubin and Strauss (1970) have studied the properties of alternating copolymers of maleic and n-alkylvinyl ethers in aqueous solutions. The monomer unit is

$$
\begin{array}{ccccccc}
 & \overset{\text{H}}{|} & & \overset{\text{H}}{|} & & \overset{\text{H}}{|} & & \overset{\text{H}}{|} \\
-\text{C} & - & \text{C} & - & \text{C} & - & \text{C}- \\
 & | & & | & & | & & | \\
 & \text{COOH} & & \text{COOH} & & \text{H} & & \text{OR}
\end{array}
\qquad (4.106)
$$

The alkyl groups were chosen to be methyl, ethyl, butyl, and hexyl. Clearly one would expect that for the small alkyl groups, the behavior of the polymer would resemble that of a typical polyelectrolyte. For the larger alkyl groups, the intramolecular HI are expected to favor (at least at low pH values) the formation of the compact structures. For this reason, these molecules provide good models for the systematic study of the effect of HI on the conformation of the polymers.

In Figure 4.32 we present some potentiometric titration data on aqueous solutions of these polymers. Here we follow the pH of the solution as a function of the degree of neutralization α of the polymer. The case $\alpha = 0$ corresponds to the unneutralized polymer where the charge–charge

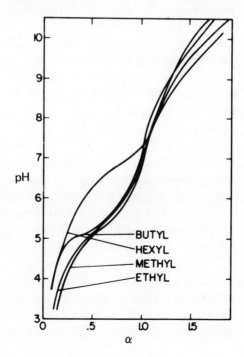

Figure 4.32. Potentiometric titration of aqueous solutions of copolymers of maleic acid and n-alkyl vinyl ethers, at 30°C. The alkyl groups are methyl, ethyl, butyl, and hexyl, as indicated. [Reproduced with changes from Dubin and Strauss (1970).]

interactions are relatively minor. The case $\alpha = 2$ corresponds to complete neutralization of the polymer (note that there are two carboxylic groups per monomer), hence the predominance of the charge–charge repulsive forces.

For the methyl and ethyl radicals, we find an almost linear dependence of the pH on α—this is considered to be the "regular" behavior of the polyelectrolyte. For the butyl and the hexyl radicals, a considerable deviation from the regular behavior is observed. We see that for the larger alkyl groups a considerable change in the pH, say from pH $= 3$ to pH $= 5$, has almost no effect on α. This clearly indicates a retention of the compact form of the polymer. The larger the alkyl group, the stronger the intramolecular HI, and the stronger is the resistance of the polymer to opening up its compact structure upon increasing the pH.

Next we turn to real biopolymers. Here HI might enter on various levels, from conformational changes of a single biopolymer, to a combination of a substrate to an enzyme, to association of subunits to form multi-subunit enzymes, etc. One of the central problems in molecular biology is the way in which a polypeptide that has just been synthesized folds itself into the "native" structure of the protein. It is known that such a process occurs spontaneously. What is less understood is the mechanism of this process. Does the polymer proceed by way of trial and error to search for its most stable conformation, or is there a well-defined path, which is dictated by the sequence of the amino acids, along which the folding process takes place? Estimates indicate that a random search for the "native" conformation would take too long a time. It seems likely, therefore, that the sequence of amino acids contains the information required to find an efficient path for folding up. In this process hydrogen bonds and hydrophobic interaction might play crucial roles.

Regarding the final (secondary and tertiary) structure of the protein, it has been established recently that in some proteins the so-called hydrophobic amino acids occupy the interior, and the polar amino acids the exterior of the protein. This arrangement is beautifully illustrated by the structure of cytochrome c (Dickerson, 1972). Clearly, this particular product is a result of the tendency of the "hydrophobic" amino acids to gather together, so as to minimize their exposure to the aqueous environment.

The next level in which we find the involvement of HI is in the so-called quaternary structure, i.e., the way subunits associate to form multisubunit proteins, such as hemoglobin, regulatory enzymes, ribosomes, etc. An interesting example, in which the concept of HI has also led to a practical application, is the case of sickle cell anemia.

It is well known that the normal hemoglobin molecule (Hb A) consists of four chains of amino acids. Two are identical α-chains (with 141 amino acid residues each) and two are β-chains (with 146 amino acid residues each). The association of these four subunits to form the Hb A molecule probably also involves HI, but this is not the point we would like to stress here.

Sickle cell anemia is a hereditary disease caused by an abnormal form of hemoglobin, referred to as Hb S. It is found that when the oxygen content of the blood is high, most of the blood cells have the normal round shape. However, at low oxygen content, a variety of forms of cells are observed, and some of them have the crescent shape—hence the name "sickle cells."

It is now known that the abnormal hemoglobin is almost identical to the normal Hb A except for *one* amino acid: a valine substitutes for glutamic acid at position 6 of the β-chain. It should be noted that the location of this amino acid is on the exterior of the molecule (Cerami and Peterson, 1975), hence one may expect that this substitution will cause an overall increase of the "hydrophobicity" of the hemoglobin molecules. This fact led to the contention that the abnormal form of hemoglobin tends to aggregate in an abnormal fashion, and that this process is the primary molecular reason for the sickle cell disease. The involvement of HI in the aggregation process has suggested also a method of treatment. The basic idea is that administration of solutes that weaken the HI might help to reverse, or perhaps prevent, the abnormal aggregation of Hb S. Indeed, some solutes such as urea and cyamate have been used as anti-sickling agents [for a review, see Murayama (1973)]. Of course, the process is more complex than merely association involving purely HI. However, this is an interesting example that demonstrates the use of the concept of HI both in the understanding of a biological phenomenon and in providing a guide for a possible molecular therapy.

There are many other more complicated processes of self-assembly of subunits to form large functional proteins. Examples are the recombination of the 21 proteins and the RNA to form the small subunit of ribosome (the 30S component), and the spontaneous reassembly of subunits and RNA to form the tobacco mosaic virus (TMV), which consists of a large number of identical subunits.

It is often claimed that the spontaneous association process indicates that the information for this process is already contained in the primary structure of the proteins. This might be true; however, it could also be the result of the structural dependence of the nonadditivity effect of the HI (see Section 4.2). In other words, if the subunits "know" how to associate because of some sort of genetic information that is carried within the sequence of the amino acids, then they should "know" to do the same thing in any medium—not necessarily aqueous solutions—and the whole question of the involvement of HI becomes irrelevant to such processes. On the other hand, if spontaneous association processes take place exclusively in aqueous solutions, then we may suspect that HI are indeed involved. This is likely to originate from the nonadditivity part of HI, an aspect about which we still know nothing.

Chapter 5

Temperature and Pressure Dependence of the Hydrophobic Interactions

5.1. INTRODUCTION

The study of the temperature and the pressure dependence of the HI is of interest for several reasons. The first is its relevance to biological processes. It is well known that proteins undergo a process of denaturation as the temperature is raised. This process disrupts their normal function and therefore threatens the entire activity, or life, of any biological system. There are indications, as will be demonstrated in Section 5.4, that the strength of the HI increases with a rise in the temperature, at least in the range of temperatures of biological interest. It seems, therefore, that this behavior of the HI contributes to the thermal stability of proteins against the tendency of denaturation.

Second, the temperature dependence of the HI, as we have already indicated in Chapters 3 and 4, is somewhat outstanding as compared with the corresponding behavior of other solvents for which we have the relevant experimental data. It is therefore of great interest to determine the reasons for such behavior on a molecular level. Finally, we know that the temperature dependence of the HI is intimately connected with the structural changes in the solvent induced by the process of aggregation of solutes. We shall devote part of this chapter to surveying this aspect of the problem.

The first part of this chapter will follow the main style of the previous chapters, namely, it will be essentially descriptive and based on classical thermodynamical concepts. However, in later sections we shall find it necessary to use some elementary concepts from statistical mechanics in order to introduce the concept of the "structure of water" and to examine its response to molecular processes occurring in aqueous systems.

In the next section we present a very detailed and slowly developed exposition of some elementary, though fundamental, concepts: the standard thermodynamic quantities of transfer. We have devoted a considerable amount of time to discussing the standard free energy of transfer in Chapter 1 and in Appendix A.1. However, when proceeding to the temperature and pressure coefficients of these quantities, there are some subtle points that should be carefully noted.

The standard thermodynamic quantities for transferring a solute between two phases are useful in the study of hydrophobic hydration, which has been discussed in Chapter 2. Furthermore, the combination of such quantities also gives the pressure and the temperature coefficients of the HI, in the sense of Chapters 3 and 4.

It is unfortunate to note that many misconceptions and misleading remarks are very commonplace in the literature on aqueous solutions. In the author's opinion it is difficult to overstate the degree of confusion that exists in this field. Part of the reason for this situation is that thermodynamics alone does not have sufficient means to interpret some of the standard partial molar quantities. One must appeal to some statistical mechanical arguments in order to gain a more penetrating insight into the meaning and the significance of these quantities. Therefore the reader is, once again, urged to read the next section carefully and patiently.

5.2. TEMPERATURE AND PRESSURE DERIVATIVES OF THE STANDARD FREE ENERGY OF TRANSFER

In this section we derive several fundamental expressions for the standard thermodynamic quantities of transfer. These are of direct interest to the study of the hydrophobic hydration as discussed in Chapter 2. Also, they serve as ingredients in the study of the temperature and pressure dependence of the HI in the sense of Chapters 3 and 4.

We start with the general expression for the chemical potential of a simple solute A in a solvent W. In this section all the expressions for the partial thermodynamic quantities are written per *molecule*, not per *mole* as is more common in thermodynamics. The Boltzmann constant k should be replaced by the gas constant R whenever we need to transform the expressions into partial *molar* quantities. For simplicity we still refer to partial *molar* quantities rather than to the more precise partial *molecular* quantities.

Thus we write

$$\mu_A = W(A \mid W) + kT \ln \varrho_A \Lambda_A^3 \qquad (5.1)$$

where $W(A \mid W)$ is the "coupling work," i.e., the work required to introduce the solute A at some fixed position in the solvent W, ϱ_A is the number density of the solute, and Λ_A^3 is the momentum partition function of A. For simplicity we have assumed that the solute particles are structureless, otherwise we should have added a term $kT \ln q_A^{-1}$, with q_A being the internal partition function of a single A molecule. Relation (5.1) is very general in the sense that it applies to any concentration of A in W. However, for most of our applications we shall use (5.1) in the limit of very dilute solutions of A in W. In that limit $W(A \mid W)$ becomes the coupling work of A against a surrounding of pure solvent W. This expression for the chemical potential is invariant to the choice of the statistical mechanical ensemble. In other words we can choose any possible set of independent thermodynamic variables to describe our system. From the practical point of view the pressure and the temperature are the most commonly used variables. Thus our system may be described by the variables P, T, N_A, N_W, where N_A and N_W are the number of solute and solvent molecules, respectively. The density ϱ_A is understood to be $\varrho_A = N_A/V$ with V the *average* volume of the system as determined by the variables P, T, N_A, N_W.

Finally we recall that the two terms in (5.1) correspond to two parts of the process of introducing a single solute particle into the system. The second term has been referred to as the *liberation free energy*[†] [see Section 1.2, Appendix A.1, and Ben-Naim (1978a)].

From now on we shall assume that A is very dilute in W. (However, some of the results obtained in this section are valid for any mixture of A and W.) Also, for completeness, we rederive a relation that we obtained in Chapter 2.

The standard chemical potential of A in W in the number density scale is defined by

$$\mu_A^{\circ W\varrho} = W(A \mid W) + kT \ln \Lambda_A^3 \qquad (5.2)$$

[†] In all of our discussions we restrict ourselves to classical systems only. In a quantum mechanical treatment, it is not clear whether the very separation of the chemical potential into two parts as in (5.1) is feasible. Fortunately, for most of the practical applications in solution chemistry we can safely apply classical statistical mechanics.

The free energy of transferring A from one solvent W to a second solvent E, keeping P and T fixed, is

$$\mu_A^E - \mu_A^W = \mu_A^{\circ E \varrho} - \mu_A^{\circ W \varrho} + kT \ln(\varrho_A^E / \varrho_A^W) \qquad (5.3)$$

For the specific case $\varrho_A^E = \varrho_A^W$ we obtain the so-called *standard free energy* of transfer

$$\Delta G\!\left(\begin{array}{c} W \to E \\ \varrho_A^W = \varrho_A^E \end{array}\right) = \mu_A^{\circ E \varrho} - \mu_A^{\circ W \varrho} = W(A \mid E) - W(A \mid W)$$

$$= \Delta G(W \to E; L) \qquad (5.4)$$

Relations (5.4) are very important and should be noted carefully. The first equality is obtained from (5.3) by substituting $\varrho_A^W = \varrho_A^E$. We thus obtain a free energy change for the process that is symbolically described by $\left(\begin{smallmatrix} W \to E \\ \varrho_A^W = \varrho_A^E \end{smallmatrix}\right)$. The second equality is obtained from (5.2) when applied to the two solvents E and W. The third equality introduces a new notation. The difference $W(A \mid E) - W(A \mid W)$ measures the difference in the coupling work of A against the two solvents E and W, respectively. Alternatively, this is the free energy change for transferring A from a fixed position in W to a fixed position in E. Since this quantity "senses" the *local* environment of A in the two solvents, we denote the corresponding free energy change by $G(W \to E; L)$, where L stresses the *local* character of this quantity. Thus we have

$$\Delta G\!\left(\begin{array}{c} W \to E \\ \varrho_A^W = \varrho_A^E \end{array}\right) = \Delta G(W \to E; L) \qquad (5.5)$$

It is very important to note that the equality (5.5) applies to the change in the Gibbs free energy for two *different* processes. On the left-hand side we have a process that can be actually carried out in the laboratory. The second process, indicated by $(W \to E; L)$, is a "thought" experiment, since we cannot actually fix the location of the solute in the solvent. However, from the theoretical point of view it is the latter process that is more significant. This is so because it provides a measure of the difference in the solvation free energy of A in the two phases, and because it does not contain any contribution due to the translational free energy of the solute. We shall see that similar significance can be ascribed to other thermodynamic quantities of transfer.

We shall often use a shorthand notation $\Delta G_A(L)$ instead of $\Delta G(W \to E; L)$ when the two phases are already specified. We shall also find it useful to denote by $\Delta G_A^{\circ}(L)$ a quantity that applies to the infinitely dilute solution of A in the two phases.

In the case where one phase, say E, is an ideal gas, we have

$$\Delta G\left(\begin{matrix} W \to G \\ \varrho_A{}^W = \varrho_A{}^G \end{matrix}\right) = -W(A \mid W) \tag{5.6}$$

which is simply a measure of the solvation free energy of A in W.

It should be stressed that the equality (5.5) for the free energy change of the two different processes is valid only for the case of the *free energy* and *only* when we choose the *number density* as our concentration unit in carrying out the standard process $\left(\begin{smallmatrix} W \to E \\ \varrho_A{}^W = \varrho_A{}^E \end{smallmatrix}\right)$.

A different, but very common, choice of concentration units is the mole fraction x_A, which for a very dilute solution is related to the density of A by

$$x_A{}^W = \varrho_A{}^W / \varrho_W{}^W \tag{5.7}$$

where $\varrho_W{}^W$ is the number density of W in pure W. Changing variables in (5.1) we can write the new standard chemical potential of A in W as

$$\mu_A{}^W = [W(A \mid W) + kT \ln \Lambda_A{}^3 \varrho_W{}^W] + kT \ln x_A{}^W$$
$$= \mu_A{}^{\circ Wx} + kT \ln x_A{}^W \tag{5.8}$$

The standard free energy change for transferring A from W to E at *equal mole fractions* in the two solvents is obtained from (5.8) when applied to the two solvents. The result is

$$\Delta G\left(\begin{matrix} W \to E \\ x_A{}^W = x_A{}^E \end{matrix}\right) = \mu_A{}^{\circ Ex} - \mu_A{}^{\circ Wx}$$
$$= W(A \mid E) - W(A \mid W) + kT \ln(\varrho_E{}^E / \varrho_W{}^W)$$
$$= \Delta G(W \to E; L) + kT \ln(\varrho_E{}^E / \varrho_W{}^W) \tag{5.9}$$

This relation should be compared with (5.5). Here we have introduced a new standard process $\left(\begin{smallmatrix} W \to E \\ x_A{}^W = x_A{}^E \end{smallmatrix}\right)$, the free energy change of which is *not equal* to any of the previously described free energy changes.

At this point we stress a very important difference between the two standard free energies defined in (5.5) and in (5.9). Whenever one of the solvents becomes rarefied, say W becomes an ideal gas, then the corresponding coupling work is zero, i.e.,

$$W(A \mid W) \xrightarrow{\varrho_W{}^W \to 0} 0 \tag{5.10}$$

As a result we get equalities of the form (5.6). This is in sharp contrast to the behavior of the quantity defined in (5.9) in the same limit. Clearly if we let $\varrho_W{}^W \to 0$, the right-hand side of (5.9) will diverge to infinity. This is a serious flaw of the standard free energy of transfer, based on the mole fraction scale. It certainly cannot be used to describe the difference in the solvation free energy of A in the two solvents.

There are many other standard processes and standard thermo-dynamical quantities in use in the literature. We shall not mention all of them here. Instead, we have chosen to discuss the one that is the simplest and the most significant and which is given in (5.5). We have also mentioned the mole fraction scale, to demonstrate its shortcomings. Unfortunately, this scale is still in use in many research articles and books.

We have elaborated in great detail on the standard free energy change. This has been done on purpose, in order to avoid some typical misconceptions that can easily "sneak in" when one proceeds to the temperature and the pressure derivatives of the standard free energy.

Having written all the necessary expressions for the chemical potential we now proceed to derive the "standard" entropy of transfer. The first and immediate question that arises is which "standard" process we should choose. We have mentioned above three different processes (out of many more possibilities). The answer to this question clearly depends on what kind of information we expect from such a quantity. Before making our choice, we shall first derive the entropy changes that correspond to the three processes indicated in (5.5) and in (5.9).

The partial molar entropy of A in the solvent is obtained from the temperature derivative of (5.1)

$$-\bar{S}_A = \left(\frac{\partial \mu_A}{\partial T}\right)_P = \frac{\partial}{\partial T}\,[W(A \mid W) + kT \ln \Lambda_A{}^3] + kT\,\frac{\partial \ln \varrho_A}{\partial T} + k \ln \varrho_A$$

$$= -\bar{S}_A{}^{\circ W \varrho} + k \ln \varrho_A \qquad (5.11)$$

where we have introduced the standard partial molar entropy of A in W in the number density scale. The standard entropy of transfer corresponding to the process indicated on the left-hand side of (5.5) is obtained from the difference of $\bar{S}_A$ in the two solvents, say E and W; the result is

$$\Delta S\left(\begin{matrix} W \to E \\ \varrho_A{}^W = \varrho_A{}^E \end{matrix}\right) = \bar{S}_A{}^{\circ E \varrho} - \bar{S}_A{}^{\circ W \varrho}$$

$$= -\frac{\partial}{\partial T}\,[W(A \mid E) - W(A \mid W)] + kT\,\frac{\partial}{\partial T}\,[\ln(V^E/V^W)]$$

$$(5.12)$$

where V^E and V^W are the volumes of the pure solvents E and W respectively. Similarly by taking the temperature derivative of (5.8) and forming the difference of the partial molar entropies for the case $x_A^E = x_A^W$ we obtain

$$\Delta S\left(\begin{array}{c} W \to E \\ x_A^W = x_A^E \end{array}\right) = \bar{S}_A^{\,\circ Ex} - \bar{S}_A^{\,\circ Wx}$$

$$= -\frac{\partial}{\partial T}\,[W(A\,|\,E) - W(A\,|\,W)] - \frac{\partial}{\partial T}\,[kT\ln(\varrho_E^E/\varrho_W^W)]$$

$$(5.13)$$

Finally, for the process indicated on the right-hand side of (5.5) we have the corresponding entropy change

$$\Delta S(W \to E; L) = -\frac{\partial}{\partial T}\,[W(A\,|\,E) - W(A\,|\,W)] \qquad (5.14)$$

In (5.12), (5.13) and (5.14) we have three "standard" entropies of transfer. Clearly, they are all different from each other [in contrast to the free energy case, where two quantities are equal to each other as shown in (5.5)]. It is also clear that the last quantity is the simplest of the three quantities. But simplicity alone cannot be used as the sole reason for preferring one quantity over the other. There exists a more fundamental reason for the choice of (5.14) as our standard quantity. A glance at (5.12) and (5.13) shows that the other two standard quantities contain the temperature derivatives of the volume of the solvent. This derivative is part of the liberation entropy of the solute. This may be seen by rewriting (5.11) in the form

$$\bar{S}_A = -\frac{\partial}{\partial T}\,[W(A\,|\,W)] - \frac{\partial}{\partial T}\,(kT\ln\varrho_A\Lambda_A^3) \qquad (5.15)$$

The first term on the right-hand side of (5.15) is the entropy change for introducing A to a fixed position in the solvent. The second term on the right-hand side is the entropy change due to the liberation of the particle from its fixed position.

Since we are interested in a standard quantity that measures the difference in the solvation entropy of A in the two solvents, we should strive to get rid of any quantity that is part of the liberation entropy. It is obvious that only $\Delta S(W \to E; L)$ defined in (5.14) has this property. Because of its fundamental importance we repeat: $\Delta S(W \to E; L)$ is the entropy change associated with the transfer of A from a fixed position in W to a fixed position in E (T and P are kept constant). This quantity has the required local character in the sense that it depends on the local environment of A in the two solvents. Furthermore, since we have shown that $\Delta G(W \to E; L)$ is the most significant standard free energy change, $\Delta S(W \to E; L)$ is its only legitimate temperature coefficient. We shall later see that none of the other standard entropy changes may be used for the temperature coefficient of the HI.

It is true that the above process may not be carried out in an actual experiment. Nevertheless it is still a measurable quantity through the temperature dependence of the partition coefficient of A between the two solvents, namely,

$$\Delta S(W \to E; L) = - \frac{\partial}{\partial T} [\Delta G(W \to E; L)]$$

$$= - \frac{\partial}{\partial T} [kT \ln(\varrho_A{}^W / \varrho_A{}^E)_{eq}] \qquad (5.16)$$

The last equality was taken from Section 1.2.

Because of the existence of considerable confusion in the literature about the choice of the proper standard quantity, we now present the more common way of choosing a standard entropy change based on thermodynamic arguments alone.

In the realm of thermodynamics we do not have explicit relations such as (5.1) or (5.8) for the chemical potential; instead we have relations of the form

$$\mu_A{}^W = \mu_A{}^{\circ W \varrho} + kT \ln \varrho_A{}^W \qquad (5.17)$$

$$\mu_A{}^W = \mu_A{}^{\circ W x} + kT \ln x_A{}^W \qquad (5.18)$$

in which the two standard chemical potentials are known to be independent of the solute concentration. No further information on their content is supplied by thermodynamics. Taking the derivatives with respect to temperature and forming the difference of the partial molar entropies of A in two solvents we obtain

$$\Delta S \binom{W \to E}{\varrho_A{}^W = \varrho_A{}^E} = - \frac{\partial}{\partial T} (\mu_A{}^{\circ E \varrho} - \mu_A{}^{\circ W \varrho}) + kT \frac{\partial}{\partial T} \left(\ln \frac{V^E}{V^W} \right) \qquad (5.19)$$

$$\Delta S \binom{W \to E}{x_A{}^W = x_A{}^E} = - \frac{\partial}{\partial T} (\mu_A{}^{\circ E x} - \mu_A{}^{\circ W x}) \qquad (5.20)$$

Thus it is apparent from the last two relations that the second is "simpler" in the sense that it contains only the derivative of the standard free energy of transfer. On the other hand, the first one also has the temperature dependence of the volume of the solvents. For this reason it is very common-place to find statements such as the following: the use of mole fractions is preferable since, in contrast to the use of number (or molar) densities, they lead to temperature-independent results.

The catch in this argument is that in *both* (5.19) and (5.20) the temperature dependence of the volume of the solvents is included. This may be seen by comparing (5.20) with (5.13). But since thermodynamics alone does not "see" the content of the standard chemical potentials in (5.20), it looks as if (5.20) is simpler than (5.19). It is clear from (5.12) and (5.13) that both contain the temperature dependence of the volume of the solvents, a quantity that we want to eliminate in order to obtain a measure of the solvation entropy of A in the two solvents. To obtain such a quantity we have simply to appeal to relation (5.16), which provides such a measure, i.e., it does not contain any part of the liberation entropy.

It is an ironic observation that the temperature dependence of ϱ_A (in contrast to x_A), which has been traditionally used as an argument *against* the use of ϱ_A in constructing standard quantities, is in fact its greatest advantage because by forming the standard quantity in (5.4) we have eliminated $kT \ln \varrho_A$ along with its temperature dependence. Hence if we adopt $\Delta G(W \to E; L)$ as our basic standard free energy of transfer, then neither $kT \ln \varrho_A$ nor its temperature (or pressure) dependence appear in any of the derivatives of $\Delta G(W \to E; L)$.

To conclude this part we stress again that thermodynamics alone may not be used to make the best choice of the standard entropy. In fact it may be slightly misleading if we look at the appearance of (5.19) and (5.20) without being concerned with what is hidden within the various standard chemical potentials. It is only with the more penetrating eyes of statistical mechanics that we could have made the choice of (5.16) as the most signif-icant standard quantity.

We now proceed to the enthalpy changes corresponding to the above-mentioned processes. The enthalpy change for any particular process is obtained from the relation

$$\Delta H = \Delta \mu + T \Delta S \tag{5.21}$$

Hence for the three processes mentioned above, we have

$$\Delta H \binom{W \to E}{\varrho_A{}^W = \varrho_A{}^E} = -T^2 \frac{\partial}{\partial T} \left[\frac{W(A \mid E) - W(A \mid W)}{T} \right]$$
$$+ kT^2 \frac{\partial}{\partial T} [\ln(V^E / V^W)] \tag{5.22}$$

$$\Delta H\!\left(\begin{matrix} W \to E \\ x_A{}^W = x_A{}^E \end{matrix}\right) = -T^2\,\frac{\partial}{\partial T}\left[\frac{W(A \mid E) - W(A \mid W)}{T}\right]$$

$$+ kT^2 \frac{\partial}{\partial T}\,[\ln(V^E/V^W)] \qquad (5.23)$$

$$\Delta H(W \to E; L) = -T^2\,\frac{\partial}{\partial T}\left[\frac{W(A \mid E) - W(A \mid W)}{T}\right] \qquad (5.24)$$

Note here that in contrast to the entropy changes, the enthalpy changes in (5.22) and (5.23) are equal to each other. They differ from the one in (5.24), which is the enthalpy change associated with the transfer of A from a fixed position in W to a fixed position in E. Clearly, as in (5.16) this enthalpy change may be obtained by direct measurement of the temperature dependence of the partition coefficient of A in the two solvents.

Next we write the corresponding expressions for the volume changes. These are

$$\Delta V\!\left(\begin{matrix} W \to E \\ \varrho_A{}^W = \varrho_A{}^E \end{matrix}\right) = \frac{\partial}{\partial P}\,[W(A \mid E) - W(A \mid W)] - kT\frac{\partial}{\partial P}\,[\ln(V^E/V^W)]$$
$$\qquad (5.25)$$

$$\Delta V\!\left(\begin{matrix} W \to E \\ x_A{}^W = x_A{}^E \end{matrix}\right) = \frac{\partial}{\partial P}\,[W(A \mid E) - W(A \mid W)] - kT\frac{\partial}{\partial P}\,[\ln(V^E/V^W)]$$
$$\qquad (5.26)$$

$$\Delta V(W \to E; L) = \frac{\partial}{\partial P}\,[W(A \mid E) - W(A \mid W)] \qquad (5.27)$$

As for the enthalpy change we find that the first two are equal to each other and that both contain the difference in the isothermal compressibilities of the two solvents, a quantity that is a residue from the liberation free energy of the solute A. The last quantity is clearly the simplest of the three.

At this point it is appropriate to note that the local quantities defined above can be used in *any* two phases, E and W. This is not the case for all the other standard quantities. As an example, consider the case where W is an ideal gas phase, G. Then

$$\Delta V(G \to E; L) = \frac{\partial}{\partial P}\,[W(A \mid E) - W(A \mid G)]$$

$$= \frac{\partial}{\partial P}\,[W(A \mid E)] \qquad (5.28)$$

This is the change in volume associated with the insertion of A into a fixed position in the liquid E. On the other hand, if we take either (5.25) or (5.26) and let the density of the phase W go to zero we obtain

$$\Delta V\left(\begin{matrix} W \to E \\ \varrho_A{}^W = \varrho_A{}^E \end{matrix}\right)$$

$$= \lim_{\varrho_W{}^W \to 0} \left\{ \frac{\partial}{\partial P} [W(A \mid E) - W(A \mid W)] - kT \frac{\partial}{\partial P} \ln(V^E/V^W) \right\}$$

$$= \frac{\partial}{\partial P} W(A \mid E) - kT \frac{\partial \ln V^E}{\partial P} + \lim_{\varrho_W{}^W \to 0} \left(kT \frac{\partial \ln V^W}{\partial P} \right) \qquad (5.29)$$

But in the limit of low density of the solvent W we have

$$\lim_{\varrho_W{}^W \to 0} \left(kT \frac{\partial \ln V^W}{\partial P} \right) = \lim_{\varrho_W{}^W \to 0} [kT(-\varkappa_T{}^W)] = -\infty \qquad (5.30)$$

where $\varkappa_T{}^W$ is the isothermal compressibility of the phase W. We see that the two quantities in (5.25) and (5.26) diverge when one of the phases becomes an ideal gas. We note that $\Delta V(G \to E; L)$ may be computed in two ways. It may be derived either by direct differentiation of the standard free energy of solution, i.e.,

$$\Delta V(G \to E; L) = \frac{\partial}{\partial P} [\Delta G(G \to E; L)] \qquad (5.31)$$

or from the measurable partial molar volume of A in E at infinite dilution. The relation between the two quantities is

$$\Delta V_A{}^0(L) \equiv \Delta V(G \to E; L) = \bar{V}_A{}^0 + kT \frac{1}{V^E} \frac{\partial V^E}{\partial P}$$

$$= \bar{V}_A{}^0 - kT\varkappa_T{}^E \qquad (5.32)$$

where $\varkappa_T{}^E$ is the isothermal compressibility of the solvent E. This relation may be obtained simply by differentiating (5.1) with respect to the pressure and comparing the result with (5.28). Also we have introduced a shorthand notation for the *local* partial molar volume of A for an infinitely dilute solution. Relation (5.32) also has an interesting interpretation. If we take a pure solvent E and add to it a single A molecule, the change of volume that we measure is $\bar{V}_A{}^0$. This quantity may be viewed as comprised of two contributions: One is due to the addition of the solute in a fixed position, which is characteristic to the solute–solvent interaction (it might be positive or negative). The second contribution is due to the release of the constraint of a fixed position, i.e., the very fact that the solute acquires a translational degree of freedom always causes a positive change of volume which amounts to $kT\varkappa_T{}^E$. This quantity is a property of the pure solvent E and is of no interest in the study of the solvation thermodynamics of A in E. We shall also see in the next section that only the *local* volume changes appear in the pressure coefficient of the HI.

For completeness, we present also the energy and heat capacity of transfer. The energy of transfer is obtained from the relation

$$\Delta E = \Delta H - P \Delta V \tag{5.33}$$

Hence we have

$$
\begin{aligned}
\Delta E\!\begin{pmatrix} W \to E \\ \varrho_A{}^W = \varrho_A{}^E \end{pmatrix} &= \Delta E\!\begin{pmatrix} W \to E \\ x_A{}^W = x_A{}^E \end{pmatrix} \\[4pt]
&= -T^2 \frac{\partial}{\partial T}\left[\frac{W(A \mid E) - W(A \mid W)}{T}\right] \\[4pt]
&\quad + kT^2 \frac{\partial}{\partial T}\left[\ln(V^E/V^W)\right] \\[4pt]
&\quad - P\frac{\partial}{\partial P}\left[W(A \mid E) - W(A \mid W)\right] \\[4pt]
&\quad - PkT\frac{\partial}{\partial P}\left[\ln(V^E/V^W)\right]
\end{aligned}
\tag{5.34}
$$

$$
\begin{aligned}
\Delta E(W \to E;\, L) &= -T^2 \frac{\partial}{\partial T}\left[\frac{W(A \mid E) - W(A \mid W)}{T}\right] \\[4pt]
&\quad - P\frac{\partial}{\partial P}\left[W(A \mid E) - W(A \mid W)\right]
\end{aligned}
\tag{5.35}
$$

The (constant pressure) partial molar heat capacity of A in E is obtained by differentiating the partial molar enthalpy of A in E, i.e.,

$$
\bar{H}_A = -T^2 \frac{\partial}{\partial T}\left[\frac{W(A \mid E)}{T}\right] + \frac{3}{2} kT + kT^2 \frac{\partial \ln V^E}{\partial T}
\tag{5.36}
$$

$$
\bar{C}_{P,A} = \left(\frac{\partial \bar{H}_A}{\partial T}\right)_P = -T \frac{\partial^2}{\partial T^2}[W(A \mid E)] + \frac{3}{2} k + \frac{\partial}{\partial T}\left(kT^2 \frac{\partial \ln V^E}{\partial T}\right)
\tag{5.37}
$$

This is the measured partial molar heat capacity of A for an infinitely dilute solution in E. Clearly it contains two terms that originate from the translational degrees of freedom of the (simple) solute A.

The corresponding quantities of transfer are

$$\Delta C_P \begin{pmatrix} W \to E \\ \varrho_A{}^W = \varrho_A{}^E \end{pmatrix} = \Delta C_P \begin{pmatrix} W \to E \\ x_A{}^W = x_A{}^E \end{pmatrix}$$

$$= -T^2 \frac{\partial^2}{\partial T^2} \left[W(A \mid E) - W(A \mid W) \right]$$

$$+ \frac{\partial}{\partial T} \left[kT^2 \frac{\partial}{\partial T} \ln(V^E/V^W) \right] \qquad (5.38)$$

$$\Delta C_P(W \to E; L) = -T^2 \frac{\partial^2}{\partial T^2} \left[W(A \mid E) - W(A \mid W) \right] \qquad (5.39)$$

Thus the *local* quantity is the one that involves only the local properties of the solute in the two solvents. The other two quantities contain the first and second derivatives of the volume of the solvent with respect to the temperature.

A particular case of interest is when the transfer is from the (ideal) gas into E, in which case we have

$$\Delta C_P(G \to E; L) = -T^2 \frac{\partial^2}{\partial T^2} \left[W(A \mid E) \right] \qquad (5.40)$$

As in (5.32) this quantity is compared with the measured partial molar heat capacity at infinite dilution, which from (5.37) is

$$\Delta \bar{C}_{P,A}^\circ = \Delta C_P(G \to E; L) + \frac{3}{2} k + \frac{\partial}{\partial T} \left(kT^2 \frac{\partial \ln V^E}{\partial T} \right) \qquad (5.41)$$

As in the case of the partial molar volume we note that the quantity $\Delta C_P(G \to E; L)$ may be obtained either by differentiating twice the experimental standard free energy of transfer $\Delta G(G \to E; L)$, or from the experimental $\bar{C}_{P,A}^\circ$ through relation (5.41). In problems involving HI it is only the *local* quantity that enters into the second derivative of the strength of the HI with respect to temperature.

This section has been devoted to the standard quantities of transferring A between two phases. Though it is true that even for this case the *local* standard quantities of transfer are the simplest and best suited for the study of the solvation properties of a solute in a solvent, one may make other choices of standard quantities if that seems more convenient. This is not the case, however, when we proceed to discuss the temperature and pressure dependence of the HI in the sense of Chapters 3 and 4. In these cases we shall find that *only* the local standard quantities are the ones to be used.

5.3. TEMPERATURE AND PRESSURE DERIVATIVES OF THE HI

In the previous section we have discussed several different standard processes for which we have derived expressions for the corresponding standard thermodynamic quantities. In this section we shall see that the choice of the process suitable for the study of the HI is dictated by the very formulation of our problem.

We shall discuss mainly the problem of the pairwise HI. All of the results of this section may be generalized for many solute particles in a straightforward manner.

The essence of the HI problem may be stated as follows: Having two solute particles in a solvent at some specified temperature and pressure, we ask what is the probability of finding these two (simple) solutes at some small separation R. We have seen in Chapter 1 that this probability is intimately connected with the change in the Gibbs free energy for the process of bringing the two solutes from fixed positions, but at an infinite distance from each other, to a small separation R. We thus write

$$\Delta G(\infty \to R) = U(R) + \delta G^{HI}(R) \qquad (5.42)$$

We see that the *process* which is relevant to the study of the HI is already determined by the very formulation of the problem of HI. We have also found an exact expression for the *indirect* part of the total free energy change, namely,

$$\delta G^{HI}(R) = \Delta\mu_R{}^\circ - 2\Delta\mu_A{}^\circ \qquad (5.43)$$

where $\Delta\mu_A{}^\circ$ is the standard free energy of transferring the solute A from a fixed position in the gas to a fixed position in the liquid W. In the notation of the previous section this is the same as

$$\Delta\mu_A{}^\circ = \Delta G(G \to W; L) \qquad (5.44)$$

Similarly, $\Delta\mu_R{}^\circ$ is the standard free energy of transferring the pair of solutes at a distance R from each other, from the gas to the liquid. In this process we require that the "dimer" be in a fixed position and orientation in the two phases.

Once we recognize the process to which the two standard free energies in (5.43) refer, it is a straightforward matter to derive the other thermodynamic quantities that correspond to the HI process. Thus, for the entropy we have

$$\delta S^{\mathrm{HI}} = - \frac{\partial \delta G^{\mathrm{HI}}}{\partial T} = - \frac{\partial}{\partial T} [\Delta \mu_R{}^{\circ} - 2\Delta \mu_A{}^{\circ}]$$

$$= \Delta S_R{}^{\circ}(L) - 2\Delta S_A{}^{\circ}(L) \qquad (5.45)$$

where for convenience we use the shorthand notation $\Delta S_A{}^{\circ}(L)$ for the local standard entropy of transferring the solute A from the gas into the liquid. With a similar notation we also have

$$\delta H^{\mathrm{HI}} = \delta G^{\mathrm{HI}} + T\delta S^{\mathrm{HI}} = \Delta H_R{}^{\circ}(L) - 2\Delta H_A{}^{\circ}(L) \qquad (5.46)$$

$$\delta V^{\mathrm{HI}} = \frac{\partial \delta G^{\mathrm{HI}}}{\partial P} = \frac{\partial}{\partial P} [\Delta \mu_R{}^{\circ} - 2\Delta \mu_A{}^{\circ}] = \Delta V_R{}^{\circ}(L) - 2\Delta V_A{}^{\circ}(L)$$

$$= V_R{}^{\circ} - 2V_A{}^{\circ} + kT\varkappa_T \qquad (5.47)$$

where in the last equality on the right-hand side of (5.47) we have used the conventional partial molar volumes of the monomer and of the "dimer." Because of the difference between $\Delta V_A{}^{\circ}(L)$ and $\bar{V}_A{}^{\circ}$ we cannot simply use the combination $\bar{V}_R{}^{\circ} - 2\bar{V}_A{}^{\circ}$ for the pressure coefficient of the HI.

Finally for the heat capacity change we write

$$\delta C_P{}^{\mathrm{HI}} = \frac{\partial \delta H^{\mathrm{HI}}}{\partial T} = \Delta C_{P,R}^{\circ}(L) - 2\Delta C_{P,A}^{\circ}(L) \qquad (5.48)$$

where again we stress that only the local partial molar heat capacities appear on the right-hand side of (5.48). In order to convert to conventional partial molar heat capacities one must use relations of the form (5.41) of Section 5.2. (But care must be exercised to use $\frac{3}{2}k$ for monatomic molecules, $3k$ for diatomic molecules, etc.)

In all of the above relations we have expressed the temperature and the pressure coefficients of the HI in terms of quantities that, in principle, are measurable. (As we noted in Chapter 3, we do not always have the relevant experimental data. For this reason various approximations were suggested to extract such quantities from experimental sources.)

We now turn to the interpretation of the temperature and the pressure coefficients of the HI in terms of the probability of finding two (or more) solute particles at a small separation.

We recall the fundamental relation between the solute–solute pair correlation function and the total free energy change defined in Sections 1.3 and 1.4:

$$g_{AA}(R) = \exp[-\Delta G(R)/kT] \qquad (5.49)$$

The function that remains after extracting the direct pair potential $U(R)$ is defined by

$$y_{AA}(R) = g_{AA}(R)\exp[+U(R)/kT] \tag{5.50}$$

Hence we have the following relation:

$$\delta G^{\text{HI}}(R) = -kT \ln y_{AA}(R) \tag{5.51}$$

The function $g_{AA}(R)$ governs the probability of finding the two solutes at a separation R. On the other hand, $y_{AA}(R)$ may be interpreted as having the same significance as $g_{AA}(R)$ but for pairs of solute particles between which the direct interaction potential has been "switched off" [i.e., we put $U(R) = 0$]. Alternatively, ratios of $y_{AA}(R)$ may be used to compare the probabilities of finding a pair of solutes at the separation R, in two solvents. The assumption is being made that $U(R)$ is independent of the solvent in which the two solutes interact.

The temperature dependence of the strength of the HI is obtained from (5.51)

$$-\delta S^{\text{HI}}(R) = \frac{\partial \delta G^{\text{HI}}(R)}{\partial T} = -k\ln y_{AA}(R) - kT\frac{\partial \ln y_{AA}(R)}{\partial T} \tag{5.52}$$

Thus the temperature coefficient of $\delta G^{\text{HI}}(R)$ has two components, which may be easily identified in terms of the following thermodynamic quantities:

$$-k\ln y_{AA}(R) = \frac{\delta G^{\text{HI}}(R)}{T} \tag{5.53}$$

$$-kT\frac{\partial \ln y_{AA}(R)}{\partial T} = -\frac{\delta H^{\text{HI}}(R)}{T} \tag{5.54}$$

It is clear from the above relations that $\delta H^{\text{HI}}(R)$ and $\delta S^{\text{HI}}(R)$ provide different and complementary information on the temperature coefficient of the HI. The overall temperature dependence of the *strength* of the HI is given by $\delta S^{\text{HI}}(R)$ as shown in (5.52). Furthermore, the average force between the two solutes is obtained by the gradient of $\Delta G(R)$ with respect to R, i.e.,

$$F(R) = -\frac{\partial}{\partial R}\Delta G(R) = -\frac{\partial \delta G^{\text{HI}}(R)}{\partial R} - \frac{\partial U(R)}{\partial R} \tag{5.55}$$

Therefore, the gradient of $\delta S^{\text{HI}}(R)$, with respect to R, gives the temperature coefficient of the average force between the two solutes.

On the other hand, $\delta H^{\text{HI}}(R)$ conveys different information. As shown in (5.54), $\delta H^{\mathrm{HI}}(R)$ measures the temperature dependence of $\ln y_{AA}(R)$. This is only part of the total temperature coefficient of $\delta G^{\mathrm{HI}}(R)$ as shown in (5.52). Since $\ln y_{AA}$ is related to the probability of finding the two solutes (the direct interaction between which has been "switched off"), we have here a measure of the temperature dependence of the probability of finding two (or more) solutes at close distance. [More properly we should examine the temperature dependence of $\varrho_A{}^2 g_{AA}(R)$ rather than $y_{AA}(R)$ if we are interested in the probability of finding two solutes at R.]

We also note that the analysis of the total temperature dependence of δG^{HI} in terms of its two contributions is of interest from the molecular point of view. As we shall see later in this chapter, the first term on the right-hand side of (5.52), though dependent on the "structure of water," is independent of the structural changes that are induced in the solvent by the HI process. The second term, on the other hand, bears the contribution from structural changes in the solvent. The discussion of this topic is postponed to Section 5.8, since it depends on having a proper definition of the concept of the "structure of water." The latter is dealt with in Section 5.7.

The pressure dependence of the function $y_{AA}(R)$ is simply obtained from (5.51), namely,

$$\frac{\partial \delta G^{\mathrm{HI}}(R)}{\partial P} = -kT \frac{\partial \ln y_{AA}(R)}{\partial P} = \delta V^{\mathrm{HI}}(R) \tag{5.56}$$

where $\delta V^{\mathrm{HI}}(R)$ is the volume change associated with the process of HI.

In a very straightforward manner we can generalize all the relations presented in this section to the HI among m solute particles. No new ideas or assumptions are involved in such a generalization.

In Sections 5.4 and 5.5 we turn to examine the available experimental or theoretical information on the temperature and pressure dependence of the HI.

5.4. EXPERIMENTAL OBSERVATIONS ON THE TEMPERATURE DEPENDENCE OF THE HI

We start with the temperature dependence of the standard free energy of solution of a simple solute A in a solvent. If we are interested only in the free energy of solvation of A, then the corresponding temperature dependence is given by the (local) standard entropy of solution, i.e.,

$$\Delta S_A{}^\circ(L) = -\frac{\partial \Delta \mu_A{}^\circ}{\partial T} \tag{5.57}$$

However, in connection with the general problem of HI we have stressed in Chapter 2 the interpretation of $\Delta\mu_A^\circ$ as a measure of the partition coefficient of A between two phases, say W and E:

$$\Delta\mu_A^\circ(W \to E) = kT \ln(\varrho_A{}^W/\varrho_A{}^E)_{eq} \qquad (5.58)$$

The partition coefficient measures the relative "phobia" of A towards the two solvents, or the relative tendency of A to prefer one solvent over the other. With this in mind, the total temperature dependence of $\Delta\mu_A^\circ$ is viewed as comprised of two components:

$$\frac{\partial \Delta\mu_A^\circ(W \to E)}{\partial T} = k \ln(\varrho_A{}^W/\varrho_A{}^E)_{eq} + kT \frac{\partial \ln(\varrho_A{}^W/\varrho_A{}^E)_{eq}}{\partial T} \qquad (5.59)$$

Thus, the standard entropy of transfer provides a measure of the *total* temperature dependence of $\Delta\mu_A^\circ(W \to E)$. However, if we are interested in the question of the relative preference of the solute for one solvent over the other, we must focus on the second term on the right-hand side of (5.59). It can easily be seen that this coefficient is related to the standard enthalpy of transfer, namely,

$$\Delta H_A^\circ(W \to E; L) = -kT^2 \frac{\partial \ln(\varrho_A{}^W/\varrho_A{}^E)_{eq}}{\partial T} \qquad (5.60)$$

Thus, both the entropy and the enthalpy of transfer are of interest; one is a measure of the temperature dependence of the difference in the solvation free energy of A in the two phases, the second measures the change in the distribution of A between two phases caused by changing the temperature. In Table 5.1 we present some numerical values for the thermodynamic quantities of transferring methane from water into various solvents at two temperatures. Note that all quantities are the local standard quantities (i.e., pertinent to the process of transferring A from a fixed position in one solvent to a fixed position in water).

The negative values of $\Delta\mu_A^\circ(W \to E; L)$ are interpreted through (5.58) in terms of the relatively larger tendency of methane to avoid water as compared with the other solvents listed in Table 5.1. In other words, methane distributes itself in such a way that it prefers the nonaqueous environment. The temperature dependence of $\Delta\mu_A^\circ$ is clearly represented by $\Delta S_A^\circ(L)$, which is always positive, i.e., $\Delta\mu_A^\circ$ increases (in absolute magnitude) as we raise the temperature. This temperature dependence has two components as shown in (5.59). It is only ΔH_A° that conveys information on the temperature dependence of the partition coefficient as shown in (5.60). A

Table 5.1

Local Standard Free Energies, Entropies, and Enthalpies of Transfer of Methane from Water into Other Solvents[a]

Other solvent E	t (°C)	$\Delta\mu_A{}^\circ(W \to E; L)$ (cal/mol)	$\Delta S_A{}^\circ(W \to E; L)$ (cal/mol deg)	$\Delta H_A{}^\circ(W \to E; L)$ (cal/mol)	$\Delta C^\circ_{P,A}(W \to E; L)$ (cal/mol deg)
Methanol	10	−1404	15.7	3039	−75
	25	−1610	11.8	1906	
Ethanol	10	−1417	15.1	2856	−60
	25	−1620	12.0	1956	
1-Propanol	10	−1402	14.0	2560	−29
	25	−1600	12.5	2125	
1-Butanol	10	−1378	15.5	3008	−86
	25	−1570	11.0	1708	
1-Pentanol	10	−1348	15.0	2897	−60
	25	−1550	11.9	1996	
1,4-Dioxane	10	−1209	17.5	3743	−60
	25	−1447	14.4	2844	
Cyclohexane	10	−1593	16.4	3048	−44
	25	−1821	14.1	2381	

[a] Based on Table 7.2 in Ben-Naim (1974).

positive value of $\Delta H_A{}^\circ$ means that the distribution of methane in favor of E relative to W *increases* with the temperature.

The above result may seem to be surprising at first glance. We expect that as the temperature increases water will become more "normal." Hence we should have found that the distribution of methane in the two liquids becomes uniform at high temperatures. This argument may be true at very high temperatures, but, as the experimental data indicate, it is not true at about room temperature. This kind of behavior will be found in other quantities that are connected with the temperature dependence of the HI. The theoretical reasons for such seemingly paradoxical behavior will be presented in Section 5.9, where we shall show how a large positive value of $\Delta H_A{}^\circ(W \rightarrow E; L)$ may be a result of the structural changes induced in the solvents.

We further note from Table 5.1 that though $\Delta H_A{}^\circ$ are positive they definitely *decrease* with the temperature. The relevant thermodynamic quantity is $\Delta C_{P,A}^{\circ}(W \rightarrow E; L)$. This means that though the distribution of methane in favor of E increases with temperature, the rate of increase becomes smaller as the temperature increases. One may therefore expect that at very high temperatures $\Delta H_A{}^\circ$ eventually becomes zero or even changes to negative values. (As, indeed, is sometimes found experimentally.)

Next we proceed to discuss the temperature dependence of the pairwise HI. Perhaps the first quantitative discussion of this topic was presented by Némethy and Scheraga (1962) and by Scheraga *et al.* (1962). These authors have computed the strength and the temperature dependence of the HI and found that it increases with increasing temperature up to about 60°C.[†]

We shall not elaborate any further on these results, since they were based on a model for water that is now considered to be obsolete. However, it must be noted that the work of Scheraga *et al.* has not only an historical value, it also can be credited with giving a strong impetus to the whole field of theoretical research on aqueous solutions in general, and on hydrophobic interactions in particular. Scheraga *et al.* also stressed the importance of

---

[†] It should be noted that these authors assigned to the *enthalpy* change the meaning of the temperature dependence of the strength of the HI. As we have seen in the previous section, the temperature dependence of the strength of the HI is represented by the entropy, not the enthalpy, change. Their calculated values of the enthalpy change indeed pass through zero at about 60°C, but the corresponding entropy change passes through zero only at about 85°C. Based on these data we can conclude that the strength of the HI ceases to increase with temperature at about 85°C and not at 60°C.

the peculiar temperature dependence of the HI for understanding the thermal stability of proteins in living systems.

The next measure of the strength of HI that we use here is the one introduced in Section 3.3. Table 5.2 presents some values of $\delta G^{HI}(\sigma_1)$, $\delta S^{HI}(\sigma_1)$, and $\delta H^{HI}(\sigma_1)$ for two methane molecules brought to a separation $\sigma_1 = 1.53$ Å.

The overall strength of the HI, as measured by the entropy change $\delta S^{HI}(\sigma_1)$ is large and positive in water, as compared with nearly zero values for the other solvents. This means that the strength of the HI (in water) *increases* with an increase in temperature (i.e., it becomes more negative). These data are for 10°C. Examining the data at higher temperatures shows that $\delta G^{HI}(\sigma_1)$ still decreases with temperature even at 80°C. The slope of $\delta G^{HI}(\sigma_1)$ against T becomes smaller at higher temperatures and it is clear that at some temperature above 80°C one would have found $\delta S^{HI}(\sigma_1) \approx 0$.

According to the discussion in Section 5.3, δH^{HI} measures the temperature coefficient of the function $y_{AA}(R)$. The best way to interpret the results of Table 5.2 is to look at the ratio of the probabilities of finding two methane molecules at the distance σ_1 in the two solvents (provided that T, P, and ϱ_A are the same in the two solvents). From Table 5.2 it follows that δH^{HI}(in water) $- \delta H^{HI}$(in nonaqueous solvent) is always positive. Hence the temperature coefficient of the ratio y_{AA}(in water)/y_{AA}(in

Table 5.2

Values of $\delta G^{HI}(\sigma_1)$, $\delta S^{HI}(\sigma_1)$, $\delta H^{HI}(\sigma_1)$ for Bringing Two Methane Molecules to the Separation $\sigma_1 = 1.533$ Å at 10°C[a]

Solvent	δG^{HI} (kcal/mol)	δS^{HI} (cal/mol deg)	δH^{HI} (kcal/mol)
Water	−1.99	12	1.40
Heavy water	−1.94	13	1.74
Methanol	−1.28	0	−1.3
Ethanol	−1.34	0	−1.3
1-Propanol	−1.39	2	−0.8
1-Butanol	−1.44	−1	−1.7
1-Pentanol	−1.49	0	−1.5
1-Hexanol	−1.51	2	−0.9
1,4-Dioxane	−1.61	3	−0.8
Cyclohexane	−1.36	1	−1.1

[a] Based on data from Yaacobi and Ben-Naim (1974).

Table 5.3

Values of $\Delta G_D{}^\circ$ (in kcal/mol) for Dimerization of Carboxylic Acids in Aqueous Solutions at Different Temperatures[a,b]

t (°C)	15°C	25°C	35°C	45°C
$\Delta G_D{}^\circ(1)$	1.65	1.70	1.75	1.80
$\Delta G_D{}^\circ(2)$	1.35	1.37	1.38	1.38
$\Delta G_D{}^\circ(3)$	1.09	1.12	1.19	1.24
$\Delta G_D{}^\circ(2) - \Delta G_D{}^\circ(1)$	−0.30	−0.33	−0.37	−0.42
$\Delta G_D{}^\circ(3) - \Delta G_D{}^\circ(1)$	−0.56	−0.58	−0.56	−0.56

[a] Based on data from Nash and Monk (1957).
[b] $\Delta G_D{}^\circ(1) = \Delta G_D{}^\circ$ (acetic acid), $\Delta G_D{}^\circ(2) = \Delta G_D{}^\circ$ (propionic acid), and $\Delta G_D{}^\circ(3) = \Delta G_D{}^\circ$ (n-butyric acid).

nonaqueous solvent) is positive (of course the data are relevant to one distance $R = 1.53$ Å only). In order to discuss the temperature dependence of the probability of finding the pair of solutes at a separation R, one must consider the temperature derivative of $\varrho_A{}^2 g_{AA}(R)$ rather than $y_{AA}(R)$ alone.

The next source of data that provides information on the temperature dependence of the HI is the dimerization constants of carboxylic acids. The method of processing such data was discussed in Section 3.7. We have already noted there that large inconsistencies exist among reports from different sources. Therefore, we present here only one set of such data, without elaborating on the reliability of the results. Table 5.3 presents data from Nash and Monk (1957) on the standard free energy of dimerization of carboxylic acids in water at different temperatures,

$$\Delta G_D{}^\circ = -kT \ln K_D \qquad (5.61)$$

where K_D is defined for the association reaction

$$2(HA) \rightleftarrows (HA)_2$$

by

$$K_D = \frac{[HA]_2}{[HA]^2} \qquad (5.62)$$

where molalities[†] are the concentration units chosen by the authors.

[†] The conversion of $\Delta G_D{}^\circ$, based on molalities, into molarities is ΔG° (molarity) = ΔG° (molality) + $RT \ln \varrho_W$, with ϱ_W the density of water in g/cm³.

The processing of the data is similar to the one described in Section 3.7. Here we do not have the values of the standard free energy of dimerization of formic acid. However, using the same arguments as in Section 3.7, we may "extract" a measure of the HI by subtracting $\Delta G_D°$ of acetic acid from $\Delta G_D°$ of the higher homologues. There are two interesting features of the results reported in the fourth and fifth rows of Table 5.3. First we note that the entries in the fifth row are systematically larger (in absolute magnitude) than the corresponding entries in the fourth row. These reflect the relatively stronger HI between two ethyl radicals, as compared with the HI between two methyl radicals. Furthermore, the values of $\Delta G_D°(2)$ − $\Delta G_D°(1)$ clearly increase with the temperature. Therefore, if we adopt this quantity as a measure of the HI between the two methyl radicals, we may conclude from these results that the strength of the HI increases with the temperature. On the other hand, this behavior is not manifested by the entries in the fifth row, which are almost temperature independent.

The above interpretation of the data in terms of HI is subject to the same reservation that we have expressed in Section 3.7 and in Appendix A.3. Thus, although part of these results are consistent with previous conclusions on the temperature dependence of the HI, they cannot serve as reliable sources of information on this matter.

Similar results on the temperature dependence of the HI were reported by Oakenfull and Fenwick (1977). In Table 5.4 we present their data on $\Delta G_{HI}°$, $\Delta H_{HI}°$, and $T \Delta S_{HI}°$. These quantities were obtained from the

Table 5.4

Values of $\Delta G_{HI}°$, $\Delta H_{HI}°$, and $T \Delta S_{HI}°$ per CH_2 Group Obtained from the Standard Free Energy of Ion-Pair Formation[a]

t (°C)	$\Delta G_{HI}°$	$\Delta H_{HI}°$	$T \Delta S_{HI}°$
5	−0.36	2.39	2.87
10	−0.42	2.15	2.63
15	−0.49	1.91	2.15
25	−0.54	1.19	1.67
35	−0.57	0.48	1.19
45	−0.58	−0.24	0.48
55	−0.62	−0.96	−0.34

[a] Based on data from Oakenfull and Fenwick (1977). All entries are in kcal/mol.

ion-pair formation of long-chain ions. The free energy $\Delta G_{\text{HI}}^{\circ}$ is defined per CH_2 group by the process discussed in Section 3.8. The corresponding entropy and enthalpy change are obtained from the temperature dependence of $\Delta G_{\text{HI}}^{\circ}$. Thus all quantities refer to the HI per CH_2 group.

From these results it is quite clear that the quantity $\Delta G_{\text{HI}}^{\circ}$, which conveys information on the HI, increases (in absolute magnitude) with temperature. Note also that at around room temperature both the corresponding enthalpy and the entropy changes are positive. This is qualitatively consistent with the results reported from other sources.

The last piece of information on the temperature dependence of the HI comes from calculations based on the scaled-particle theory. In principle if we write the HI among m solute particles in the form (see Section 4.6)

$$\delta G_m{}^{\text{HI}} = \Delta \mu_{\text{agg}}^{\circ} - m \, \Delta \mu_M{}^{\circ} \tag{5.63}$$

then the corresponding entropy change is

$$\delta S_m{}^{\text{HI}} = \Delta S_{\text{agg}}^{\circ}(L) - m \, \Delta S_M{}^{\circ}(L) \tag{5.64}$$

where here one must use the local standard entropy of solution of the monomers M and of the aggregate (agg). The results reported in Section 4.6 indicate that both the entropy and enthalpy changes associated with the HI of the aggregation process are large and positive (compared with the corresponding quantities in nonaqueous solutions). This means that the strength of HI among m solute particles as well as the probability of aggregation increases with increasing temperature. The appropriate results have already been described in Section 4.6.

Finally, we estimate the change in the heat capacity involved in the process of HI. We recall from Section 5.2 the relationship between the partial molar heat capacity of a solute A at infinite dilution in W, and the local partial molar heat capacity. For monatomic solutes we have

$$\bar{C}_{P,A}^{\circ} = \Delta C_P[G \to W; L] + \frac{3}{2} R + \frac{\partial}{\partial T} \left(RT^2 \frac{\partial \ln V^W}{\partial T} \right) \tag{5.65}$$

where R is the gas constant. For diatomic molecules having also a rotational degree of freedom, the term $\frac{3}{2}R$ is replaced by $3R$. We have estimated the last term on the right-hand side of (5.65) for water at around room temperature to be about 2.1 cal/mol deg.

Taking the quantity $\delta G^{\text{HI}}(\sigma_1)$ as a measure of the HI between two methane molecules, one obtains for the heat capacity change

$$\delta C_P{}^{\mathrm{HI}} = \Delta C_{P,\mathrm{ethane}}(G \to W; L) - 2\Delta C_{P,\mathrm{methane}}(G \to W; L)$$

$$= \bar{C}^{\circ}_{P,\mathrm{ethane}} - 2\bar{C}^{\circ}_{P,\mathrm{methane}} + 2.1$$

$$\cong 80 - 2 \times 60 + 2.1 = -38 \text{ cal/mol deg} \qquad (5.66)$$

We can conclude that though the strength of the HI increases with the temperature, its rate of increase decreases with temperature.

We conclude this section with a short reflection on the relevance of these data to biochemical processes. Consider the equilibrium between two forms of a protein, say the native and the denatured forms. We know that in most cases proteins are denatured at high temperatures. This may be symbolically written as

$$\text{native} \xrightarrow{\ T\ } \text{denatured} \qquad (5.67)$$

Suppose that we follow only the nonpolar groups and ignore all other changes that occur in the transformation (5.67). We find that some nonpolar groups are transferred from the interior of the protein to an essentially aqueous environment. We also find that two (or more) nonpolar groups that had been close to each other in the native form are now separated.

From our studies on model systems we found that both the extent of the "phobia" for water and the strength of the HI *increases* with temperature. Therefore, we should expect that these two factors will operate to shift the reaction (5.67) *leftwards* as we increase the temperature.

Thus if we ignore all other factors and only consider the hydrophobic hydrations and interactions, our prediction will tell us that an increase in temperature would stabilize the native form. The fact that the opposite process is observed is a strong indication that the HI is not the major or the dominant factor involved in the process discussed above. The most we can say at present is that the HI does contribute to thermal stability of the native form. However, we cannot say how efficient the HI is in comparison with all the other factors that contribute to the same process.

5.5. EXPERIMENTAL INFORMATION ON THE PRESSURE DEPENDENCE OF THE HI

In this section we present a sample of experimental data that bear direct relevance to the pressure dependence of the HI. As we have done in the previous section we start with the pressure dependence of the standard free energy of transfer of a solute A from one phase W to another phase E:

$$\Delta\mu_A^\circ(W \to E) = kT \ln(\varrho_A^W/\varrho_A^E)_{eq} \tag{5.68}$$

Hence

$$\Delta V_A^\circ(W \to E; L) = \frac{\partial \Delta\mu_A^\circ(W \to E)}{\partial P} = \frac{\partial}{\partial P} [kT \ln(\varrho_A^W/\varrho_A^E)_{eq}] \tag{5.69}$$

Here, in contrast to the temperature dependence of $\Delta\mu_A^\circ$, we have only one component for the pressure dependence of the standard free energy of transfer. Thus $\Delta V_A^\circ(W \to E; L)$, the *local* standard volume of transfer, is both a measure of the pressure coefficient of the *local* free energy of transfer and a measure of the pressure coefficient of the distribution coefficient of A between the two phases [compare with the two *different* measures involved in the case of the temperature dependence of $\Delta\mu_A^\circ$; see (5.59)].

If one phase is an ideal gas, then we have the following relation between the local standard volume of transfer and the conventional partial molar volume of A at infinite dilution in the solvent E:

$$\Delta V_A^\circ(L) \equiv \Delta V_A^\circ(G \to E; L) = \bar{V}_A^\circ - kT\varkappa_T^E \tag{5.70}$$

where $\varkappa_T^E$ is the isothermal compressibility of the phase E. As we have noted in Section 5.2, $\bar{V}_A^\circ$ is a directly measurable quantity. On the other hand $\Delta V_A^\circ(G \to E; L)$ may be obtained from the pressure dependence of the partition coefficient as shown in (5.69); the two quantities differ by the amount $kT\varkappa_T^E$. We shall see below that only the *local* standard volume of transfer is required in the expression for the pressure coefficient of the HI.

In Table 5.5 we present some values of the conventional partial molar volume and the corresponding *local* standard volume of transfer, for methane and ethane in water, benzene, carbon tetrachloride, and hexane at 25°C.

We see that both the conventional and the local partial molar volumes of methane and ethane in water are somewhat smaller than the corresponding values in the nonaqueous solutes. This means that if we take the pressure derivative of the partition coefficient of, say, methane between water and benzene, we obtain

$$\Delta V_A^\circ(\text{water} \to \text{benzene}; L) = \frac{\partial}{\partial P} [kT \ln(\varrho_A^W/\varrho_A^B)_{eq}] = 18.44 \text{ cm}^3/\text{mol}$$

At room temperature and at atmospheric pressure the value of $\Delta\mu_A^\circ(W \to B)$ is negative; this means that the solutes favor the benzene more than the aqueous environment. The positive pressure coefficient

Table 5.5

Conventional $(\bar{V}_A^\circ)$ and Local $[\Delta V_A^\circ(L)]$ Partial Molar Volumes of Methane and Ethane in Some Solvents at 25°C[a]

Solvent	$RT\varkappa_T$ for the solvent	Methane		Ethane	
		$\bar{V}_A^\circ$	$\Delta V_A^\circ(L)$	$\bar{V}_A^\circ$	$\Delta V_A^\circ(L)$
Water	1.13	37.3	36.17	51.2	50.07
Carbon tetrachloride	2.64	51.7	49.06	66.0	63.36
n-Hexane	3.97	60.0	56.03	69.3	65.33
Benzene	2.39	57.0	54.61	73.0	70.61

[a] All entries are in cm³/mol. Data are from Masterton (1954) and Horiuti (1931).

indicates that the "phobia" of methane for water relative to that for benzene *decreases* with pressure. (Note that this pressure effect has an opposite sign to the corresponding temperature effect.)

The next quantity that we use as a measure of the strength of the HI is $\delta G^{\mathrm{HI}}(\sigma_1)$ introduced in Section 3.3. The pressure coefficient of this quantity is

$$\delta V^{\mathrm{HI}}(\sigma_1) = \frac{\partial \delta G^{\mathrm{HI}}(\sigma_1)}{\partial P} = \Delta V_{\mathrm{Et}}^\circ(L) - 2\Delta V_{\mathrm{Me}}^\circ(L) \qquad (5.71)$$

where only the *local* quantities can be used on the right-hand side of (5.71). Thus, for the four solvents listed in Table 5.5 we have

$$\delta V^{\mathrm{HI}} = \begin{cases} -22.27 \text{ cm}^3/\text{mol} & \text{in water} \\ -34.76 \text{ cm}^3/\text{mol} & \text{in carbon tetrachloride} \\ -46.73 \text{ cm}^3/\text{mol} & \text{in } n\text{-hexane} \\ -38.61 \text{ cm}^3/\text{mol} & \text{in benzene} \end{cases} \qquad (5.72)$$

We see that the volume change associated with the process of bringing two methane molecules to a small separation is *negative* in all solvents. Thus the strength of the HI, as measured by the quantity $\delta G^{\mathrm{HI}}(\sigma_1)$, *increases* with an increase in pressure. This conclusion is valid for all the solvents listed above, with a somewhat smaller value for δV^{HI} in water.[†]

[†] Sometimes one finds the following argument leading to the conclusion that δV^{HI} should be positive. One looks at the volume change when, say, a methane molecule is

Table 5.6

Conventional Partial Molar Volumes $\bar{V}_A{}^\circ$ for Ammonium and Diazonium
Salts in Water at 25°C[a]

Solute[a]	$\bar{V}_A{}^\circ$ (cm³/mol)	Solute	$\bar{V}_A{}^\circ$ (cm³/mol)
M_1	300.5	D_1	577.9
M_2	241.6	D_2	418.5
M_3	279.3	D_3	494.2

[a] The symbols for the solutes are defined in Figure 5.1. From Jolicoeur and Boileau (1974).

A very similar conclusion may be reached from other experimental sources. First we present some interesting data reported by Jolicoeur and Boileau (1974). In Table 5.6 we present the conventional partial molar volumes of certain long-chain ammonium and diazonium salts.

Viewing the D_i (see Figure 5.1) solutes as "dimers" of the monomers M_i, we may compute the hydrophobic volume of the "dimerization" process [in the same sense as $\delta V^{\mathrm{HI}}(\sigma_1)$]. To do that we must first transform the values in Table 5.6 into local quantities which, in water, requires the subtraction of $RT\varkappa_T = 1.13$ cm³/mol. Also we divide by the number of "bonds" formed in each "dimer" to obtain, for the three "reactions," the results

$$\delta V^{\mathrm{HI}} = V_{D_1}^\circ(L) - 2V_{M_1}^\circ(L) = -21.97 \text{ cm}^3/\text{mol}$$

$$\delta V^{\mathrm{HI}} = \tfrac{1}{3}\big(V_{D_2}^\circ(L) - 2V_{M_2}^\circ(L)\big) = -21.19 \text{ cm}^3/\text{mol} \tag{5.73}$$

$$\delta V^{\mathrm{HI}} = \tfrac{1}{3}\big(V_{D_3}^\circ(L) - 2V_{M_3}^\circ(L)\big) = -21.09 \text{ cm}^3/\text{mol}$$

The values of δV^{HI} in (5.73) though obtained for quite complex molecules

---

transferred from benzene into water. This volume change is negative (see Table 5.5). Now if the process of HI is viewed as being a "partial reversal of the transfer process" from nonaqueous to aqueous environment, one would expect the sign of δV^{HI} to be positive. We believe that the fundamental flaw in this argument rests in the view that the HI process is a "partial reversal of the transfer process" between two liquids. As we have discussed in Section 4.6, the identification of the aggregation process with the transfer process may be appropriate in the limit of many solute particles forming a compact droplet. In this case it is true that the thermodynamic quantities associated with the two processes should have opposite signs. This is not true for the dimerization process, however (see also the example given in Section 3.4).

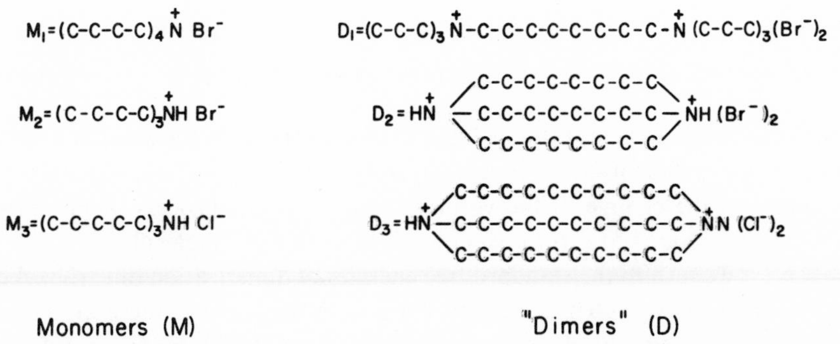

$M_I = (C-C-C-C)_4 \overset{+}{N} Br^-$ $D_I = (C-C-C)_3 \overset{+}{N}-C-C-C-C-C-C-C-C-C-\overset{+}{N}(C-C-C)_3(Br^-)_2$

$M_2 = (C-C-C-C)_3 \overset{+}{N}H Br^-$ $D_2 = H\overset{+}{N} \begin{matrix} C-C-C-C-C-C-C-C \\ -C-C-C-C-C-C-C-C-\overset{+}{N}H (Br^-)_2 \\ C-C-C-C-C-C-C-C \end{matrix}$

$M_3 = (C-C-C-C)_3 \overset{+}{N}H Cl^-$ $D_3 = H\overset{+}{N} \begin{matrix} C-C-C-C-C-C-C-C-C \\ -C-C-C-C-C-C-C-C-C-\overset{+}{N}N (Cl^-)_2 \\ C-C-C-C-C-C-C-C-C \end{matrix}$

Monomers (M)   "Dimers" (D)

Figure 5.1. Schematic formulas for the ammonium salts (monomers) and the diazonium salts ("dimers"). Each methyl or methylene group is represented by the letter "C."

are very similar to the value of $\delta V^{HI}(\sigma_1)$ in water shown in (5.72). This indicates that in all of these volume changes the crucial process is the formation of the C–C bond. Note also that all the values in (5.73), per bond formation, are very close to each other. Fortunately, the value of $RT\varkappa_T$ for water at 25°C is quite small (Table 5.5). Thus, if one uses $\bar{V}_A^\circ$ rather than $\Delta V_A^\circ(L)$, the results would not be much different from the ones reported in (5.73). This would not be the case for other solvents in which $RT\varkappa_T$ is large.

Some further results of a similar kind are listed in Table 5.7. The mono- and dialcohols are schematically shown on the left-hand side. The

Table 5.7

Volume Change Associated with the "Dimer" Formation of α,w-Alcohols from Monoalcohols in Water at 25°C[a]

Monomer	"Dimer"	δV^{HI}
C—OH	HO—C—C—OH	−19.87
C_2—OH	HO—C_4—OH	−20.57
C_3—OH	HO—C_6—OH	−19.47
C_4—OH	HO—C_8—OH	−19.47
C_5—OH	HO—C_{10}—OH	−19.27

[a] Values of δV^{HI} (cm³/mol) computed from data of Edwards *et al.* (1977). Similar data have also been reported by Jolicoeur and Lacroix (1976).

volume change for the formation of a "dimer" is defined by

$$\delta V^{\mathrm{HI}} = \Delta V^{\circ}_{\text{"dimer"}}(L) - 2\Delta V^{\circ}_{\text{monomer}}(L) \tag{5.74}$$

The results reported in Table 5.7 are similar to the ones reported in (5.73) and to the value of δV^{HI}(in water) given in (5.72). All correspond essentially to the change in the volume induced by the process of bringing two methyl groups to form a C–C bond in a "dimer" molecule.

Next we turn to the standard free energies of dimerization of carboxylic acids. As we have noted in Section 5.4 and in Section 3.7, there are large discrepancies between data from different sources. Therefore, we have selected what we believe is the most complete set of data that is suitable for demonstrating one aspect of the behavior of $\Delta G_D{}^{\circ}$, such as its temperature or pressure dependence. In Table 5.8 we present values of the standard free energy of dimerization defined by

$$\Delta G_D{}^{\circ} = -kT \ln K_D \tag{5.75}$$

where K_D is the dimerization constant (the concentrations used are in molarities). We have also processed these data according to the method suggested by the authors and which was outlined in Section 3.7. Namely, we define

$$\Delta[\Delta G_D{}^{\circ}(\mathrm{R})] = \Delta G_D{}^{\circ}(\mathrm{RCOOH}) - \Delta G_D{}^{\circ}(\mathrm{HCOOH}) \tag{5.76}$$

This quantity is presumed to give a measure of the HI between the two alkyl groups in the dimers. (For more details, see also Section 3.7 and Appendix A.3.)

The results reported in Table 5.8 do not show a clear-cut pressure dependence of $\Delta[\Delta G_D{}^{\circ}(\mathrm{R})]$. For $\mathrm{R} = \mathrm{CH_3}$ it looks as if initially this quantity increases with pressure and then decreases. For the other two alkyl radicals the pressure coefficient is clearly positive. These results, if they do convey information on HI, are inconsistent with the results reported on the pressure dependence of $\delta G^{\mathrm{HI}}(\sigma_1)$.

However, it is not clear whether the above method of processing the experimental data is really the best one to obtain information on the pressure dependence of the HI. We have expressed some reservations about this method in Section 3.7 and in Appendix A.3.

The flaw in this method becomes more conspicuous if we write the statistical mechanical expression for $\Delta G_D{}^{\circ}$ (see Appendix A.3):

$$\Delta G_D{}^{\circ} = U(D) - kT \ln\left\{\frac{\langle\exp[-\beta B(D)]\rangle_0}{\langle\exp[-\beta B(M)]\rangle_0{}^2}\right\} + kT \ln(q^2_{\mathrm{int},M}/q_{\mathrm{int},D}) \tag{5.77}$$

Table 5.8

Values of $\Delta G_D{}^\circ$ and $\Delta[\Delta G_D{}^\circ(R)]$ for Carboxylic Acids at 30°C and at Different Pressures[a]

Pressure (kg/cm²)	$\Delta G_D{}^\circ$ (kcal/mol)				$\Delta[\Delta G_D{}^\circ(R)]$ (kcal/mol)		
	Formic acid	Acetic acid	Propionic acid	n-Butyric acid	CH_3	CH_3CH_2	$CH_3CH_2CH_2$
1 atm	1.64	1.14	1.02	0.94	−0.50	−0.62	−0.70
860	1.38	0.88	0.84	0.84	−0.50	−0.54	−0.54
1650	1.19	0.65	0.70	0.74	−0.54	−0.49	−0.45
2560	1.07	0.24	0.53	0.94	−0.83	−0.54	−0.13
3410	0.85	0.25	0.62	1.18	−0.60	−0.23	+0.33
4250	0.70	0.32	0.85	1.39	−0.38	+0.15	+0.69
5020	0.48	0.48	1.05	1.50	0.00	+0.57	+1.02
5910	0.36	0.62	1.24	1.72	+0.26	+0.88	+1.36

[a] Based on data from Suzuki et al. (1973).

where $U(D)$ is the direct work of formation of the dimer, $\beta = (kT)^{-1}$, $B(D)$ and $B(M)$ are the binding energies of the dimer and the monomer, respectively, and q_{int} is the internal partition function of either a monomer or a dimer. We may assume that both $U(D)$ and the internal partition functions are pressure independent, hence

$$\Delta V_D{}^\circ = \frac{\partial \Delta G_D{}^\circ}{\partial P} = \frac{\partial}{\partial P} \left\{ -kT \ln \frac{\langle \exp[-\beta B(D)] \rangle_0}{\langle \exp[-\beta B(M)] \rangle_0{}^2} \right\}$$

$$= \frac{\partial}{\partial P} (\delta G_D{}^{HI}) \qquad (5.78)$$

Thus we see that the pressure coefficient of $\Delta G_D{}^\circ$ is equal to the pressure coefficient of the indirect part of the work, $\delta G_D{}^{HI}$ (this is in contrast to the case of the temperature coefficient of $\Delta G_D{}^\circ$). The quantity $\delta G_D{}^{HI}$ does contain contributions from the formation of a hydrogen bond and from the alkyl–alkyl hydrophobic interaction. However, as pointed out in Appendix A.3 there is no obvious way of separating these two contributions to $\delta G_D{}^{HI}$. Therefore, it is impossible to determine which quantity better represents the pressure coefficient of $\Delta G_D{}^\circ$. We believe that $\Delta V_D{}^\circ$ itself is at least as good a measure of the pressure coefficient of the HI as are the differences $\Delta V_D{}^\circ(\text{RCOOH}) - \Delta V_D{}^\circ(\text{HCOOH})$. Estimates of the initial slope of $\Delta G_D{}^\circ$ as a function of pressure, made by Suzuki et al. (1973), are shown in Table 5.9.

Thus, if we use $\Delta V_D{}^\circ$ as our measure of the pressure coefficient of the HI we get negative values that are consistent with similar results that were obtained from other sources.

We present one further series of results on the association of rhodamine B (RB) and methylene blue (MB) in aqueous solutions. Suzuki and Tsuchiya

Table 5.9

Standard Volume $\Delta V_D{}^\circ$ of Dimerization of Carboxylic Acids, at 1 atm and 30°C[a]

Acid	$\Delta V_D{}^\circ$ (cm³/mol)	Acid	$\Delta V_D{}^\circ$ (cm³/mol)
Formic acid	−14.0	Propionic acid	−8.8
Acetic acid	−13.0	n-Butyric acid	−6.2

[a] From Suzuki et al. (1973).

Rhodamine B(RB)

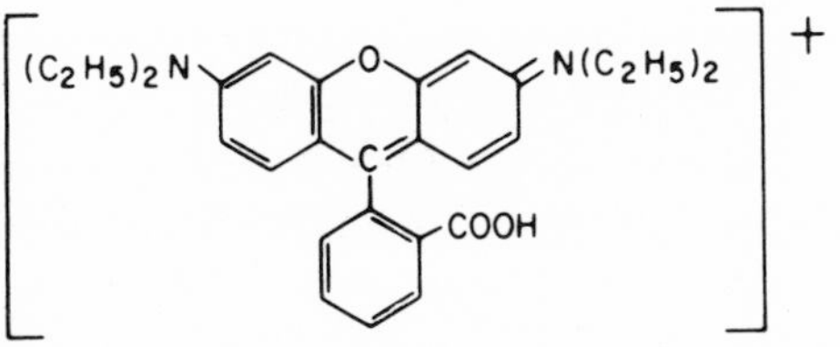

Methylene
Blue (MB)

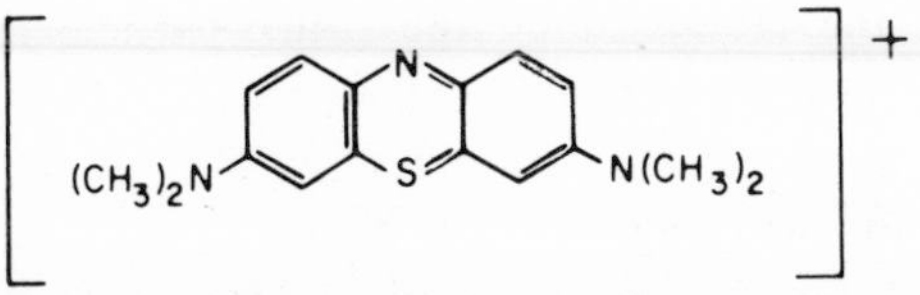

Figure 5.2. Chemical formulas of rhodamine B and methylene blue.

(1971) reported values of the equilibrium constant for the dimerization of RB and MB (Figure 5.2) as a function of pressure at 20°C. Their results were converted into standard free energies of dimerization and are shown in Table 5.10.

From Table 5.10 it is clear that the pressure coefficient of ΔG_D° is negative. Estimates made by the authors for the standard volume of the association are

$$\Delta V_D^\circ(\text{RB}) = -10.4 \text{ cm}^3/\text{mol}$$
$$\Delta V_D^\circ(\text{MB}) = -10.6 \text{ cm}^3/\text{mol} \tag{5.79}$$

Table 5.10

Standard Free Energy of Dimerization $\Delta G_D^\circ = -kT \ln K_D$ [a]

Pressure (atm)	$\Delta G_D^\circ(\text{RB})$ (kcal/mol)	$\Delta G_D^\circ(\text{MB})$ (kcal/mol)
1	3.08	4.36
900	2.95	4.11
1800	2.75	3.98
2700	2.54	3.78
3600	2.27	3.43
4500	1.94	3.18

[a] K_D is the equilibrium constant in (liter/mol) for rhodamine B (RB) at 30°C and methylene blue (MB) at 20°C as a function of pressure. Data from Suzuki and Tsuchiya (1971).

which are roughly of the same order of magnitude as the values reported in Table 5.9 for the carboxylic acids.

Of course the "driving force" for the formation of the dimer between RB or MB involves several factors, only one of which is the HI. Therefore, the above-cited values do not convey information on the pressure coefficient of the HI alone.

Finally, we recall the results of the computation of the pressure dependence of the HI based on the scaled-particle theory. In Section 4.6 we presented values of δG_m^{HI} for the process of association of m solute molecules as a function of the pressure. The pressure coefficient obtained by that method was negative, i.e., the change of volume associated with the process of aggregation is negative. This result is consistent with most of the other results reported in this section, namely, that the strength of the HI increases with an increase in pressure.

We conclude this section with a comment on the relevance of the knowledge gathered from model systems to biological processes. As in Section 5.4, we look at the pressure effect on the process of denaturation of biopolymers, which we schematically write as

$$\text{native} \xrightarrow{P} \text{denatured} \tag{5.80}$$

If we focus attention only on the nonpolar groups (see Figure 1.1), we find from the study of the model systems that on the one hand the transfer of nonpolar groups from the interior of the protein to the aqueous environment should be enhanced with an increase in pressure. On the other hand, the HI between two (or more) nonpolar groups increases with pressure. Hence the latter effect would tend to shift the reaction (5.80) leftward. Thus, in contrast to the case of the temperature effect on the hydrophobic hydration and on the hydrophobic interaction, we find here that the two pressure coefficients will have opposite effects on the conformational change of a protein. This fact, together with the recognition of all the other factors involved in the process (5.80), renders all our information on the pressure coefficient of the HI only remotely relevant to the pressure coefficient of real biological processes.

Comment

The experimental data from which one may extract information on the pressure dependence of the HI are clearly fragmentary, and their interpretation is inconclusive. We have stressed our reservations about the method of processing standard free energies of association of carboxylic acid to obtain information on HI. The same reservations hold for the temperature and pressure dependence of such quantities. In spite of some inconsistencies we believe that the pressure coefficient of the HI is negative, both for the case of "dimers" and for larger aggregates. Of course much more experimental work must be done before the clouds over this topic are dispersed. We believe that the most promising information could be obtained from the partial molar volumes of molecules of the type discussed in Section 4.7. The processing of the data according to the method described there would give information on the pressure coefficient of the intramolecular hydrophobic interactions, a quantity which seems to be the best representative of the contribution of the HI to processes in real biopolymers.

5.6. FORMAL STATISTICAL MECHANICAL EXPRESSIONS FOR THE ENTROPY, ENTHALPY, AND VOLUME CHANGES ASSOCIATED WITH THE HI

In this section we diverge from the style that prevailed in the main part of the book. From here on we shall appeal more to the statistical mechanical, rather than to the thermodynamical language. This will enable us to obtain a deeper insight into the content of the temperature and the pressure coefficients of the HI.

We start with the very general expression for the chemical potential of a simple solute A in a very dilute solution in a solvent W:

$$\mu_A = kT \ln(\varrho_A \Lambda_A^3 q_A^{-1}) + W(A \mid W) \tag{5.81}$$

This expression is derived in Appendix A.1 in the T, V, N ensemble. However, it can be easily shown that the same formal expression is also obtained in the T, P, N ensemble. The latter is the more useful set of independent variables for practical applications. Here $\varrho_A = N_A/V$, with V the average volume of the pure solvent, specified by the variables T, P, N. Λ_A^3 is the momentum partition function of A. q_A is the internal partition function of a single A molecule. For simplicity we assume that A is a structureless particle, hence we put $q_A = 1$. $W(A \mid W)$ is the coupling work of A against the pure solvent W. This may be written as an average, in the T, P, N ensemble, of the form (for more details see Ben-Naim, 1974)

$$W(A \mid W) = -kT \ln \langle \exp(-\beta B_A) \rangle_0 \tag{5.82}$$

where $\beta = (kT)^{-1}$ and B_A is the total binding energy of A with the solvent at some specific configuration:

$$B_A = \sum_{i=1}^{N} U(\mathbf{R}_A, \mathbf{X}_i) \qquad (5.83)$$

$\mathbf{R}_A$ is the position vector of the solute A and $\mathbf{X}_i$ comprises the set of all coordinates required to specify the configuration of the ith solvent molecule. The average $\langle \ \rangle_0$ is defined as

$$\langle \exp(-\beta B_A) \rangle_0 = \frac{\int dV \int d\mathbf{X}^N \exp[-\beta U(\mathbf{X}^N) - \beta B_A - \beta PV]}{\int dV \int d\mathbf{X}^N \exp[-\beta U(\mathbf{X}^N) - \beta PV]}$$

$$= \int dV \int d\mathbf{X}^N P^\circ(\mathbf{X}^N, V) \exp(-\beta B_A) \qquad (5.84)$$

Here $P^\circ(\mathbf{X}^N, V)$ is the fundamental distribution function in the T, P, N ensemble. The circle superscript on $P^\circ(\mathbf{X}^N, V)$ stresses the fact that this distribution function is pertinent to the *pure* solvent, i.e., *before* we have introduced the solute A into the system.

The standard free energy of solution of A is given by

$$\Delta\mu_A^\circ = \Delta G^\circ(G \to W; L) = -kT \ln\langle \exp(-\beta B_A) \rangle_0 \qquad (5.85)$$

The HI between two solutes A separated by a distance R may be written in several equivalent forms

$$\delta G^{\mathrm{HI}}(R) = -kT \ln \frac{\langle \exp[-\beta B(R)] \rangle_0}{\langle \exp[-\beta B(\infty)] \rangle_0}$$

$$= -kT \ln \frac{\langle \exp(-\beta B_D) \rangle_0}{\langle \exp(-\beta B_A) \rangle_0^2} = \Delta\mu_D^\circ - 2\Delta\mu_A^\circ \qquad (5.86)$$

where $B(R)$ and $B(\infty)$ are the binding energies of the two solutes at a distance R and at infinite distance from each other, respectively. In the latter case the average in the denominator may be factorized into a product of two averages each of which is pertinent to the binding energy of one solute A. We have also denoted the "dimer," consisting of two solutes at a distance R, by D.

We may now proceed to compute the temperature and the pressure derivatives of $\delta G^{\mathrm{HI}}(R)$. We first derive the corresponding expressions for the standard entropy, enthalpy, and volume of solution. By taking the temperature and pressure derivatives of (5.85) we obtain the following results:

$$-\Delta S_A{}^{\circ}(L) = \frac{\partial \Delta \mu_A{}^{\circ}}{\partial T} = -k \ln \langle \exp(-\beta B_A) \rangle_0$$

$$- \frac{1}{T} [\langle B_A \rangle_A + \langle U(N) \rangle_A - \langle U(N) \rangle_0 + P \langle V \rangle_A - P \langle V \rangle_0] \tag{5.87}$$

$$\Delta H_A{}^{\circ}(L) = \langle B_A \rangle_A + \langle U(N) \rangle_A - \langle U(N) \rangle_0 + P \langle V \rangle_A - P \langle V \rangle_0 \tag{5.88}$$

$$\Delta V_A{}^{\circ}(L) = \frac{\partial \Delta \mu_A{}^{\circ}}{\partial P} = \langle V \rangle_A - \langle V \rangle_0 \tag{5.89}$$

$$\Delta E_A{}^{\circ}(L) = \Delta H_A{}^{\circ}(L) - P \, \Delta V_A{}^{\circ}(L)$$

$$= \langle B_A \rangle_A + \langle U(N) \rangle_A - \langle U(N) \rangle_0 \tag{5.90}$$

In the above expressions we have introduced a new notation for the conditional average, which is an average over all the possible configurations of the solvent molecules, given that a solute A is at some fixed position $\mathbf{R}_A$. Alternatively, this is an average in the T, P, N ensemble which is subjected to an "external" field of force produced by the solute A at $\mathbf{R}_A$. For example,

$$\langle U(N) \rangle_A = \frac{\int dV \int d\mathbf{X}^N \exp[-\beta U(\mathbf{X}^N) - \beta B_A - \beta PV] U(\mathbf{X}^N)}{\int dV \int d\mathbf{X}^N \exp[-\beta U(\mathbf{X}^N) - \beta B_A - \beta PV]}$$

$$= \int dV \int d\mathbf{X}^N P(\mathbf{X}^N, V/\mathbf{R}_A) U(\mathbf{X}^N) \tag{5.91}$$

where $P(\mathbf{X}^N, V/\mathbf{R}_A)$ is the conditional distribution function of finding a configuration $\mathbf{X}^N$ and a volume V, given that a solute A is at $\mathbf{R}_A$.

We also stress that all the thermodynamic standard quantities listed in (5.87) to (5.90) are *local* standard quantities, i.e., they all refer to the process of transferring a simple solute A from a fixed position in the ideal gas phase to a fixed position in the solvent W.

A glance at equation (5.85) shows that, in contrast to relations (5.87) to (5.90), it contains only an average with the distribution function $P^{\circ}(\mathbf{X}^N, V)$ of the *pure* solvent. In the other relations we also have conditional averages. This is an important observation, and as we shall later see, it leads to an interesting interpretation in terms of the structural changes in the solvent induced by the process of introducing A to a fixed position in W.

From the local standard quantities we can easily construct the thermodynamic quantities associated with the HI. Alternatively, we may obtain these quantities by direct differentiation of $\delta G^{\mathrm{HI}}(R)$ in (5.86) with respect to temperature and pressure. The results are

$$\delta S^{\mathrm{HI}}(R) = \Delta S_D{}^\circ(L) - 2\Delta S_A{}^\circ(L)$$

$$= k \ln\left\{\frac{\langle \exp[-\beta B(R)]\rangle_0}{\langle \exp[-\beta B(\infty)]\rangle_0}\right\} - \frac{1}{T}\left[\langle B(R)\rangle_R - \langle B(\infty)\rangle_\infty\right.$$

$$\left. + \langle U(N)\rangle_R - \langle U(N)\rangle_\infty + P\langle V\rangle_R - P\langle V\rangle_\infty\right] \qquad (5.92)$$

$$\delta H^{\mathrm{HI}}(R) = \Delta H_D{}^\circ(L) - 2\Delta H_A{}^\circ(L)$$

$$= \langle B(R)\rangle_R - \langle B(\infty)\rangle_\infty + \langle U(N)\rangle_R - \langle U(N)\rangle_\infty$$

$$+ P\langle V\rangle_R - P\langle V\rangle_\infty \qquad (5.93)$$

$$\delta V^{\mathrm{HI}}(R) = \langle V\rangle_R - \langle V\rangle_\infty \qquad (5.94)$$

$$\delta E^{\mathrm{HI}}(R) = \langle B(R)\rangle_R - \langle B(\infty)\rangle_\infty + \langle U(N)\rangle_R - \langle U(N)\rangle_\infty \qquad (5.95)$$

where $\langle\ \rangle_R$ and $\langle\ \rangle_\infty$ stand for the conditional averages over all the configurations of the solvent molecules, given that two solutes are at fixed positions at a distance R or at infinite distance, respectively. Thus, for example, the energy associated with the HI consists of two terms. $\langle B(R)\rangle_R - \langle B(\infty)\rangle_\infty$ is the change of the binding energy of the pair of solutes when they are brought from infinity to the distance R. The averages are conditional averages with different conditions in each case. The second term $\langle U(N)\rangle_R - \langle U(N)\rangle_\infty$ is the change of the average potential energy of interaction among the solvent molecules caused by the process of bringing the two solvents from infinity to R. This quantity will be reinterpreted in subsequent sections in terms of structural changes in the solvent induced by the process of the HI. In order to construct such an interpretation we must first define what we mean by the structure of the solvent—we shall discuss this matter for the particular case of water in Section 5.7. A more general definition of the "structure" of a solvent and structural changes in the solvent may be found in Ben-Naim (1974).

5.7. DEFINITION OF THE STRUCTURE OF WATER (SOW)

The concept of the structure of water (SOW) has been used in the literature ever since scientists were concerned with the peculiar properties of liquid water and aqueous solutions. At the time when the mixture model approach to the theory of water was prevalent in the literature, it was relatively easy to construct a definition of the SOW. Basically, water was viewed as a mixture of several components, one of which was identified as the more structured species. Hence, its concentration could have served as a proper measure of the structure of the liquid. In this way a variety of concepts such as the "degree of icelikeness," the "degree of crystallinity," or the amount of "icebergs" had appeared in many discussions and interpretations of phenomena in aqueous systems. All these conveyed more or less the same meaning, namely, the extent to which the structure of liquid water resembles, or deviates from, the structure of solid (ordinary) ice.

In this section we present one possible definition of the SOW which is based on a particular choice of the water–water pair potential. We shall also see that this definition is in conformity with the current ideas on the SOW as has been expressed in various qualitative ways in the literature. It may also be reinterpreted in terms of a mixture model approach to liquid water.

Before presenting this definition, it is worthwhile reflecting on the more fundamental question: in constructing a definition for the SOW, what kind of definition are we looking for?

In the well-known book by Eisenberg and Kauzmann (1969), a considerable amount of space is devoted to elaboration on the concept of the SOW. However, in no place do they offer a definition of this concept that may be assigned a numerical value (which, in turn, may be either measured experimentally or computed by one of the simulation techniques). Another common way of invoking the concept of the SOW is to follow the tradition in the theory of simple fluids. Here one views the pair (or higher-order) correlation function as conveying structural information on the liquid. More precisely, it provides information on the average mode of packing of the molecules in the liquid state. The same is obviously true for liquid water. However, with this kind of definition it is impossible to compare two states of the liquid and conclude which is the more "structured" one. This leads us to look for a definition of the SOW that is a quantity that can be assigned a numerical value. Of course it should conform to the intuitive notion, and perhaps to the traditional ideas, of this concept. The main feature of this definition is that given two states of the system we could determine which state is the more structured of the two. For instance, if we start with pure water at 25°C and add a small amount of alcohol we would like to be able to say whether the SOW has been increased or decreased, rather than merely noting the fact that the SOW has been changed. We stress that the definition presented below is specifically tailored for water and may not be applied to other fluids.

In the spirit of the above requirement we now turn to construct our definition. For convenience and simplicity we assume that the total potential function of N water molecules may be written as a sum of pairwise potential functions, namely,

$$U(\mathbf{X}^N) = \sum_{i<j} U(\mathbf{X}_i, \mathbf{X}_j) \qquad (5.96)$$

This assumption is certainly incorrect for water [see also Ben-Naim (1974)] and in fact it is not even essential for the very definition of the SOW (see below). However, for later application it will be found useful to adopt this assumption.

Furthermore, the pair potential $U(\mathbf{X}_i, \mathbf{X}_j)$ for any pair of water molecules at the configuration $\mathbf{X}_i, \mathbf{X}_j$ is presumed to have the following general form:

$$U(\mathbf{X}_i, \mathbf{X}_j) = U(R_{ij}) + U_{el}(\mathbf{X}_i, \mathbf{X}_j) + \varepsilon_{HB} G(\mathbf{X}_i, \mathbf{X}_j) \qquad (5.97)$$

The first term on the right-hand side of (5.97) is a spherically symmetric contribution to the total interaction between the two molecules. This part may conveniently be chosen to have the Lennard-Jones form. Its main function is to account for the strong repulsive forces which are operative at very short distances, say, $R_{ij} \lesssim 2.5$ Å. The electrostatic part U_{el} may include the interaction between a few electric multipoles. Its main function is to account for the long-range interaction between two water molecules. Of course, at very large distances, only the dipole–dipole interaction will contribute to the total interaction. The third term on the right-hand side of (5.97) is the problematic, and presently the least known, part of the potential. We know that at a distance of about $R \approx 2.8$ Å and at some particular configuration of the pair of molecules a hydrogen bond is formed. However, there is no reliable information on the analytical form of this part of the potential. (In fact, it is not clear whether the hydrogen bond may be completely described as a function of only the configuration of the pair of molecules.) There have been some suggestions for an analytical form of a model potential [see Ben-Naim (1974, 1978c)], but for our purposes we shall not need to describe any details of such a function. Instead, we focus on two important features that such a potential function is supposed to have. In the first place, it includes an energy parameter ε_{HB} which we may refer to as the hydrogen bond energy. Secondly, the function $G(\mathbf{X}_i, \mathbf{X}_j)$ is essentially a geometrical stipulation on the relative configuration of the pair of water molecules. Namely, this function attains a maximum value of unity whenever the two molecules are in the configuration most favorable to form a hydrogen bond, and its value drops to zero when the configuration deviates considerably from the one required for the formation of a hydrogen bond. Recently, Ben-Naim (1974) suggested an explicit analytical form of such a function, but in the context of the present treatment, we shall be satisfied with the following definition:

$$G(\mathbf{X}_i, \mathbf{X}_j) = \begin{cases} 1 & \text{if } i \text{ and } j \text{ are in a configuration favorable for} \\ & \text{a hydrogen-bond formation} \\ 0 & \text{for all other configurations} \end{cases} \qquad (5.98)$$

Of course, it is implicitly assumed that we are given a rule, by the use of which we can determine which configurations are favorable or unfavorable for the formation of a hydrogen bond. Once such a rule is given, then the function $G(\mathbf{X}_i, \mathbf{X}_j)$ becomes well defined. For simplicity, we have described a discontinuous behavior of this function, i.e., it changes abruptly from zero to unity. Of course, a continuous function would be closer to the real case.

For any given configuration of the N water molecules, we may now define, for each molecule, the function

$$\psi_i(\mathbf{X}^N) = \sum_{\substack{j=1 \\ j \neq i}}^{N} G(\mathbf{X}_i, \mathbf{X}_j) \qquad (5.99)$$

Since $G(\mathbf{X}_i, \mathbf{X}_j)$ contributes unity to the sum on the right-hand side of (5.99) whenever the jth molecule is hydrogen-bonded to the ith molecule, ψ_i measures the number of hydrogen bonds in which the ith molecule participates when the system is at the configuration $\mathbf{X}^N$. Based on what we know about the behavior of water molecules, we expect that ψ_i may attain only one of the five possible integral values 0, 1, 2, 3, 4. Clearly, if $\psi_i(\mathbf{X}^N) = 4$, we may say that the local structure around the ith molecule at the configuration $\mathbf{X}^N$ is similar to that of ice. If $\psi_i(\mathbf{X}^N) = 0$, we have a local structure that bears no resemblance to that of ice. Hence we can use $\psi_i(\mathbf{X}^N)$ to serve as a measure of the *local* structure around the ith molecule at the given configuration $\mathbf{X}^N$ of the whole system.

Next we define the average value of $\psi_i(\mathbf{X}^N)$ in the T, P, N ensemble

$$\langle \psi \rangle_0 = \int dV \int d\mathbf{X}^N P(\mathbf{X}^N, V) \psi_i(\mathbf{X}^N) \qquad (5.100)$$

Since all molecules in the system are equivalent, the integral gives the same numerical value independently of the index i, hence we have denoted the average in (5.100) by $\langle \psi \rangle_0$.

Clearly the average number of hydrogen bonds in the system may be obtained from $\langle \psi \rangle_0$ by

$$\langle G \rangle_0 = \frac{N}{2} \langle \psi \rangle_0 \tag{5.101}$$

The division by 2 is required since in $N\langle \psi \rangle_0$ we count each hydrogen bond twice. From (5.99), (5.100), and (5.101) we also obtain the relation

$$\langle G \rangle_0 = \frac{N}{2} \int dV \int d\mathbf{X}^N P(\mathbf{X}^N, V) \sum_{\substack{j=1 \\ j \neq i}}^{N} G(\mathbf{X}_i, \mathbf{X}_j)$$

$$= \frac{1}{2} \int dV \int d\mathbf{X}^N P(\mathbf{X}^N, V) \sum_{i=1}^{N} \sum_{\substack{j=1 \\ j \neq i}}^{N} G(\mathbf{X}_i, \mathbf{X}_j)$$

$$= \int dV \int d\mathbf{X}^N P(\mathbf{X}^N, V) \sum_{i=1}^{N} \sum_{\substack{j=i \\ i < j}}^{N} G(\mathbf{X}_i, \mathbf{X}_j) \tag{5.102}$$

For the second form on the right-hand side of (5.102) we have used the equivalency of the molecules, i.e., instead of N times a single integral, we have written a sum of N (equal) integrals differing by the index i. In the third form, we have absorbed the factor $1/2$ into the integrand (by taking the summation over all pairs i and j with $i < j$).

From relation (5.101) it is clear that we may use either $\langle \psi \rangle_0$ or $\langle G \rangle_0$ as a convenient measure of the SOW. In ordinary ice $\langle \psi \rangle_0$ attains its maximum value of 4 [or $\langle G \rangle_0$ equal to $2N$]. This is considered to be an upper limit to the structure of liquid water. On the other hand, in a completely random orientation, say at a very high temperature and pressure, $\langle \psi \rangle_0$ is expected to reach its lowest value of zero. At any intermediate state we can use $\langle \psi \rangle_0$ as a measure of the degree of the structure of the system relative to these two extreme cases.

Perhaps it is worth noting that the function $G(\mathbf{X}_i, \mathbf{X}_j)$ has been introduced in (5.97) as a part of the pair potential of water. However, once we have such a function with the property described in (5.98), the definition of the SOW as given in (5.102) is not dependent on any assumption on the total potential energy of the system. In this sense the definition (5.102) is quite general and depends only on our geometrical criterion according to which we determine whether a pair is hydrogen-bonded or not.

Although we have not used the mixture model approach in the construction of our definition, it is instructive to show that the same result (5.102) may be obtained through the (exact) formalism of the mixture model approach.

To show that, we define five species in our system. Molecules engaged in K hydrogen bonds are referred to as K-cules. The average number (in the T, P, N ensemble) of K-cules is

$$\langle N_K \rangle = \int dV \int d\mathbf{X}^N P(\mathbf{X}^N, V) \sum_{i=1}^{N} \delta[\psi_i(\mathbf{X}^N) - K] \qquad (5.103)$$

where $\delta(A - B)$ is the Kronecker delta function

$$\delta(A - B) = \begin{cases} 1 & \text{if } A = B \\ 0 & \text{if } A \neq B \end{cases} \qquad (5.104)$$

The summation in the integrand of (5.103) means that we scan through the system, and each molecule that participates in exactly K hydrogen bonds contributes unity to the sum. Hence $\sum_{i=1}^{N} \delta[\psi_i(\mathbf{X}^N) - K]$ "counts" the number of molecules in the system, each of which is engaged in K hydrogen bonds. $\langle N_K \rangle$ is thus the average number of such molecules. We now define the mole fraction of such molecules by

$$x_K = \langle N_K \rangle / N \qquad (5.105)$$

Thus the vector $(x_0, x_1, x_2, x_3, x_4)$ defines the "composition" of the system when viewed as a mixture of five species. This is the exact version of the mixture model approach to the theory of liquid water.

Clearly, we may define the SOW through the mixture model approach by the quantity $N \sum_{K=0}^{4} K x_K / 2$, which can easily be shown to be equal to the quantity $\langle G \rangle_0$ defined in (5.102). It is worth noting that in earlier theories of water two-component mixture models were commonplace, in which case one assumed the existence of a nonbonded and a fully hydrogen-bonded species. Hence the above measure of the SOW reduces to

$$N \sum_{K=0}^{4} \frac{K x_K}{2} \approx \frac{N}{2} [4 x_4 + 0 \cdot x_0] = 2N x_4 \qquad (5.106)$$

i.e., the concentration of the "icelike" form is the measure of the SOW.

Once we have agreed upon a definition of the structure of water we can turn to the following questions that are relevant to the process of HI (or to any process occurring in aqueous solutions). (1) For any specific process, say the dimerization of two simple solutes, how much structural change is involved in the solvent? (2) Knowing the effect of the process on the SOW, to what extent do these structural changes affect the thermodynamics of the process?

The next sections will be devoted to an elaboration of the answers to these questions. We shall see that in general exact answers cannot be given. Therefore, we shall appeal in Section 5.10 to a simple solvable model where exact answers may be given to these questions.

5.8. HOW MUCH STRUCTURAL CHANGE IN THE SOLVENT?

The first question raised at the end of the previous section has a very long history. It is sufficient to mention the terms "structure breaker" and "structure promoter" solutes to remind those who are familiar with properties of aqueous solutions of an immense literature on this subject. Yet, in spite of the large number of articles written on it, this question is still unanswered. In most cases only a qualitative definition of the structure of water has been adopted, and experimental observations were interpreted, in a very speculative manner, in terms of the structural changes in the solvent.

We shall present here one answer to this question which makes use of the definition of the SOW given in the previous section. The answer is evidently a very approximate one. However, having no better one at present, we feel that its presentation here is well justified. We shall deal with two simple processes: one is the transfer of a simple solute from a fixed position in the gas to a fixed position in the pure water. The second is the HI process, namely, the process of bringing two simple solutes from infinite separation to a small separation R. (All processes are carried out at a constant temperature T and pressure P.)

We start with the simplest process, i.e., the transfer of a solute A from the gas to the liquid. The standard free energy of solution is

$$\Delta\mu_A{}^\circ = -kT \ln\langle \exp(-\beta B_A)\rangle_0 \tag{5.107}$$

where the average is taken in the T, P, N ensemble. We now assume that the total potential energy of the system is pairwise additive. Namely, for pure water we write

$$U(\mathbf{X}^N) = \sum_{i<j} U(\mathbf{X}_i, \mathbf{X}_j) \tag{5.108}$$

We assume that each pair potential has the form (5.97) which we rewrite as

$$U(\mathbf{X}_i, \mathbf{X}_j) = U^1(\mathbf{X}_i, \mathbf{X}_j) + \varepsilon_{\mathrm{HB}}G(\mathbf{X}_i, \mathbf{X}_j) \tag{5.109}$$

Hence the total potential energy is written as

$$U(\mathbf{X}^N) = U^1(\mathbf{X}^N) + \varepsilon_{HB}G(\mathbf{X}^N) \qquad (5.110)$$

where

$$G(\mathbf{X}^N) = \sum_{i<j} G(\mathbf{X}_i, \mathbf{X}_j) \qquad (5.111)$$

Thus in $U^1(\mathbf{X}^N)$ we have lumped together the repulsive and the electrostatic interactions among all the molecules.

We now take the derivative of $\Delta\mu_A{}^\circ$ with respect to ε_{HB} to obtain

$$\left(\frac{\partial\Delta\mu_A{}^\circ}{\partial\varepsilon_{HB}}\right)_{T,P} = \int dV \int d\mathbf{X}^N P(\mathbf{X}^N, V/\mathbf{R}_A)G(\mathbf{X}^N)$$

$$- \int dV \int d\mathbf{X}^N P^\circ(\mathbf{X}^N, V)G(\mathbf{X}^N)$$

$$= \langle G \rangle_A - \langle G \rangle_0 \qquad (5.112)$$

The quantity on the right-hand side is exactly the change in the average structure of the water induced by the process of placing a solute at a fixed position $\mathbf{R}_A$. As we noted before, the particular point $\mathbf{R}_A$ is of no importance since we are dealing with a homogeneous fluid. However, it is important to stress that the change of the structure corresponds to the process of inserting A at a *fixed* position. The change in the SOW might be different if we simply add a (free) solute A to the liquid.

Within the assumptions made in this section Equation (5.112) is an exact relation between the structural changes in the solvent induced by adding a single solute to a fixed position and the derivative of the standard free energy of solution of A with respect to the parameter ε_{HB}.

We now turn to an approximate relation based on (5.112). Suppose we view light and heavy water as two liquids that fulfill the assumptions made above. Furthermore, we assume that the two liquids have the same pair potential as (5.109) except that the hydrogen bond energy is slightly stronger in D_2O compared to H_2O, i.e.,

$$\varepsilon_{HB}(D_2O) < \varepsilon_{HB}(H_2O) \qquad (5.113)$$

We also assume that the difference $\varepsilon_{HB}(D_2O) - \varepsilon_{HB}(H_2O)$ is small compared with the values of the hydrogen bond energies themselves. In such a case we may write the first-order expansion of the standard free energy of solution as

$$\Delta\mu_A{}^\circ(D_2O) - \Delta\mu_A{}^\circ(H_2O) = [\varepsilon_{HB}(D_2O) - \varepsilon_{HB}(H_2O)](\langle G \rangle_A - \langle G \rangle_0) \qquad (5.114)$$

This relation may be used in two different ways. First, suppose we have two liquids that fulfill all the assumptions made above, and for which $\varepsilon_{HB}(D_2O) - \varepsilon_{HB}(H_2O) < 0$. Then from the experimental measurements of $\Delta\mu_A{}^\circ$ in these two liquids, we can estimate the structural change that is induced by the addition of A to a fixed position in H_2O. Some numerical results of this kind are given below.

The second possible application of the same relation is the following. Suppose that we know from some other sources that addition of A to H_2O causes an increase in the structure of water, i.e., it is given that $\langle G \rangle_A - \langle G \rangle_0 > 0$. Then we can predict that $\Delta\mu_A{}^\circ(D_2O) < \Delta\mu_A{}^\circ(H_2O)$, or equivalently the solubility of A is larger in D_2O than in H_2O. This kind of reasoning may be used to resolve the puzzling finding that the "phobia" of, say, methane, for H_2O is greater than for D_2O (see Section 2.3).

As an illustration of the application of relation (5.114) we take the values of

$$\varepsilon_{HB}(H_2O) = -3.57 \text{ kcal/mol}$$
$$\varepsilon_{HB}(D_2O) = -3.80 \text{ kcal/mol}$$

(5.115)

as estimated by Némethy and Scheraga (1964). We use the experimental results of the standard free energies of solution of argon, methane, and ethane to estimate the structural change in the solvent induced by placing such solutes at a fixed position in the liquid. Table 5.11 presents such numerical results.

The values shown in this table are all positive. This is consistent with similar conclusions that have been reached by many other authors, i.e., these solutes increase the structure of the solvent, and therefore may justifiably be referred to as "structure makers" or "structure promoters."

Table 5.11

Values of $\Delta\langle G \rangle_A \equiv \langle G \rangle_A - \langle G \rangle_0$ for Argon, Methane, and Ethane in Water at Different Temperatures[a]

t (°C)	5	10	15	20	25
Argon	0.28	0.25	0.23	0.22	0.20
Methane	0.21	0.19	0.17	0.15	0.13
Ethane	0.19	0.14	0.12	0.12	0.14

[a] From Ben-Naim (1975).

It should be stressed, however, that because of the extreme approximation that has been used, the exact magnitude of these results should not be taken too seriously.

Next we proceed to estimate how much structural change is involved in the process of HI. The only process for which we may make such an estimation is the one discussed in Section 3.3. Namely, the one for which the indirect part of the free energy change is given by

$$\delta G^{\mathrm{HI}}(\sigma_1) = \Delta \mu_{\mathrm{Et}}^{\circ} - 2\Delta \mu_{\mathrm{Me}}^{\circ} \tag{5.116}$$

where $\Delta \mu_{\mathrm{Et}}^{\circ}$ and $2\Delta \mu_{\mathrm{Me}}^{\circ}$ are the standard free energies of solution of ethane and methane, respectively. The corresponding isotope effect on the strength of the HI, as measured by the quantity (5.116), is given by

$$\delta G^{\mathrm{HI}}(D_2O) - \delta G^{\mathrm{HI}}(H_2O)$$
$$= \Delta \mu_{\mathrm{Et}}^{\circ}(D_2O) - \Delta \mu_{\mathrm{Et}}^{\circ}(H_2O) - 2[\Delta \mu_{\mathrm{Me}}^{\circ}(D_2O) - \Delta \mu_{\mathrm{Me}}^{\circ}(H_2O)]$$
$$\cong [\varepsilon_{\mathrm{HB}}(D_2O) - \varepsilon_{\mathrm{HB}}(H_2O)]\{\langle G \rangle_{\mathrm{Et}} - \langle G \rangle_0 - 2[\langle G \rangle_{\mathrm{Me}} - \langle G \rangle_0]\} \tag{5.117}$$

[Note that the letter G is used with two different meanings in (5.117).]

Let us denote by $\langle G \rangle_{\mathrm{2Me}} - \langle G \rangle_0$ the structural change in the solvent induced by the process of placing *two* methane molecules at *fixed* positions and at infinite separation from each other. Since the structural change occurs in the close vicinity of each of the solutes, the total amount of structural change produced by this process is exactly twice the structural change produced by placing *one* methane molecule at a fixed position in the liquid. Hence we have

$$\langle G \rangle_{\mathrm{2Me}} - \langle G \rangle_0 = 2[\langle G \rangle_{\mathrm{Me}} - \langle G \rangle_0] \tag{5.118}$$

Substituting in (5.117) we obtain

$$\delta G^{\mathrm{HI}}(D_2O) - \delta G^{\mathrm{HI}}(H_2O)$$
$$\approx [\varepsilon_{\mathrm{HB}}(D_2O) - \varepsilon_{\mathrm{HB}}(H_2O)][\langle G \rangle_{\mathrm{Et}} - \langle G \rangle_{\mathrm{2Me}}] \tag{5.119}$$

Thus from the isotope effect on the quantity δG^{HI} we may estimate the structural change induced by the process of bringing two methane molecules from infinite separation to a small separation $R \approx \sigma_1$ (see Section 3.3 for details).

Table 5.12 presents some values of $\langle G \rangle_{\mathrm{Et}} - \langle G \rangle_{\mathrm{2Me}}$ estimated from relation (5.119) [using the values in (5.115) for the hydrogen-bond parameters for H_2O and D_2O].

We see that the HI process induces a net breakdown of the SOW. This is in agreement with the qualitative conclusions that were reached before (Némethy and Scheraga, 1962; Ben-Naim, 1974). It must be stressed

Table 5.12

The Amount of Structural Change Induced by the Process of Bringing Two Methane Molecules to a Separation $R = \sigma_1 = 1.53$ Å from Fixed Positions at Infinite Separation[a]

t (°C)	5	10	15	20	25
$\langle G \rangle_{\text{Et}} - \langle G \rangle_{\text{2Me}}$	−0.23	−0.23	−0.22	−0.18	−0.12

[a] From Ben-Naim (1975).

again that the values given in Table 5.12 are based on a very approximate procedure of computation, and that they are relevant to one particular process as described above. Nevertheless, we believe that the *sign* of these results is correct.

The following intuitively appealing argument is often expressed in the literature. We know that the HI in water is stronger than in ethanol. We also attribute this phenomenon to the peculiar structure of water. If we also believe the D_2O is a more "structured" solvent, then we should expect that the HI in D_2O will be stronger than in H_2O. This reasoning is incorrect, however, since we do not know how the HI changes with the degree of structure of the solvent. In fact, from relation (5.119) it follows that if the process of HI induces a net structural breakdown, and if $[\varepsilon_{\text{HB}}(D_2O) - \varepsilon_{\text{HA}}(H_2O)] < 0$, which is equivalent to the statement that D_2O is more structured than H_2O (see below), then it follows that the HI in D_2O will be weaker than in H_2O. The same kind of reasoning may be applied to the extent of "phobia" of simple solutes for H_2O and D_2O. Note also that all the above relations are valid for small deviations from the structure of pure water, and that these relations may not be extended to discuss the different behavior of water as compared with, say, ethanol.

We conclude this section by noting that within the assumption of this section, the parameter ε_{HB} may serve as an indicator to the degree of structure of water. To show this, we differentiate $\langle G \rangle_0$ as defined in Section 5.7 to obtain

$$\langle G \rangle_0 = \frac{\int dV \int d\mathbf{X}^N \exp[-\beta U(\mathbf{X}^N) - \beta PV]G(\mathbf{X}^N)}{\int dV \int d\mathbf{X}^N \exp[-\beta U(\mathbf{X}^N) - \beta PV]} \qquad (5.120)$$

$$\left[\frac{\partial \langle G \rangle_0}{\partial (-\varepsilon_{\text{HB}})} \right]_{T,P,N} = \beta(\langle G^2 \rangle_0 - \langle G \rangle_0^2)$$
$$= \beta[\langle (\langle G \rangle_0 - G)^2 \rangle_0] \geq 0 \qquad (5.121)$$

which means that the average structure of the system, as measured by the quantity $\langle G \rangle_0$, is a monotonic increasing function of the positive parameter $(-\varepsilon_{\text{HB}})$. Applying this result to H_2O and D_2O, we may conclude from (5.115) that D_2O is more structured than H_2O. This conclusion has been reached by many authors, using other more qualitative arguments. We may now use (5.121) to rewrite (5.112) in a different form:

$$
\left(\frac{\partial \Delta \mu_A{}^\circ}{\partial \langle G \rangle_0} \right)_{T,P,N} = \left[\frac{\partial \Delta \mu_A{}^\circ}{\partial (-\varepsilon_{\text{HB}})} \right]_{T,P,N} \left[\frac{\partial (-\varepsilon_{\text{HB}})}{\partial \langle G \rangle_0} \right]_{T,P,N}
$$

$$
= \frac{-kT(\langle G \rangle_A - \langle G \rangle_0)}{\langle (\langle G \rangle_0 - G)^2 \rangle_0} \tag{5.122}
$$

The last equality is an interesting one: we recall that it is based on relations (5.112) and (5.121), which depend on the definition of the parameter ε_{HB} introduced in (5.109). However, in (5.122) we have eliminated the explicit dependence on ε_{HB}. Therefore, we may estimate the changes in $\Delta \mu_A{}^\circ$ by considering directly the change in the structure of the system. As an example, if we know that a solute causes an increase in the SOW, in the sense that $\langle G \rangle_A - \langle G \rangle_0$ is positive, then we can predict that an increase in the structure would lead to a decrease in $\Delta \mu_A{}^\circ$, or increase in the solubility of A. We may hope that some kind of a general relation of the type (5.122) might be derived for other definitions of the SOW which do not depend on the drastic approximations used in establishing this particular relation.

The analog of (5.122) for the HI process is

$$
\left[\frac{\partial \delta G^{\text{HI}}(\sigma_1)}{\partial \langle G \rangle_0} \right]_{T,P,N} = \left(\frac{\partial \Delta \mu_{\text{Et}}^\circ}{\partial \langle G \rangle_0} \right)_{T,P,N} - 2 \left(\frac{\partial \Delta \mu_{\text{Me}}^\circ}{\partial \langle G \rangle_0} \right)_{T,P,N}
$$

$$
= \frac{-kT(\langle G \rangle_{\text{Et}} - \langle G \rangle_{2\text{Me}})}{\langle (\langle G \rangle_0 - G)^2 \rangle_0} \tag{5.123}
$$

Thus, if we know that the HI process causes a net breakdown of the structure, i.e., $\langle G \rangle_{\text{Et}} - \langle G \rangle_{2\text{Me}} < 0$, then we conclude from (5.123) that the strength of the HI will decrease with an increase in the structure of the water. This conclusion is at least consistent with some experimental findings discussed in Chapters 2 and 3 of this book.

Comment

This section is based on a very crude approximation of the total interaction energy among water molecules. Therefore, the precise numerical figures are not of great significance. However, we believe that the *signs* of the structural changes induced by the processes of solubility and HI are correct. Effort expended in improving these estimates is well justified since our understanding of the hydrophobic hydration and interaction phenomena might crucially depend on our understanding of the structural changes in the solvent induced by these processes.

5.9. EFFECT OF STRUCTURAL CHANGES IN THE SOLVENT (SCIS) ON THE THERMODYNAMICS OF SOLUBILITY AND HI

In Section 5.6 we derived formal statistical mechanical expressions for various thermodynamic quantities associated with the process of dissolution and HI. We now turn to reinterpreting the same results from a different point of view. The motivation for doing so stems from the historical recognition of the importance of the SOW in determining the outstanding properties of aqueous solutions.

We start with a thermodynamic approach to the problem. Let N_K represent the average number of water molecules engaged in K hydrogen bonds. (Here we use the simpler notation N_K for $\langle N_K \rangle$ defined in Section 5.7.) For any given system of N_A solute molecules and N solvent molecules the chemical potential of the solute is given by

$$\mu_A = \left(\frac{\partial G}{\partial N_A}\right)_{T,P,N} \tag{5.124}$$

However, the same system may be viewed as a mixture of six species with composition $N_A, N_0, \ldots, N_4$. With this set of variables we may rewrite the various partial molar quantities of A as follows:

$$\mu_A = \left(\frac{\partial G}{\partial N_A}\right)_{eq} = \left(\frac{\partial G}{\partial N_A}\right)_{N_0,\cdots,N_4} + \sum_{i=0}^{4}\left(\frac{\partial G}{\partial N_i}\right)\left(\frac{\partial N_i}{\partial N_A}\right) \tag{5.125}$$

$$\bar{S}_A = \left(\frac{\partial S}{\partial N_A}\right)_{eq} = \left(\frac{\partial S}{\partial N_A}\right)_{N_0,\cdots,N_4} + \sum_{i=0}^{4}\left(\frac{\partial S}{\partial N_i}\right)\left(\frac{\partial N_i}{\partial N_A}\right) \tag{5.126}$$

$$\bar{V}_A = \left(\frac{\partial V}{\partial N_A}\right)_{eq} = \left(\frac{\partial V}{\partial N_A}\right)_{N_0,\cdots,N_4} + \sum_{i=0}^{4}\left(\frac{\partial V}{\partial N_i}\right)\left(\frac{\partial N_i}{\partial N_A}\right) \tag{5.127}$$

In each of these equations the first equality is the thermodynamic definition of the partial molar quantity. This is indicated by "eq" to stress that "chemical" equilibrium among all the species of the solvent is maintained. The second equality employs the new set of variables T, P, N_A, N_0, $\ldots$, N_4. If we view the vector $(N_0, \ldots, N_4)$ as representing the "structure of the water" in the sense discussed in Section 5.7, then we may say that the derivatives of the thermodynamic functions at constant $(N_0, \ldots, N_4)$ are also derivatives at constant *structure* of the system. The sum over i on the right-hand side of (5.125), (5.126), and (5.127) will represent the contribution of structural changes in the solvent to the corresponding partial molar quantities. In other words, each of the partial molar quantities may be viewed as consisting of two parts. First there is the contribution due to insertion of A into the system keeping the vector $(N_0, \ldots, N_4)$ fixed. In this part the structure of the system is kept constant. The second contribution involves structural changes in the solvent which in turn carries with it some change in the corresponding thermodynamic quantity.

Among all the partial molar quantities of the solute, the chemical potential has a unique behavior. This follows from the condition of chemical equilibrium among all the species in the system, i.e.,

$$\mu_0 = \mu_1 = \mu_2 = \mu_3 = \mu_4 \tag{5.128}$$

which when applied to the second term on the right-hand side of (5.125) yields

$$\sum_{i=0}^{4} \mu_i \frac{\partial N_i}{\partial N_A} = \mu_0 \sum_{i=0}^{4} \frac{\partial N_i}{\partial N_A} = \mu_0 \frac{\partial}{\partial N_A}\left(\sum_{i=0}^{4} N_i\right) = 0 \tag{5.129}$$

The last equality follows from the conservation of the total number of solvent species in the system. Thus we see that structural changes in the solvent, in the above sense, do not contribute to the chemical potential of the solute A. On the other hand, all other partial molar quantities may be affected by the structural changes in the solvent. It is of particular interest to note that the contribution to the partial entropy and the enthalpy of a solute A compensate each other in the following sense:

For each species we write

$$\mu_i = \bar{H}_i - T\bar{S}_i, \qquad i = 0, \ldots, 4 \tag{5.130}$$

Substituting in (5.129) we obtain

$$\sum_{i=0}^{4} (\bar{H}_i - T\bar{S}_i)\left(\frac{\partial N_i}{\partial N_A}\right) = 0 \tag{5.131}$$

or equivalently

$$\sum_{i=0}^{4} \bar{H}_i \left(\frac{\partial N_i}{\partial N_A} \right) = T \sum_{i=0}^{4} \bar{S}_i \left(\frac{\partial N_i}{\partial N_A} \right) \tag{5.132}$$

where on the left-hand side we have the contribution of the SCIS to $\bar{H}_A$, and on the right-hand side the contribution of the SCIS to $T\bar{S}_A$. We shall see below that this compensation is also valid in a more general sense.

We now rewrite equations (5.125) and (5.126) in a more compact form:

$$\mu_A = \mu_A{}^* \tag{5.133}$$

$$\bar{S}_A = \bar{S}_A{}^* + \sum_{i=0}^{4} \bar{S}_i \left(\frac{\partial N_i}{\partial N_A} \right) \tag{5.134}$$

where the asterisk indicates a partial molar quantity evaluated in a system in which the equilibrium among all the species is "frozen in." The second term on the right-hand side of (5.134) may be referred to as the relaxation term, i.e., this is the contribution to $\bar{S}_A$ due to SCIS, induced by the addition of dN_A. It is instructive at this point to demonstrate that the contribution due to SCIS as written in (5.134) may also be expressed in terms of the SCIS caused by changing the temperature. This may be easily obtained by taking the temperature derivative of the chemical potential:

$$-\bar{S}_A = \left(\frac{\partial \mu_A}{\partial T} \right)_{\text{eq}} = \left(\frac{\partial \mu_A}{\partial T} \right)^* + \sum_{i=0}^{4} \left(\frac{\partial \mu_A}{\partial N_i} \right) \left(\frac{\partial N_i}{\partial T} \right) \tag{5.135}$$

Note that $\bar{S}_A$ is obtained from μ_A as a derivative along the "equilibrium line" among all the species. We now show that the second term on the right-hand side of (5.135) is identical with the second term on the right-hand side of (5.134). To prove that, we use the following two derivatives of the chemical potential in the equilibrated and in the "frozen-in" systems:

$$\left(\frac{\partial \mu_i}{\partial N_A} \right)_{\text{eq}} = \left(\frac{\partial \mu_i}{\partial N_A} \right)^* + \sum_{j} \left(\frac{\partial \mu_i}{\partial N_j} \right)^* \left(\frac{\partial N_j}{\partial N_A} \right) \qquad \text{for } i = 0, \ldots, 4 \tag{5.136}$$

$$\left(\frac{\partial \mu_j}{\partial T} \right)_{\text{eq}} = \left(\frac{\partial \mu_j}{\partial T} \right)^* + \sum_{i} \left(\frac{\partial \mu_j}{\partial N_i} \right)^* \left(\frac{\partial N_i}{\partial T} \right) \qquad \text{for } j = 0, \ldots, 4 \tag{5.137}$$

Note that because of the equilibrium condition (5.128), the value of the derivative on the left-hand side of (5.136) is independent of the index i. The same comment applies to the index j in (5.137).

We start from the second term on the right-hand side of (5.135), and by standard manipulation, using also the identity (5.129), we obtain

$$\sum_{i=0}^{4}\left(\frac{\partial \mu_A}{\partial N_i}\right)^{*}\left(\frac{\partial N_i}{\partial T}\right) = \sum_{i}\left(\frac{\partial \mu_i}{\partial N_A}\right)^{*}\left(\frac{\partial N_i}{\partial T}\right)$$

$$= \sum_{i}\left[\left(\frac{\partial \mu_i}{\partial N_A}\right)_{eq} - \sum_{j}\left(\frac{\partial \mu_i}{\partial N_j}\right)^{*}\left(\frac{\partial N_j}{\partial N_A}\right)\right]\left(\frac{\partial N_i}{\partial T}\right)$$

$$= -\sum_{i}\sum_{j}\left(\frac{\partial \mu_i}{\partial N_j}\right)^{*}\left(\frac{\partial N_j}{\partial N_A}\right)\left(\frac{\partial N_i}{\partial T}\right)$$

$$= -\sum_{j}\left(\frac{\partial N_j}{\partial N_A}\right)\sum_{i}\left(\frac{\partial \mu_j}{\partial N_i}\right)^{*}\left(\frac{\partial N_i}{\partial T}\right)$$

$$= -\sum_{j}\left(\frac{\partial N_j}{\partial N_A}\right)\left[\left(\frac{\partial \mu_j}{\partial T}\right)_{eq} - \left(\frac{\partial \mu_j}{\partial T}\right)^{*}\right]$$

$$= \sum_{j}\left(\frac{\partial \mu_j}{\partial T}\right)^{*}\left(\frac{\partial N_j}{\partial N_A}\right)$$

$$= -\sum_{j}\bar{S}_j \frac{\partial N_j}{\partial N_A} \tag{5.138}$$

Thus the identification of the two relaxation terms in (5.134) and in (5.135) has been proved.

Two comments are now in order. First, in writing equations (5.125), (5.126), and (5.127) we have not restricted ourselves to dilute solutions. However, care must be exercised in the identification of the SCIS that are involved in each system. For instance, if we start from a system described by the variables $(T, P, N_A, N_0, \ldots, N_4)$ the "structure" of this system is different from the "structure" of pure water (i.e., when $N_A = 0$). However, the SCIS that we were concerned with is the change in the structure brought about by the addition of dN_A molecules (or moles) of A to the system. In other words, we were interested in the infinitesimal SCIS for the process

$$(T, P, N_A, N_0, \ldots, N_4) \rightarrow (T, P, N_A + dN_A, N_0', \ldots, N_4') \tag{5.139}$$

Secondly, all the above considerations may be applied to the case when we add one solute A to the pure solvent. In this case we obtain the partial molar (or molecular) quantities of A in the liquid ideal dilute solutions. Furthermore, we could have added the solute A with or without restricting it to a fixed position. In each case the extent of SCIS might be different, but the general conclusion stated after equation (5.129) is valid for both cases. This comment permits us to apply the same conclusion to any of the *standard* thermodynamic quantities of transferring A between two phases. Thus $\Delta\mu_A{}^\circ$ will gain no contribution from SCIS whereas $\Delta S_A{}^\circ$ may, in principle, gain such a contribution. (For an explicit example, see Section 5.10.)

With the last comment in mind we may proceed to discuss the thermodynamics of the HI process, or any other standard process in aqueous solution. These may be expressed in terms of standard thermodynamics of solution. Hence, the same conclusion stated after Equation (5.129) is also valid for these processes. For instance, we may write for the HI process the two exact relations

$$\delta G^{\text{HI}} = \Delta\mu_D{}^\circ - 2\Delta\mu_M{}^\circ \tag{5.140}$$

$$\delta S^{\text{HI}} = \Delta S_D{}^\circ(L) - 2\Delta S_M{}^\circ(L) \tag{5.141}$$

where D and M denote the "dimer" and the monomer, respectively, and we stress that in (5.141) one must use the *local* standard entropy of solution of D and of M.

Since δG^{HI}, δS^{HI}, δH^{HI}, etc. are simple combinations of standard thermodynamic quantities of solution we may conclude that SCIS induced by the process of HI will not have any effect on δG^{HI} whereas they may affect all other thermodynamic quantities associated with this process.

Before generalizing the above conclusion to include other possible definitions of the SOW, it is appropriate to present a concrete example and cite a very common conclusion which contains a false argument.

Consider the process of bringing two methane molecules from fixed positions, but at infinite separation, to the small separation $R = \sigma_1 = 1.53$ Å. For this process we have the following thermodynamic results (see Section 3.3):

$$\delta G^{\text{HI}} = -1.99 \text{ kcal/mol} \tag{5.142}$$

$$T\delta S^{\text{HI}} = 3.39 \text{ kcal/mol} \tag{5.143}$$

$$\delta H^{\text{HI}} = 1.40 \text{ kcal/mol} \tag{5.144}$$

It is clear from the general relation

$$\delta G^{\text{HI}} = \delta H^{\text{HI}} - T\delta S^{\text{HI}} \tag{5.145}$$

that $T\delta S^{\text{HI}}$ dominates the right-hand side of (5.145). Hence, one correctly concludes that the process is "entropy driven." This simply means that the absolute magnitude of $T\delta S^{\text{HI}}$ is large compared with δH^{HI}.

Next, one interprets the large and positive value of $T\delta S^{\mathrm{HI}}$ as arising from the breakdown of the SOW induced by the process. Although this is a qualitative statement, it is probably correct, and we have indeed estimated in Section 5.8 the magnitude of the SCIS for such a process.

From the above consideration it is very tempting to conclude that the large and negative value of δG^{HI} in water is explainable in terms of SCIS. However, the flaw in this argument is that whatever the contribution of the SCIS to $T\delta S^{\mathrm{HI}}$, exactly the same contribution is also contained in δH^{HI}. Hence, when we construct the combination (5.145), the two contributions compensate each other.

We have demonstrated the compensation effect with the help of the (exact) mixture model approach to liquid water. We now proceed to show that the same conclusion may also be reached for the definition of the SOW as presented in Section 5.7.

To do that we use expression (5.110) in the general expression for $\Delta H_A°(L)$ and $\Delta S_A°(L)$ of Section 5.6 [see (5.87) and (5.88)]:

$$\langle U(N)\rangle_A - \langle U(N)\rangle_0 = \langle U^1(N)\rangle_A - \langle U^1(N)\rangle_0$$
$$+ \varepsilon_{\mathrm{HB}}[\langle G\rangle_A - \langle G\rangle_0] \qquad (5.146)$$

where the second term on the right-hand side of (5.146) is identified as the contribution of the conventional SCIS to the enthalpy of solution of the solute A. Similarly, the same term divided by T appears in the expression for $\Delta S_A°(L)$. Hence we conclude that SCIS, in the sense of the definition presented in Section 5.7, do not contribute either to $\Delta \mu_A°$ or to δG^{HI}. This statement may be somewhat generalized to include other possible definitions of the SOW. For more details the reader is referred to Ben-Naim (1974, 1978b).

A few concluding comments are now in order. We have discussed in this section the overall SCIS that is induced by a process. We did not examine the question of the locality of the SCIS. It is, however, clear that the structure of the solvent is affected in the close vicinity of the solutes involved. The solvent molecules at a large distance from the center of the solute would not "feel" the effect of the solute; therefore the local structure in those regions will be the same as in a pure solvent.

Another point that should be clarified is the distinction between the "structural changes in the solvent" and the "structure of the solvent around the solute." This section has been concerned with the former concept only. It is only with this concept in mind that we have constructed the definition of the SOW—a quantity that may be given a numerical value, both for pure water and for aqueous solutions. The concept of the "structure of the solvent around the solute" is clearly relevant to both $\Delta\mu_A{}^\circ$ and δG^{HI}. Perhaps the simplest way of seeing that is through the well-known expression for $\Delta\mu_A{}^\circ$

$$\Delta\mu_A{}^\circ = \varrho_W \int_0^1 d\xi \int d\mathbf{X}_W U(\mathbf{R}_A, \mathbf{X}_W) g_{AW}(\mathbf{R}_A, \mathbf{X}_W, \xi) \qquad (5.147)$$

where ξ is the coupling parameter and g_{AW} is the solute–solvent pair correlation function. [For more details, see Ben-Naim (1974).] If we consider g_{AW} as conveying information on the "structure of the solvent around the solute," then it is clear that the change of this function, as we increase the coupling parameter ξ, will determine the value of $\Delta\mu_A{}^\circ$. However, this kind of reasoning is distinctly different from the SCIS that we were concerned with in this section, i.e., with a measure that may be assigned numerical values for any two states of the system.

5.10. A SIMPLE SOLVABLE MODEL

In the previous section we have discussed, in very general terms, the relation between some thermodynamic quantities associated with processes in aqueous solutions and the structural changes in the solvent (SCIS) induced by these processes. Except for the very approximate calculation made in Section 5.8, it is very difficult to estimate the amount of SCIS involved even in the simplest processes. This is a very typical situation in many problems in the field of solution thermodynamics.

For this reason it is sometimes very useful to study simplified models, which, though caricatures of the real system, do contain some features in common with the real systems. We devote this last section to an elaboration of such a model. The main virtue of this section is a didactic one. It shows how one can construct very simple models that contain the most essential elements of the problem that we are concerned with. Yet at the same time they are simple enough that an exact statistical mechanical analysis of its behavior is feasible. In this model every question pertaining to the equilibrium properties of the system may be given an exact answer.

In a sense what we do is to "extract" our problem from the complex system where it has been raised, and to embed it in a simple model where it can be solved. After having solved the problem one must make a clear-cut distinction between those results that are valid only for the particular model, and those that have more general validity. [A similar treatment of the partial molar heat capacities of solutes in water has been presented by Ben-Naim (1970).]

More specifically, our attention will be focused on the interrelation between the thermodynamics of certain processes and the analog of the SCIS in the system. In addition, the study of this kind of a model is helpful in gaining a deeper insight into the content and meaning of some thermodynamic quantities.

The physical model that will be described and solved in the following sections contains the following ingredients. We have two processes, analogs to the solubility and the HI that occur in a "solvent," which may respond to the process by changing its "structure." We first solve for the partition function of our system without explicit reference to the "structure" of the system. Next we repeat the same calculation, but now from the point of view of the mixture model approach. In doing so, we illustrate the equivalence between the two approaches to the same system. [This topic is discussed in more general terms in Ben-Naim (1974).] The two main questions that are of interest to us in this chapter and that were raised at the end of Section 5.7 are discussed at great length, i.e., how much SCIS is involved in a certain process and how these SCIS affect the thermodynamics of the processes. We shall also demonstrate a general relation between the probability of an event and the free energy change for the creation of such an event. This relation has been cited, without proof, in Section 1.4.

Finally, a word of caution is perhaps in order. The model worked out below is neither a model for water nor for aqueous solutions. It is specifically designed for the study of the questions raised above, and nothing more.

5.10.1. The Model and Its Solution

Consider a system of M independent, identical, and localized adsorbing sites, each of which may attain one of two states L or H with corresponding energy levels E_L and E_H, and we assume that $E_L < E_H$. Each site can adsorb a single gas molecule G; the adsorption energies are U_L and U_H, according to whether the site is in the state L or H.

Thus in essence we have a system of two states in chemical equilibrium,

$$L \rightleftharpoons H \qquad (5.148)$$

and a third agent G interacts with both L and H in the most elementary fashion. As a result of this interaction a shift in the equilibrium composition of L and H may be expected upon adsorption of G. This model may be viewed as a prototype of the equilibrium between two states of a molecule (say, helix-coil transformation in biopolymers) and we are interested in the effect of an interacting agent G on this equilibrium.

We have chosen localized sites to simplify slightly the mathematical treatment (i.e., we do not consider any translational or rotational degrees of freedom of the sites), but this assumption is not essential for the results of this section.

Let us use the notation

$$Q_L = \exp(-\beta E_L), \qquad Q_H = \exp(-\beta E_H) \tag{5.149}$$

where $\beta = (kT)^{-1}$, with k the Boltzmann constant and T the absolute temperature. The canonical partition function of the empty system, i.e., of M sites at temperature T, is (Hill, 1960)

$$Q(M, T) = (Q_L + Q_H)^M = \sum_{M_L=0}^{M} \frac{M!}{M_L! \, M_H!} Q_L^{M_L} Q_H^{M_H}$$

$$= \sum_{M_L} Q^*(M_L, M_H, T) \tag{5.150}$$

where M_L and M_H are the number of sites in the L and H states, respectively, and $Q^*(M_L, M_H, T)$ may be interpreted as the partition function of the same system with fixed values of M_L and M_H.

The equilibrium concentrations of L and M are obtained from the condition

$$\frac{\partial \ln Q^*}{\partial M_L} = 0 \tag{5.151}$$

which leads to the solution

$$\frac{\bar{M}_H}{\bar{M}_L} = \frac{Q_H}{Q_L} = K \tag{5.152}$$

which is the equilibrium condition for the "reaction" (5.148). We use a bar over M_L and M_H to denote the values of M_L and M_H that maximize Q^* in (5.150).

We also introduce the *mole fractions* of sites in the two states by

$$x_L{}^\circ = \frac{\bar{M}_L}{M} = \frac{Q_L}{Q_L + Q_H} = \frac{1}{1 + K}$$

$$x_H{}^\circ = \frac{\bar{M}_H}{M} = \frac{Q_H}{Q_L + Q_H} = \frac{K}{1 + K} \tag{5.153}$$

These are also the probabilities of finding a specific site in the L or H state for the *empty* system.

We next turn to the case where, in addition to the M sites, we also have N adsorbed molecules distributed over the sites. For simplicity we assign no internal degrees of freedom to these molecules: they are characterized solely by their adsorption energies U_L and U_H.

The canonical partition function for such a system is

$$Q(N, M, T) = \sum_{\substack{M_L+M_H=M \\ N_L+N_H=N}} \binom{M}{M_L}\binom{M_L}{N_L}\binom{M_H}{N_H} Q_L{}^{M_L} Q_H{}^{M_H} q_L{}^{N_L} q_H{}^{N_H} \quad (5.154)$$

where we introduced the notation $q_L = \exp(-\beta U_L)$ and $q_H = \exp(-\beta U_H)$ and N_L and N_H are the number of molecules adsorbed on L and H sites, respectively. The summation in (5.154) extends over all possible values of M_L, M_H, N_L, N_H with the obvious conditions

$$M_L + M_H = M, \quad N_L + N_H = N, \quad N_L \leq M_L, \quad N_H \leq M_H \quad (5.155)$$

Clearly the auxiliary parameters N_L, N_H, M_L, M_H serve as intermediate quantities which determine the "energy levels" of the system, namely,

$$E(N_L, N_H, M_L, M_H) = M_L E_L + M_H E_H + N_L U_L + N_H U_H \quad (5.156)$$

The degeneracy of this energy level is given by the product of the three combinatorial factors in (5.154).

The summation in (5.154) cannot be carried out to obtain a closed form of the partition function. This is easily achieved, however, by transforming to an open system with respect to the gas molecules, i.e., we define the grand partition function by

$$\begin{aligned}
\Xi(\lambda, M, T) &= \sum_{N=0}^{M} \lambda^N Q(N, M, T) \\
&= \sum_{M_L=0}^{M} \sum_{N_L=0}^{M_L} \sum_{N_H=0}^{M_H} \lambda^{N_L} \lambda^{N_H} \binom{M}{M_L}\binom{M_L}{N_L}\binom{M_H}{N_H} Q_L{}^{M_L} Q_H{}^{M_H} q_L{}^{N_L} q_H{}^{N_H} \\
&= \sum_{M_L=0}^{M} \binom{M}{M_L} Q_L{}^{M_L}(1 + \lambda q_L)^{M_L} Q_H{}^{(M-M_L)}(1 + \lambda q_H)^{(M-M_L)} \\
&= (Q_L + Q_H + \lambda q_L Q_L + \lambda q_H Q_H)^M = \xi^M \quad (5.157)
\end{aligned}$$

where $\lambda = \exp(\beta\mu)$ is the absolute activity of the gas and μ is its chemical potential. The last form of the partition function is typical of a system of independent sites, where ξ may be viewed as the grand partition function of a single site. The four terms of ξ correspond to the four possible states of a site, i.e., empty L, empty H, occupied L, and occupied H. The corresponding probabilities of finding the system in each of these states are given by

$$P(L, 0) = Q_L/\xi, \qquad P(H, 0) = Q_H/\xi$$
$$P(L, G) = \lambda q_L Q_L/\xi, \qquad P(H, G) = \lambda q_H Q_H/\xi \tag{5.158}$$

For later applications it will also be useful to introduce the conditional probability $P(L/G)$ of finding a site in state L when it is known to be occupied by a gas molecules, i.e.,

$$P(L/G) = \frac{P(L, G)}{P(G)} = \frac{\lambda q_L Q_L/\xi}{(\lambda q_L Q_L + \lambda q_H Q_H)/\xi} = \frac{q_L Q_L}{q_L Q_L + q_H Q_H} \tag{5.159}$$

where $P(G)$ is the probability of finding an occupied site. Similarly, the conditional probability of finding a site in state H, given that it is occupied, is

$$P(H/G) = \frac{q_H Q_H}{q_L Q_L + q_H Q_H} \tag{5.160}$$

The average number of gas molecules in the system is given by the standard relation

$$\bar{N} = \lambda \frac{\partial \ln \Xi}{\partial \lambda} = \frac{\lambda M}{\xi} (q_L Q_L + q_H Q_H) \tag{5.161}$$

Let $x = \bar{N}/M$ be the average fraction of occupied sites. From (5.158) we obtain

$$x = P(L, G) + P(H, G) \tag{5.162}$$

Elimination of λ from (5.161) gives

$$\lambda = \left(\frac{x}{1 - x}\right) \frac{Q_L + Q_H}{q_L Q_L + q_H Q_H} \tag{5.163}$$

from which we can obtain all the partial thermodynamic quantities of the adsorbed gas. First the chemical potential is

$$\mu = kT \ln\left(\frac{x}{1 - x}\right) - kT \ln\left(\frac{q_L Q_L + q_H Q_H}{Q_L + Q_H}\right) \tag{5.164}$$

Using the distribution of the two states in the empty system given in (5.153) we may rewrite (5.164) as

$$\mu = kT \ln\left(\frac{x}{1-x}\right) - kT \ln(q_L x_L{}^\circ + q_H x_H{}^\circ)$$

$$= kT \ln\left(\frac{x}{1-x}\right) - kT \ln\langle \exp(-\beta B_G)\rangle_0 \qquad (5.165)$$

where B_G represents the "binding energy" of the gas to the site. This may attain one of the two possible values U_L and U_H. The notation $\langle \ \rangle_0$ signifies an average, using the probability distribution (5.153), of a system containing no gas molecules. It is important to realize that for *any* x the second term of (5.165) is always an average over a probability distribution taken in the *empty* system. In most of the discussions in this section we shall be interested only in the limit of the dilute system, where $x \ll 1$, so that the first term on the right-hand side of (5.165) is $kT \ln x$.

The partial entropy of the gas is obtained from (5.165) by differentiation with respect to the temperature:

$$\bar{S} = -\left(\frac{\partial \mu}{\partial T}\right)_{M,N} = -k \ln\left(\frac{x}{1-x}\right) + k \ln\langle \exp(-\beta B_G)\rangle_0$$

$$+ \frac{1}{T}\left\{\frac{q_L Q_L(E_L + U_L) + q_H Q_H(E_H + U_H)}{q_L Q_L + q_H Q_H} - \frac{Q_L E_L + Q_H E_H}{Q_L + Q_H}\right\}$$
$$(5.166)$$

Using the conditional probabilities of (5.159) and (5.160), relation (5.166) may be rewritten as

$$\bar{S} = -k \ln\left(\frac{x}{1-x}\right) + k \ln\langle \exp(-\beta B_G)\rangle_0 + \frac{1}{T}[\langle E + B_G\rangle_G - \langle E\rangle_0]$$
$$(5.167)$$

where the symbol $\langle \ \rangle_G$ signifies a conditional average taken with the distribution given in (5.159) and (5.160).

The partial energy of the gas is given by

$$\bar{E} = \mu + T\bar{S} = \langle E + B_G\rangle_G - \langle E\rangle_0 = \langle B_G\rangle_G + (\langle E\rangle_G - \langle E\rangle_0) \qquad (5.168)$$

The three quantities μ, $\bar{S}$, and $\bar{E}$ pertaining to the gas adsorbed in the system were computed here through the exact and standard formalism of statistical mechanics. We have not, so far, introduced any notions such as "structure" of the system or "structural changes" in the system. We note, however, an important observation that has already been made, namely, whereas $\bar{S}$ and $\bar{E}$ are expressed in terms of conditional averages, the chemical potential of the gas contains only an average with the distribution (5.153) pertaining to the empty system.

In order to gain a deeper insight into the reason for this difference and at the same time to proceed with the analogy to the problem of aqueous solutions, we shall turn next to a different formulation of the same quantities using the mixture model approach. The results will be completely equivalent to those obtained above, but we shall gain a new interpretation of the various expressions obtained for the partial thermodynamic quantities of the gas.

5.10.2. A Mixture Model Approach to the Same System

We now develop an equivalent approach to the same system but we shall be using a different terminology. This terminology is similar in a certain sense to the one frequently used for aqueous solutions. The gas G will be referred to as a "solute" and the adsorbing system as the "solvent." In the solvent we distinguish between two "species" L and H. In this rather primitive model, the "structure" of the solvent will be simply the equilibrium composition of the solvent, i.e., the mole fraction of L (or H) molecules (or sites) x_L (or x_H, with $x_L + x_H = 1$). In this section we shall be interested in the effect of the "*solubility*" (adsorption) of the gas on the *structure* of the solvent; later we shall consider an analog of the hydrophobic interaction process and its relation to structural changes in the solvent.

To gain more detailed information on the effect of the gas G on the structure of the system we turn back to the canonical partition function (5.154), which we shall rewrite as

$$Q(N, M, T) = \sum_{M_L + M_H = M} Q^*(N, M_L, M_H, T)$$

$$= \sum_{M_L + M_H = M} \sum_{N_L + N_H = N} Q^{**}(N_L, N_H, M_L, M_H, T) \qquad (5.169)$$

where

$$Q^{**}(N_L, N_H, M_L, M_H, T) = \binom{M}{M_L}\binom{M_L}{N_L}\binom{M_H}{N_H} Q_L{}^{M_L} Q_H{}^{M_H} q_L{}^{N_L} q_H{}^{N_H} \qquad (5.170)$$

is the partition function of the completely *frozen-in system* (FIS), i.e., this is the partition function of a system with *any* fixed (and arbitrarily chosen) values of N_L, N_H, M_L, and M_H. Once we sum all possible values of N_L and N_H with the condition $N_L + N_H = N$, but keeping M_L and M_H fixed, we get $Q^*(N, M_L, M_H, T)$, which is the partition function of the *partially equilibrated system* (PES), i.e., we have equilibrium with respect to the gas molecules on the sites; the equilibrium values of N_L and N_H, denoted by $\bar{N}_L$ and $\bar{N}_H$, are obtainable from the condition that Q^{**} be maximum with respect to N_L and N_H subject to the restriction $N_L + N_H = N$ (M_L and M_H being constants). This procedure leads to the equilibrium constant for the PES, namely,

$$\frac{\bar{N}_H(M_L - \bar{N}_L)}{\bar{N}_L(M_H - \bar{N}_H)} = \frac{q_H}{q_L} \equiv h \qquad (5.171)$$

This can be easily solved for $\bar{N}_H$ in terms of the other parameters of the system; the result is

$$\bar{N}_H = \{-(N - M_L - hN - hM_H)$$
$$\pm [(N-M_L-hN-hM_H)^2-4(h-1)hNM_H]^{1/2}\}[2(h-1)]^{-1} \quad (5.172)$$

For later use, we shall be interested only in the limiting behavior when $N \ll M$ (very dilute solutions), thus, taking only the linear dependence of $\bar{N}_L$ (and $\bar{N}_H$) on N, we obtain

$$\bar{N}_L = \frac{NM_L}{M_L + hM_H}, \qquad \bar{N}_H = \frac{hNM_H}{M_L + hM_H} \qquad \text{(PES, } N \ll M)$$
$$(5.173)$$

We have stressed in (5.173) that this result is for the PES.

Next we turn to the completely equilibrated system (CES) which is obtained by letting M_L and M_H reach their equilibrium values, which we denote by $\bar{M}_L$ and $\bar{M}_H$, respectively. These are obtained by taking the maximum of $Q^*(N, M_L, M_H, T)$ with respect to M_L and M_H subject to the condition $M_L + M_H = M$. This procedure leads to a new equilibrium condition

$$\frac{\bar{M}_H - \bar{N}_H}{\bar{M}_L - \bar{N}_L} = \frac{Q_H}{Q_L} = K \qquad \text{(CES)} \qquad (5.174)$$

From (5.171) and (5.174), when evaluated at the condition of CES, we can solve for $\bar{N}_L$, $\bar{N}_H$, $\bar{M}_L$, and $\bar{M}_H$ in terms of the molecular parameters h, K and the macroscopic parameters N, M, and T; the result is

$$\bar{N}_L = \frac{N}{1 + hK}, \qquad \bar{N}_H = \frac{hKN}{1 + hK} \qquad (5.175)$$

$$\bar{M}_L = M\frac{1 + hK - xK(h - 1)}{(1 + K)(1 + hK)}, \qquad \bar{M}_H = M\frac{K + hK^2 + xK(h - 1)}{(1 + K)(1 + hK)}$$
$$(5.176)$$

where $x = N/M$. From (5.176) we obtain the mole fractions of L and H at CES for any x.

Clearly for $x = 0$ relations (5.176) reduce to the one corresponding to the empty system. Also if $h = 1$ [see (5.171)], which means that $U_L = U_H$, then again (5.176) reduces to (5.153) since in this case the adsorbed gas cannot affect the equilibrium composition of the system. The difference between x_L and $x_L{}^\circ$ gives the total shift in the equilibrium composition of L due to the addition of solute at concentration x. Of particular importance to the partial thermodynamic quantities of the gas is the *differential shift*, or the stabilization of (say) the L form per one solute molecule

$$\left(\frac{\partial \bar{M}_L}{\partial N}\right)_{T,M} = \left(\frac{\partial x_L}{\partial x}\right)_T = \frac{-K(h-1)}{(1+K)(1+hK)} \tag{5.177}$$

Thus we have an exact answer to the question, how much structural change in the "solvent" is induced by addition of a "solute" to the system?

5.10.3. Partial Molecular Thermodynamic Quantities of the Gas in the Mixture-Model Formalism

In this subsection we evaluate the chemical potential and the partial molecular entropy and enthalpy of the gas, using the so-called mixture-model approach. The final results will be exactly the same as in (5.165), (5.167), and (5.168), thus proving the equivalence of the two methods. However, the more important aspect of this approach is the split of each thermodynamic quantity into two terms: one corresponds to a PES and the second is a relaxation term for the CES. We first calculate the chemical potential in the PES, i.e., when M_L and M_H are fixed; this will be denoted by $\mu^*(M_L, M_H)$. Next, we substitute $\bar{M}_L$ and $\bar{M}_H$ for M_L and M_H to get $\mu^*(\bar{M}_L, \bar{M}_H)$ and this will turn out to be exactly the same as (5.165). Carrying out a similar procedure for S^* and E^* we shall see that these are only *parts* of $\bar{S}$ and $\bar{E}$, respectively.

For the sake of mathematical simplicity we shall restrict ourselves to a "very dilute solution," i.e., we assume that $N \ll M$ (or $x \ll 1$).

The chemical potential of the gas in any PES may be obtained by differentiating Q^* with respect to N, i.e.,

$$\mu^*(M_L, M_H) = \left(\frac{\partial A^*}{\partial N}\right)_{T,M_L,M_H} = \left(\frac{-kT\partial \ln Q^*}{\partial N}\right)_{T,M_L,M_H} \tag{5.178}$$

Evaluating μ^* in the PES, taking the limit $x \ll 1$, and then substituting $\bar{M}_L$ and $\bar{M}_H$ for M_L and M_H, we get, after some elementary algebra, the final result

$$\mu^*(\bar{M}_L, \bar{M}_H) = kT \ln x - kT \ln\left(\frac{q_L Q_L + q_H Q_H}{Q_L + Q_H}\right) \qquad (5.179)$$

which is exactly equal to μ of (5.164) taken at the limit $x \ll 1$.

Since there is a subtle point here that has sometimes been overlooked in the past and therefore led to misunderstandings, we shall further elaborate on the conceptual difference between the meaning of μ and μ^*. The quantity μ is defined as the derivative of the free energy with respect to N in a CES, i.e., we add dN molecules of gas while any chemical equilibrium that exists in the system must respond by changing the equilibrium concentrations. The quantity μ^*, on the other hand, is evaluated in a PES, i.e., M_L and M_H are *fixed*. Even though in the final expression we substitute the values of $\bar{M}_L$ and $\bar{M}_H$ of the CES, these are still *fixed* quantities, and in the very definition of μ^* in (5.178) we do not allow the equilibrium composition of the system to respond to the addition of dN. The fact that (5.179) equals (5.164) will become clearer later on when we examine the corresponding compensation effect in this system. At this moment, however, it will be instructive to demonstrate that if we undertake the same procedure as above for S^* and E^* we do *not* regenerate $\bar{S}$ and $\bar{E}$ of (5.167) and (5.168), respectively. We define S^* for any PES by

$$S^*(M_L, M_H) = -\left(\frac{\partial \mu^*}{\partial T}\right)_{N, M_L, M_H} \qquad (5.180)$$

Differentiating (5.178) with respect to T (*before* substituting $\bar{M}_L$ and $\bar{M}_H$), taking the limit $x \ll 1$, and evaluating $S^*(\bar{M}_L, \bar{M}_H)$, we get, after lengthy algebra, the result

$$S^*(\bar{M}_L, \bar{M}_H) = -k \ln x + k \ln\langle\exp(-\beta B_G)\rangle_0 + (1/T)\langle B_G \rangle_G \qquad (5.181)$$

This should be compared with $\bar{S}$ in (5.167) (evaluated at $x \ll 1$). We observe that the two quantities differ from each other by the term $(1/T)(\langle E \rangle_G - \langle E \rangle_0)$. Similarly for the partial energy E^* we have

$$E^*(\bar{M}_L, \bar{M}_H) = \mu^* + TS^* = \langle B_G \rangle_G \qquad (5.182)$$

which should be compared with (5.168), and we note that the difference is just $(\langle E \rangle_G - \langle E \rangle_0)$. Using the mixture model formalism as in Section

5.9, we obtain

$$\mu = \mu^* + (\mu_L - \mu_H)\left(\frac{\partial \bar{M}_L}{\partial N}\right)_{T,M} \tag{5.183}$$

$$\bar{S} = S^* + (\bar{S}_L - \bar{S}_H)\left(\frac{\partial \bar{M}_L}{\partial N}\right)_{T,M} \tag{5.184}$$

$$\bar{E} = E^* + (\bar{E}_L - \bar{E}_H)\left(\frac{\partial \bar{M}_L}{\partial N}\right)_{T,M} \tag{5.185}$$

where all the quantities here are evaluated in the (N, M, T) ensemble. Because of the existence of chemical equilibrium, $\mu_L - \mu_H = 0$, the second term on the right-hand side of (5.183) drops out. This is true independently of whether $\partial \bar{M}_L/\partial N$ is zero or nonzero. In fact, we have shown that in general this quantity is nonzero (except for the extreme cases of $h = 1$ or $K = 0$ [see (5.177)]). On the other hand, $\bar{S}_L - \bar{S}_H$ as well as $\bar{E}_L - \bar{E}_H$ do not have to be zero, and therefore both $\bar{S}$ and $\bar{E}$ may get contributions due to structural changes in the solvent. We can, in our particular model, evaluate exactly all the quantities in (5.183) to (5.185) by using standard statistical mechanical relations similar to the ones used above. The results of particular importance in the present context are

$$(\mu_L - \mu_H) = 0 \tag{5.186}$$

$$(\bar{S}_L - \bar{S}_H)\left(\frac{\partial \bar{M}_L}{\partial N}\right)_{T,M} = \frac{E_L - E_H}{T}\,\frac{K(1 - h)}{(1 + K)(1 + hK)} \tag{5.187}$$

$$(\bar{E}_L - \bar{E}_H)\left(\frac{\partial \bar{M}_L}{\partial N}\right)_{T,M} = (E_L - E_H)\,\frac{K(1 - h)}{(1 + K)(1 + hK)} \tag{5.188}$$

Note that $\bar{E}_i$ and $\bar{S}_i$ are *partial* (molecular) quantities, whereas E_L and E_H are the molecular quantities of the model. We see from the last three relations that if $E_L \neq E_H$ and if $K \neq 0$ and $h \neq 1$ that both $\bar{S}$ and $\bar{E}$ get finite contributions from structural changes in the solvent. In any case, the chemical potential does not get such a contribution. These are exact results for our particular model, but the conclusion has a more general validity beyond this model, as has been discussed in Section 5.9. We note also that the difference between $\bar{S}$ in (5.167) and S^* in (5.181) can now be rewritten as

$$\frac{1}{T}\,(\langle E \rangle_G - \langle E \rangle_0) = (\bar{S}_L - \bar{S}_H)\left(\frac{\partial \bar{M}_L}{\partial N}\right)_{T,M} \tag{5.189}$$

and similarly

$$\langle E \rangle_G - \langle E \rangle_0 = (\bar{E}_L - \bar{E}_H)\left(\frac{\partial \bar{M}_L}{\partial N}\right)_{T,M} \tag{5.190}$$

Finally we note the *exact* compensation between the contribution of structural changes in the solvent to $T\bar{S}$ and to $\bar{E}$.

The question is often raised: how could it be that a process causes a structural change in the solvent, yet this change does not affect the standard free energy change of the process? An exact answer may be given to this question, and for the particular process of solubility within this specific model. To do this we reformulate our problem in a slightly different way. Consider an empty system (pure solvent) with composition $\bar{M}_L^\circ$ and $\bar{M}_H^\circ$. Now add one solute (gas) molecule to this system. It is well known that the standard chemical potential for the infinitely dilute solution may be obtained by adding a single solute to a macroscopic pure solvent. The addition of the solute may cause some structural change in the solvent, and as a matter of fact, we know from (5.177) exactly how much structural change is brought about by the addition of one solute molecule; this quantity is in general nonzero (except for the cases $h = 1$ or $K = 0$). The new composition will be denoted by $\bar{M}_L$ and $\bar{M}_H$, respectively. The chemical potential of the solute may be obtained as follows (see also Section 1.2):

$$
\begin{aligned}
\mu(\bar{M}_L, \bar{M}_H) &= \lim_{dN \to 0} \left[\frac{A(M, dN) - A(M)}{dN} \right] \\
&= \lim_{M \to \infty} [A(M, N = 1) - A(M)] \\
&= \lim_{M \to \infty} [A(\bar{M}_L, \bar{M}_H, N = 1) - A(\bar{M}_L^\circ, \bar{M}_H^\circ)] \\
&= \lim_{M \to \infty} [A(\bar{M}_L^\circ, \bar{M}_H^\circ, N = 1) - A(\bar{M}_L^\circ, \bar{M}_H^\circ)] \\
&\quad + \lim_{M \to \infty} [A(\bar{M}_L, \bar{M}_H, N = 1) - A(\bar{M}_L^\circ, \bar{M}_H^\circ, N = 1)] \\
&= \mu^*(\bar{M}_L^\circ, \bar{M}_H^\circ) + \Delta\mu^r
\end{aligned}
\tag{5.191}
$$

where μ^* is the chemical potential of the solute in the PES and $\Delta\mu^r$ is the contribution to μ from structural changes in the solvent. Expanding $A(\bar{M}_L^\circ, \bar{M}_H^\circ, N = 1)$ about the equilibrium point $(\bar{M}_L, \bar{M}_H, N = 1)$ we obtain

$$
\begin{aligned}
-\Delta\mu^r &= [(\bar{M}_L^\circ - \bar{M}_L)\mu_L(\bar{M}_L, \bar{M}_H, N = 1) \\
&\quad + (\bar{M}_H^\circ - \bar{M}_H)\mu_H(\bar{M}_L, \bar{M}_H, N = 1)] \\
&\quad + \tfrac{1}{2}(\mu_{LL} - 2\mu_{HL} + \mu_{HH})(\bar{M}_L^\circ - \bar{M}_L)^2 + \cdots
\end{aligned}
\tag{5.192}
$$

where $\mu_{ij} = \partial A / (\partial \bar{M}_i \, \partial \bar{M}_j)$. Now since at the point $(\bar{M}_L, \bar{M}_H, N = 1)$ a chemical equilibrium exists, we have $\mu_L = \mu_H$ and obviously $\bar{M}_L{}^\circ - \bar{M}_L = -(\bar{M}_H{}^\circ - \bar{M}_H)$. Therefore, the first term on the right-hand side of (5.192) is zero. In addition, the second term will be as small as we wish in the limit of $M \to \infty$ as required in (5.191). This is so because $(\bar{M}_L{}^\circ - \bar{M}_L)^2$ is finite but $(\mu_{LL} - 2\mu_{HL} + \mu_{HH})$ is a quantity that tends to zero as M^{-1}. More explicitly, in this model we have

$$(\mu_{LL} - 2\mu_{HL} + \mu_{HH}) = \frac{kT}{Mx_L{}^\circ x_H{}^\circ} \xrightarrow{M \to \infty} 0 \tag{5.193}$$

Similarly all higher-order terms in the expansion (5.192) will tend to zero even faster than the second term. We therefore conclude that in the limit of a macroscopic system, $\Delta\mu^r \to 0$ and hence $\mu = \mu^*$, which is the same result we obtained before. If we had carried out the same procedure for, say, the entropy, we would obtain instead of (5.191) and (5.192) the relations

$$\bar{S}(\bar{M}_L, \bar{M}_H) = S^*(\bar{M}_L{}^\circ, \bar{M}_H{}^\circ) + \Delta S^r \tag{5.194}$$

where

$$\begin{aligned} -\Delta S^r = {} & (\bar{S}_L - \bar{S}_H)(\bar{M}_L{}^\circ - \bar{M}_L) \\ & + \tfrac{1}{2}(S_{LL} - 2S_{LH} + S_{HH})(\bar{M}_L{}^\circ - \bar{M}_L)^2 + \cdots \end{aligned} \tag{5.195}$$

where we use the notation

$$S_{ij} = \frac{\partial S}{\partial \bar{M}_i \, \partial \bar{M}_j}$$

Here again the second term and all higher-order ones in (5.195) will tend to zero as $M \to \infty$, but in contrast to (5.192) ΔS^r still contains the first-order term in (5.195), which is nonzero.

Thus we see that although $\Delta\mu^r$ tends to zero in the limit of a macroscopic system, $(\partial \bar{M}_L / \partial N)$ remains finite. This may be seen in yet another way through the identity (Ben-Naim, 1974)

$$\left(\frac{\partial \bar{M}_L}{\partial N}\right)_{M,T} = -\left(\frac{\partial(\mu_L - \mu_H)}{\partial N}\right)_{T,M_L,M_H} (\mu_{LL} - 2\mu_{LH} + \mu_{HH})^{-1} \tag{5.196}$$

The two factors on the right-hand side of (5.196) tend to zero as M^{-1}. In particular, in our model we have

$$\left(\frac{\partial(\mu_L - \mu_H)}{\partial N}\right)_{T,M_L,N_H} = \frac{-kT(1 - h)}{M(x_L{}^\circ + hx_H{}^\circ)} \tag{5.197}$$

Thus, although the leading term in (5.192), which is determined by (5.193), tends to zero as $M \to \infty$, the ratio of the two quantities (5.197) and (5.193) is finite in this limit. This answers the question posed above.

A very similar consideration holds for the process of hydrophobic interaction, the analog of which in our model will be treated in Section 5.10.4.

5.10.4. The Analog of the Hydrophobic Interaction (HI) Process

We now extend our model in such a way that it enables us to examine a process similar to the HI. Most of the assumptions of the model remain unchanged except one, namely, we allow each site (either L or H) to be able to accommodate at most two solute particles. We shall then consider the process of bringing two solutes from two different, but singly occupied, sites to a single site. The free energy, entropy, and energy changes of this process may be computed. We shall find that any structural change that may occur in the "solvent" as a result of the HI process does not affect the free energy of the process, whereas it may well affect other thermodynamic quantities such as the entropy or the energy change of the same process.

For singly occupied sites the binding energies are as before U_L and U_H. For doubly occupied sites we assume that the binding energy of the dimer d is

$$U_L{}^d = 2U_L + \varepsilon$$
$$U_H{}^d = 2U_H + \varepsilon \tag{5.198}$$

where $\varepsilon > 0$ is introduced merely to account for the fact that in a dimer the binding energy of each solute with the site is less than U_L (or U_H) because part of its surface is not exposed to the "solvent." We also introduce the direct interaction between the two solutes on a site U_{12}.

The process we consider is the following. We start with a system with fixed M, T and $N = 2$, and with the two solutes on different sites $i \neq j$. The process, analogous to the HI, consists of bringing the two solutes from fixed, but *different*, sites $i \neq j$, to a *fixed single* site i. In a real solution we consider the process of bringing two solutes from fixed positions but at infinite separation to a small separation $R_{12} \simeq \sigma$, where σ is of the order of a molecular diameter of the solute. Here, because of the independence of the sites, we can take as our initial state *any* two different sites, since

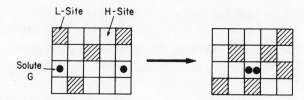

Figure 5.3. Schematic illustration of the analog of the hydrophobic interaction in the model system.

there is no solute–solute interaction between solutes on different sites. The process is schematically shown in Figure 5.3.

Using a thermodynamical cycle drawn schematically in Figure 5.4 we obtain the free energy change for the required process

$$\Delta A^d = \Delta \mu_d{}^\circ - 2\Delta \mu_m{}^\circ + U_{12} \tag{5.199}$$

where $\Delta \mu_d{}^\circ$ is the standard free energy of "solution" of the dimer—more specifically, this is the free energy of transferring the dimer, as a single entity, from a *fixed* position and (orientation) in the gaseous phase to a fixed site in the adsorption model. Similarly, $\Delta \mu_m{}^\circ$ is the standard free energy of "solution" of the monomer (m), i.e., a single gas molecule as was treated in Sections 5.10.1–5.10.3.

In this model, $\Delta \mu_d{}^\circ$ and $\Delta \mu_m{}^\circ$ are easily computable, and we find

$$\Delta \mu_m{}^\circ = -kT \ln \langle \exp(-\beta B_m) \rangle_0$$
$$= -kT \ln [x_L{}^\circ \exp(-\beta U_L) + x_H{}^\circ \exp(-\beta U_H)] \tag{5.200}$$

$$\Delta \mu_d{}^\circ = -kT \ln \langle \exp(-\beta B_d) \rangle_0$$
$$= -kT \ln \{x_L{}^\circ \exp[-\beta(2U_L + \varepsilon)] + x_H{}^\circ \exp[-\beta(2U_H + \varepsilon)]\} \tag{5.201}$$

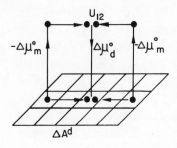

Figure 5.4. A thermodynamic cycle corresponding to relation (5.199). The free energy change ΔA^d corresponds to the process of bringing two solute molecules from *fixed* but different sites ($i \neq j$) to a single site. Instead one can transfer the two solutes to two fixed positions (but with infinite separation from each other) in the gaseous phase, then form the dimer d, and finally transfer the dimer as a single entity to a fixed site. The equality of the free energy changes along the two routes is given in (5.199).

and hence

$$
\begin{aligned}
\Delta A^d &= U_{12} - kT \ln\left[\frac{\langle \exp(-\beta B_d)\rangle_0}{\langle \exp(-\beta B_m)\rangle_0{}^2}\right] \\
&= U_{12} + \varepsilon - kT \ln\left[\frac{(x_L{}^\circ \exp(-\beta 2U_L) + x_H{}^\circ \exp(-\beta 2U_H)}{[x_L{}^\circ \exp(-\beta U_L) + x_H{}^\circ \exp(-\beta U_H)]^2}\right] \\
&= U_{12} + \varepsilon - kT \ln\left\{\frac{[x_L{}^\circ + (1 - x_L{}^\circ)h^2]}{[x_L{}^\circ + (1 - x_L{}^\circ)h]^2}\right\} \qquad (5.202)
\end{aligned}
$$

We note, as in (5.165), that the averages appearing in ΔA^d are taken over the distribution $x_L{}^\circ$ and $x_H{}^\circ$ of the pure system. This is again the essential reason ΔA^d does not get contributions from structural changes in the solvent due to the HI process.

It is interesting to note that if $U_L = U_H$, then the third term on the right-hand side of (5.202) would be zero and all the free energy change would be due to direct interaction between the two solutes U_{12} and the change in the binding energy of the solute pairs in the two states. Thus the third term arises from the ratio of the probabilities of finding the two solutes at a single site (L or H) and at two different sites (L or H) (see also Section 5.10.5).

The change of the entropy and energy for the HI is obtained from

$$
\begin{aligned}
\Delta S^d &= -\left(\frac{\partial \Delta A^d}{\partial T}\right)_{M,T} \\
&= k \ln\left[\frac{\langle \exp(-\beta B_d)\rangle_0}{\langle \exp(-\beta B_m)\rangle_0{}^2}\right] + kT \frac{\partial}{\partial T} \ln\langle \exp(-\beta B_d)\rangle_0 \\
&\quad - kT \frac{\partial}{\partial T}[\ln\langle \exp(-\beta B_m)\rangle_0{}^2] \\
&= k \ln\left[\frac{\langle \exp(-\beta B_d)\rangle_0}{\langle \exp(-\beta B_m)\rangle_0{}^2}\right] + \frac{1}{T}(\langle E + B_d\rangle_d - \langle E\rangle_0) \\
&\quad - \frac{2}{T}(\langle E + B_m\rangle_m - \langle E\rangle_0) \\
&= k \ln\left[\frac{\langle \exp(-\beta B_d)\rangle_0}{\langle \exp(-\beta B_m)\rangle_0{}^2}\right] + (\bar{S}_L - \bar{S}_H)\left(\frac{\partial \bar{M}_L}{\partial N_d} - 2\frac{\partial \bar{M}_L}{\partial N_m}\right) \qquad (5.203)
\end{aligned}
$$

$$
\Delta E^d = \Delta A^d + T \Delta S^d = U_{12} + (\bar{E}_L - \bar{E}_H)\left(\frac{\partial \bar{M}_L}{\partial N_d} - 2\frac{\partial \bar{M}_L}{\partial N_m}\right) \qquad (5.204)
$$

In the fourth line on the right-hand side of (5.203), we have exploited the fact that the derivative of $\langle \exp(-\beta B_i)\rangle$ has already been computed in

(5.167), and in the sixth line on the right-hand side of (5.203) we used the reinterpretation given in (5.189) in terms of the structural changes in the solvent. Similar arguments apply to (5.204).

The conclusion of this section is very similar to the one we have arrived at in our examination of the solubility process. Here we see again that structural changes in the "solvent" induced by the HI process do not contribute to the free energy change of the process. On the other hand, the entropy and the energy change of the process may well be affected. The only point left to be examined is that the structural change itself is nonzero. Clearly $(\partial \bar{M}_L / \partial N_m)$ is the same as the quantity defined in (5.177). The corresponding term for the dimer is

$$\left(\frac{\partial \bar{M}_L}{\partial N_d} \right) = \frac{-K(h_d - 1)}{(1 + K)(1 + h_d K)} \tag{5.205}$$

where

$$h_d = \frac{\exp[-\beta(2U_H + \varepsilon)]}{\exp[-\beta(2U_L + \varepsilon)]} = h^2$$

Hence, the difference between the structural change in the solvent induced by the dimer and the two separate solutes is

$$\left(\frac{\partial \bar{M}_L}{\partial N_d} \right) - 2\left(\frac{\partial \bar{M}_L}{\partial N_m} \right) = \frac{-K}{1 + K} \left(\frac{h^2 - 1}{1 + h^2 K} - 2 \frac{h - 1}{1 + hK} \right) \tag{5.206}$$

which in general would not be zero. There is one subtle point that should be mentioned, and that is to show that the quantity on the left-hand side of (5.206) is indeed the structural change in the solvent due to the HI process. To show this we note that for macroscopic systems we may write the two derivatives in (5.206) as

$$\left(\frac{\partial \bar{M}_L}{\partial N_d} \right) = \bar{M}_L(d) - \bar{M}_L(0), \qquad \left(\frac{\partial \bar{M}_L}{\partial N_m} \right) = \bar{M}_L(m) - \bar{M}_L(0) \tag{5.207}$$

where $\bar{M}_L(0)$, $\bar{M}_L(m)$, and $\bar{M}_L(d)$ are the values of $\bar{M}_L$ when the system is empty, contains one m, and contains one d, respectively. Clearly the structural change due to the addition of two m's at two different sites is twice the structural change due to the addition of a single m. We write this symbolically as

$$\bar{M}_L(m, m) - \bar{M}_L(0) = 2[\bar{M}_L(m) - \bar{M}_L(0)] \tag{5.208}$$

$\bar{M}_L(m, m)$ is the value of $\bar{M}_L$ for a system with two m's at different sites (for real solutions we require that the two solutes be at fixed positions but at infinite separation). Thus we obtain

$$\frac{\partial \bar{M}_L}{\partial N_d} - 2\frac{\partial \bar{M}_L}{\partial N_m} = \bar{M}_L(d) - M_L(0) - 2[\bar{M}_L(m - \bar{M}_L(0)]$$

$$= M_L(d) - \bar{M}_L(m, m) \qquad (5.209)$$

which is the required form for the SCIS due to the HI process. The argument given here is similar to the one presented in Section 5.8.

5.10.5. Probability Approach to the Problem of HI in the Present Model

In Section 5.10.4 we presented the process of HI for exactly two "solute" particles in a "solvent." The connection between ΔA^d and the thermodynamics of solution was established through the thermodynamic cycle in Figure 5.4. It is instructive to present here a different derivation of the same result using a probability point of view. This procedure not only provides a new insight into the meaning of ΔA^d but also is more general since we do not have to limit ourselves to extremely dilute solutions.

For the model presented in Section 5.10.4 we can immediately write down the grand partition function (with respect to the solute), namely,

$$\Xi(\lambda, M, T) = (Q_L + Q_H + \lambda q_L Q_L + \lambda q_H Q_H + \lambda^2 q_L{}^2 Q_L \eta + \lambda^2 q_H{}^2 Q_H \eta)^M$$

$$= \xi^M \qquad (5.210)$$

where $\eta = \exp[-\beta(\varepsilon + U_{12})]$ and ξ may be viewed as the grand partition function of a single site. The six terms correspond to the six possible states of the site: empty (L or H), singly occupied (L or H), doubly occupied (L or H).

The probability that a specific site, say i, will be singly occupied is

$$P^{(1)}(i) = \frac{\lambda q_L Q_L + \lambda q_H Q_H}{\xi} \qquad (5.211)$$

The probability that two *specific* but different sites, say i and j with $i \neq j$, will be singly occupied is

$$P^{(2)}(i \neq j) = P(i)P(j) = \left(\frac{\lambda q_L Q_L + \lambda q_H Q_H}{\xi}\right)^2 \qquad (5.212)$$

This relation is clearly due to the assumption of independence of the sites.

The probability of finding a *specific* site being doubly occupied is

$$P^{(2)}(i = j) = \frac{(\lambda^2 q_L{}^2 Q_L + \lambda^2 q_H{}^2 Q_H)\eta}{\xi} \qquad (5.213)$$

Hence the correlation function, which is the analog of the radical distribution function in liquid solutions, is

$$
\begin{aligned}
g^{(2)}(i = j) &= \frac{P^{(2)}(i = j)}{P^{(2)}(i \neq j)} = \frac{(\lambda^2 q_L{}^2 Q_L + \lambda^2 q_H{}^2 Q_H)\xi\eta}{(\lambda q_L Q_L + \lambda q_H Q_H)^2} \\
&= \exp[-\beta(\varepsilon + U_{12})] \frac{q_L{}^2 x_L{}^\circ + q_H{}^2 x_H{}^\circ}{(q_L x_L{}^\circ + q_H x_H{}^\circ)^2} \\
&\quad \times [1 + \lambda(q_L x_L{}^\circ + q_H x_H{}^\circ) + \lambda^2 \eta(q_L{}^2 x_L{}^\circ + q_H{}^2 x_H{}^\circ)] \qquad (5.214)
\end{aligned}
$$

[Actually, in the definition of $g^{(2)}(i = j)$ one should use in the denominator of (5.214) $p^{(1)}(i)p^{(1)}(i)$ rather than $p^{(2)}(i \neq j)$; but because of the independence of the two sites i and j we could also use $p^{(2)}(i \neq j) = p^{(1)}(i)p^{(1)}(j)$.]

The last relation is valid for any activity of the solute λ. In particular, for very dilute solutions we regain the result of Section 5.10.4, namely, for $\lambda \to 0$ we obtain

$$
\begin{aligned}
g^{(2)}(i = j) &= \frac{\langle \exp(-\beta B_d) \rangle_0}{\langle \exp(-\beta B_m) \rangle_0{}^2} \exp(-\beta U_{12}) \\
&= \exp(-\beta \, \Delta A^d) \qquad (5.215)
\end{aligned}
$$

The last equality has been established here by comparison with (5.202) for our specific model and in the limit $\lambda \to 0$. However, the connection between $g^{(2)}$ and the free energy change ΔA^d is valid for more general cases (see also Section 1.4). This concludes our treatment of this model.

Comment

Working with simplified models for very complex systems is both rewarding and dangerous. It is rewarding because it provides a clear insight into the meaning and content of various thermodynamic quantities—an insight that is not easy to gain from the general formalism. It is dangerous, however, if one does not exercise extreme care in distinguishing between those results that are characteristic of the simple model and those results that have a more general validity.

Appendixes

A.1. THE FUNDAMENTAL EQUATION FOR THE CHEMICAL POTENTIAL IN A TWO-COMPONENT SYSTEM

This appendix provides a statistical mechanical basis for the expressions cited in Chapter 1. It is the author's conviction that a minimal knowledge of statistical mechanics is essential for a proper interpretation of the various standard chemical potentials and of the corresponding quantities for the transfer of a solute between two phases. The reader will soon find that indeed very little statistical mechanics is needed for the full understanding of all the expressions given in Chapter 1.

The fundamental quantity in statistical mechanics is the partition function. This function is the master key to obtaining all the thermodynamic information on the system. For simplicity we shall treat here a system of simple spherical molecules obeying the laws of classical statistical mechanics. Also for simplicity of presentation we shall use the so-called canonical ensemble, but all the results obtained are of more general validity.

Consider a system of N_A molecules of type A and N_B molecules of type B, contained in a volume V at a given temperature T. For such a system the classical partition function is

$$Q(T, V, N_A, N_B)$$
$$= \frac{q_A{}^{N_A} q_B{}^{N_B}}{N_A! \, N_B! \, \Lambda_A{}^{3N_A} \Lambda_B{}^{3N_B}} \int \cdots \int_V d\mathbf{R}^{N_A} d\mathbf{R}^{N_B} \exp[-\beta U(\mathbf{R}^{N_A}, \mathbf{R}^{N_B})]$$
$$= \frac{q_A{}^{N_A} q_B{}^{N_B} Z(N_A, N_B)}{N_A! \, N_B! \, \Lambda_A{}^{3N_A} \Lambda_B{}^{3N_B}} \tag{A.1}$$

Here $\beta = (kT)^{-1}$, with k the Boltzmann constant. q_A is the internal partition function of a single molecule A, and it is assumed that the internal properties of a single molecule are independent of the type of environment that surrounds the molecule. We shall not need the explicit form of this function because in the thermodynamic quantities that will be of interest to us these functions cancel out. $\Lambda_A{}^3$ is the momentum partition function of an A molecule. It results from the integration over all possible momenta of the molecule, and again we shall not need its explicit form for our thermodynamic quantities. It is important, however, to know its source and its meaning for a proper interpretation of the general expression of the chemical potential, as derived below.

The quantity $U(\mathbf{R}^{N_A}, \mathbf{R}^{N_B})$ is a shorthand notation for the total potential energy of interaction of all the molecules of the system at a specific configuration $\mathbf{R}^{N_A}, \mathbf{R}^{N_B}$. The last symbol simply stands for the set of all the position vectors of the $N_A + N_B$ molecules. In the following we shall use a simpler notation, $U(N_A, N_B)$, for this quantity. The integration in (A.1) is carried out over all possible locations of the molecules in the system. This integral is often referred to as the configuration partition function and is denoted by $Z(N_A, N_B)$.

With this rather brief presentation of the partition function we proceed to state the fundamental connection between thermodynamics and statistical mechanics, which for our particular choice of the independent variables, T, V, N_A, and N_B, is

$$A(T, V, N_A, N_B) = -kT \ln Q(T, V, N_A, N_B) \qquad (A.2)$$

where $A(T, V, N_A, N_B)$ is the Helmholtz free energy of the system. The reader who is not familiar with the fundamentals of statistical mechanics, and who is not interested in any further details on this subject, may safely *adopt* the two relations (A.1) and (A.2) as the starting point (or the axioms) for all that follows below. (In fact, in any theoretical structure of a physical theory, one always has to accept some basic postulates or axioms on which the theory is constructed. The suggestion made above is a practical one for a theory that leads to the results of this appendix, and in fact for the content of most of the book.)

The chemical potential of, say, component A is defined by

$$\mu_A = \left(\frac{\partial A}{\partial N_A} \right)_{T,V,N_B} = -\left(kT \frac{\partial \ln Q}{\partial N_A} \right)_{T,V,N_B} \qquad (A.3)$$

Because of the extensive character of the Helmholtz free energy one can replace this derivative with respect to N_A by a difference, namely,

$$\mu_A = A(T, V, N_A + 1, N_B) - A(T, V, N_A, N_B) \tag{A.4}$$

provided we take the proper "thermodynamic limit," i.e., that N_A, N_B, and V are very large, but $\varrho_A = N_A/V$ and $\varrho_B = N_B/V$ are constants.

Writing the partition function (A.1) once for N_A and once for $N_A + 1$, and then taking the ratio of the two quantities, we obtain

$$\mu_A = -kT \ln \frac{Q(T, V, N_A + 1, N_B)}{Q(T, V, N_A, N_B)}$$

$$= -kT \ln \frac{q_A Z(N_A + 1, N_B)}{(N_A + 1)\Lambda_A{}^3 Z(N_A, N_B)} \tag{A.5}$$

Now we write the total potential energy of the system of $N_A + 1$ and N_B particles as

$$U(N_A + 1, N_B) = U(N_A, N_B) + B_A(\mathbf{R}_A) \tag{A.6}$$

where $\mathbf{R}_A$ is the location of the added molecule of type A; all the locations of the remaining particles are not specified explicitly, but they are implicitly contained in the arguments of U. $B_A(\mathbf{R}_A)$ is now referred to as the total *binding energy* of one A molecule at $\mathbf{R}_A$ to all the other molecules of the system at some specified configuration.

Since we can choose the origin of our coordinate system at any point, without affecting the properties of the system, one can measure the distances of all the molecules relative to the origin chosen at $\mathbf{R}_A$. Thus by transforming to coordinates relative to $\mathbf{R}_A$ one can easily show [for details see Ben-Naim (1974), Section 3.5] that integration over $\mathbf{R}_A$ may be performed to obtain

$$Z(N_A + 1, N_B) = \int \cdots \int_V d\mathbf{R}^{N_A+1} d\mathbf{R}^{N_B} \exp[-\beta U(N_A, N_B) - \beta B_A(\mathbf{R}_A)]$$

$$= V \int \cdots \int d\mathbf{R}^{N_A} d\mathbf{R}^{N_B} \exp[-\beta U(N_A, N_B) - \beta B_A(\mathbf{R}_A)] \tag{A.7}$$

where in the second form on right-hand side of (A.7) the integration is over all locations of the $N_A + N_B$ molecules only. The added molecule is now at a *fixed* position, which we still denote by $\mathbf{R}_A$. It is also important to realize that the whole integral is not dependent on the particular choice of $\mathbf{R}_A$, since all the points in the (macroscopic) liquid are presumed to be equivalent.

Taking $\varrho_A = (N_A + 1)/V \approx N_A/V$ as the number density of the A molecules, we write (A.5) as

$$\mu_A = kT \ln(\varrho_A \varLambda_A{}^3 q_A{}^{-1}) - kT \ln\langle\exp[-\beta B_A(\mathbf{R}_A)]\rangle \qquad \text{(A.8)}$$

The symbol $\langle\ \rangle$ stands for an average over all the locations of the $N_A + N_B$ molecules, i.e.,

$$\langle\exp[-\beta B_A(\mathbf{R}_A)]\rangle = \frac{\int \cdots \int d\mathbf{R}^{N_A}\, d\mathbf{R}^{N_B} \exp[-\beta U(N_A, N_B) - \beta B_A(\mathbf{R}_A)]}{\int \cdots \int d\mathbf{R}^{N_A}\, d\mathbf{R}^{N_B} \exp[-\beta U(N_A, N_B)]}$$

$$= \int \cdots \int d\mathbf{R}^{N_A}\, d\mathbf{R}^{N_B} P(\mathbf{R}^{N_A}, \mathbf{R}^{N_B})\, \exp[-\beta B_A(\mathbf{R}_A)]$$

$$\text{(A.9)}$$

Note that the average is taken using the distribution function $P(\mathbf{R}^{N_A}, \mathbf{R}^{N_B})$ of a system of $N_A + N_B$ particles; this distribution is given by the well-known Boltzmann relation

$$P(\mathbf{R}^{N_A}, \mathbf{R}^{N_B}) = \frac{\exp[-\beta U(N_A, N_B)]}{\int \cdots \int d\mathbf{R}^{N_A}\, d\mathbf{R}^{N_B} \exp[-\beta U(N_A, N_B)]} \qquad \text{(A.10)}$$

The most important result for our purposes is the expression for the chemical potential in (A.8). This form is sufficient for most of our applications. However, for the purpose of interpretation of certain quantities, it is convenient at this stage to introduce an auxiliary quantity, which we shall call the pseudo-chemical-potential. This is defined as in (A.4) but with the additional restriction that the added molecule be placed at a *fixed* position, say $\mathbf{R}_A$, in the solvent, i.e.,

$$\tilde{\mu}_A = A(T, V, N_A + 1, N_B, \mathbf{R}_A) - A(T, V, N_A, N_B) \qquad \text{(A.11)}$$

(In Section 1.3 a similar quantity has been defined in the so-called T, P, N ensemble, which is more useful for practical purposes.) Repeating almost exactly the steps leading from (A.4) to (A.8) for μ_A, we obtain for $\tilde{\mu}_A$ the final relation

$$\tilde{\mu}_A = kT \ln q_A{}^{-1} - kT \ln\langle\exp[-\beta B_A(\mathbf{R}_A)]\rangle \qquad \text{(A.12)}$$

Comparing this with (A.8) shows that the difference between μ_A and $\tilde{\mu}_A$ is in the term $kT \ln \varrho_A \varLambda_A{}^3$. Combining (A.8) and (A.12) we obtain

$$\mu_A = \tilde{\mu}_A + kT \ln \varrho_A \varLambda_A{}^3 \qquad \text{(A.13)}$$

This relation has a very simple interpretation. The free energy change for adding a new A particle to the system is split into two parts corresponding to two consecutive steps: First we place the molecule at a *fixed* position, the corresponding work being $\bar{\mu}_A$, and second we release the constraint of a fixed position, which leads to an additional change in free energy—by the amount $kT \ln \varrho_A \Lambda_A{}^3$. This term originates from the acquisition of translational degrees of freedom of the newly added particle [for more details on this see Ben-Naim (1974), Sections 3.4 and 3.5], and may be referred to as the "liberation" free energy change. We note here that the same expression is valid in the T, P, N ensemble where ϱ_A is understood to be the average density of the A molecules.

It is convenient to denote the second term on the right-hand side of (A.12) by

$$W(A \mid A + B; x_A) = -kT \ln \langle \exp[-\beta B_A] \rangle \qquad (A.14)$$

where $W(A \mid \cdots)$ stands for the work of coupling an A molecule, located at some fixed position $\mathbf{R}_A$, to the entire environment, the composition of which is specified on the right-hand side of the vertical line. In our particular example this is a mixture of A and B with composition $x_A = N_A/(N_A+N_B)$. We have also used the shorter notation B_A for $B_A(\mathbf{R}_A)$.

Note that in the very general case μ_A contains the typical term $kT \ln \varrho_A$ with ϱ_A the *number density* of A. This is true for any mixture of A and B. We now specify some particular cases.

(1) *The Pure Component A.* For this case let $\varrho_A{}^p$ be the number density of pure A and $\mu_A{}^p$ its chemical potential; we have

$$\begin{aligned} \mu_A{}^p &= \bar{\mu}_A + kT \ln \varrho_A{}^p \Lambda_A{}^3 \\ &= W(A \mid \text{pure } A) + kT \ln(\varrho_A{}^p \Lambda_A{}^3 q_A{}^{-1}) \end{aligned} \qquad (A.15)$$

(2) *The Symmetrical Ideal Solution.* Suppose that A and B are very similar to each other, say two isotopes of the same molecule. In such a case we should have for any composition x_A

$$W(A \mid A + B; x_A) = W(A \mid \text{pure } A) \qquad (A.16)$$

provided that the number density $\varrho_A{}^p$ of the pure A is equal to the total density of the mixture $\varrho_A{}^p = \varrho_A + \varrho_B$. The meaning of (A.16) is self-evident: the coupling work of one A against an environment composed of a mixture of two almost identical components is almost equal to the coupling work of A against pure A, provided that the total density around the particle is the same in the two cases. In this case we have the so-called *symmetrical ideal* solution, and the chemical potential of A has the form

$$
\begin{aligned}
\mu_A &= W(A \mid A + B; x_A) + kT \ln[\varrho_A \Lambda_A{}^3 q_A{}^{-1}] \\
&= W(A \mid A) + kT \ln(\varrho_A \Lambda_A{}^3 q_A{}^{-1}) \\
&= \mu_A{}^p - kT \ln[(\varrho_A + \varrho_B)\Lambda_A{}^3 q_A{}^{-1}] + kT \ln[\varrho_A \Lambda_A{}^3 q_A{}^{-1}] \\
&= \mu_A{}^p + kT \ln x_A
\end{aligned}
\tag{A.17}
$$

where we have used the equality $\varrho_A{}^p = \varrho_A + \varrho_B$ and introduced the mole fraction x_A. The last relation is well known for *symmetrical ideal* solutions. We see that in this case the natural composition variable is the mole fraction. We stress, however, that the more general concentration variable is the number density ϱ_A as appears in (A.13).

We also note that the sufficient and necessary conditions for a solution to be symmetrical ideal are less stringent than the condition given above. For a more detailed discussion of this aspect the reader is referred to Ben-Naim (1974), Chapter 4.

(3) *Very Dilute Solution of A in B.* If A is extremely dilute in B, then from (A.8) and (A.14) we write

$$
\mu_A = kT \ln(\varrho_A \Lambda_A{}^3 q_A{}^{-1}) + W(A \mid B)
\tag{A.18}
$$

where now the coupling work of A is against a surrounding that is almost pure B. [From the molecular point of view it is sufficient to consider the case of a solution of *one* A molecule in pure B, in which case $W(A \mid B)$ is the coupling work of A against a strictly pure B.]

We now summarize the results obtained in this appendix. The general expression for the chemical potential of A in any mixture of composition x_A is

$$
\mu_A = kT \ln(\varrho_A \Lambda_A{}^3 q_A{}^{-1}) + W(A \mid A + B; x_A)
\tag{A.19}
$$

For pure A

$$
\mu_A{}^p = kT \ln(\varrho_A{}^p \Lambda_A{}^3 q_A{}^{-1}) + W(A \mid A)
\tag{A.20}
$$

For symmetrical ideal solutions

$$
\mu_A = \mu_A{}^p + kT \ln x_A = [kT \ln(\varrho_A \Lambda_A{}^3 q_A{}^{-1}) + W(A \mid A)] + kT \ln x_A
\tag{A.21}
$$

and for extremely dilute solution of A in B

$$
\mu_A = \mu_A{}^{\circ\varrho} + kT \ln \varrho_A = [W(A \mid B) + kT \ln \Lambda_A{}^3 q_A{}^{-1}] + kT \ln \varrho_A
\tag{A.22}
$$

In (A.21) and (A.22) we have written the chemical potential of A in the more conventional form. It is very important to realize that whereas $\mu_A{}^p$ in (A.21) is a *proper* chemical potential of A in a real system (i.e., pure A), the symbol $\mu_A{}^{o\varrho}$ is not a chemical potential of A in any real system—since it lacks the typical $kT \ln \varrho_A$ term which is present in every expression of the chemical potential (see also Ben-Naim, 1978a).

In concluding it is instructive to demonstrate how by an improper notation one can easily be led to misinterpretations.

For very dilute solutions of A in B one can write the mole fraction of A as

$$x_A = \frac{\varrho_A}{\varrho_A + \varrho_B} \sim \frac{\varrho_A}{\varrho_B} \qquad (A.23)$$

Replacing ϱ_A by $\varrho_B\, x_A$ in (A.22) we obtain

$$
\begin{aligned}
\mu_A &= \mu_A{}^{o\varrho} + kT \ln \varrho_B x_A \\
&= (\mu_A{}^{o\varrho} + kT \ln \varrho_B) + kT \ln x_A \\
&= [W(A \mid B) + kT \ln \varrho_B \Lambda_A{}^3 q_A{}^{-1}] + kT \ln x_A \\
&= \mu_A{}^{ox} + kT \ln x_A
\end{aligned}
\qquad (A.24)
$$

where we have denoted by $\mu_A{}^{ox}$ the standard chemical potential in the mole-fraction scale. Comparing (A.21) with (A.24) shows that the two expressions have the form

$$\mu_A = \begin{Bmatrix} \mu_A{}^p \\ \mu_A{}^{ox} \end{Bmatrix} + kT \ln x_A \qquad (A.25)$$

Now it is clear that the quantity $\mu_A{}^p$ is fundamentally different from $\mu_A{}^{ox}$. The first contains the term $kT \ln \varrho_A$ and the coupling work $W(A \mid A)$. The second contains the term $kT \ln \varrho_B$ and the coupling work $W(A \mid B)$. It is this difference that makes $\mu_A{}^p$ a *chemical potential* of A whereas $\mu_A{}^{ox}$ is merely a quantity that is independent of x_A. Unfortunately, in many textbooks and research articles these two quantities are denoted by the same symbol, and as a result the same meaning is assigned to both.

A.2. CONDITIONAL STANDARD FREE ENERGY OF TRANSFER

Consider a solute of the type $A—B$, where A and B are two groups of the same molecule. As a typical example, treated in Section 2.3, $A—B$ may be alanine, A being the CH_3 group, and B the $—CH(NH_2)COOH$

group. For simplicity we shall disregard internal degrees of freedom of each group and assume that the configuration of the molecule A—B may be described by the centers of the groups A and B.

The standard free energy of solution (i.e., the standard free energy of transfer from the gas to the liquid) of A—B is given by

$$\exp(-\beta\,\Delta\mu^{\circ}_{A-B}) = \frac{\int \cdots \int d\mathbf{X}^{N} \exp[-\beta U_{N}(\mathbf{X}^{N}) - \beta U(\mathbf{X}^{N}\mid A\text{—}B)]}{\int \cdots \int d\mathbf{X}^{N} \exp[-\beta U_{N}(\mathbf{X}^{N})]} \tag{A.26}$$

where $\beta = (kT)^{-1}$ and $U_{N}(\mathbf{X}^{N})$ is the total potential energy of interaction of the solvent molecules at a specified configuration $\mathbf{X}^{N} = \mathbf{X}_{1}, \ldots, \mathbf{X}_{N}$. $U(\mathbf{X}^{N}\mid A\text{—}B)$ is the total binding energy of a solute molecule, at some specific configuration, denoted symbolically by A—B, to the solvent at the configuration $\mathbf{X}^{N}$.

As for the total potential energy of the solvent we do not require the assumption of pairwise additivity. However, for the total binding energy we assume that it may be split into two parts:

$$U(\mathbf{X}^{N}\mid A\text{—}B) = U(\mathbf{X}^{N}\mid A) + U(\mathbf{X}^{N}\mid B) \tag{A.27}$$

This assumption is certainly valid for a "hard solute," i.e., a solute with no attractive part, in which case $U(\mathbf{X}^{N}\mid A\text{—}B)$ becomes infinity whenever a solvent molecule penetrates into the excluded volume determined by the presence of A—B. This excluded volume can be viewed as a superposition of the two excluded volumes of the groups A and B. If A and B are the two methylene groups of $CH_{3}CH_{3}$, then (A.27) is a reasonably good approximation. However, in general, this approximation is not valid. As an example consider the case where A is an ionic group, say $A = COO^{-}$ and $B = CH_{3}$, and the solvent consists of one water molecule W at some specified configuration. Let us denote by $U(W\mid A)$ and $U(W\mid B)$ the interaction energies of W with A and B separately. In general we do not have an equality of the form

$$U(W\mid A\text{—}B) = U(W\mid A) + U(W\mid B) \tag{A.28}$$

The reason is that the presence of A may polarize the water molecule to such an extent that it affects the interaction between W and B. Thus (A.28) should be modified to

$$U(W\mid A\text{—}B) = U(W\mid A) + U^{*}(W\mid B) \tag{A.29}$$

The meaning of (A.29) is the following. The total work of bringing a water molecule from infinity to some close configuration relative to A—B is split into two parts: First we bring A and W to the final configuration, the corresponding work being $U(W \mid A)$. Next we bring the group B to its final position, the corresponding work being $U^*(W \mid B)$; but this work is different from the work one would have done in the second step, had A not been placed in its position.

Clearly, if there are N solvent molecules, the same considerations apply to each solvent molecule. The point we would like to demonstrate now is that a simple split of the standard free energy of solution of A—B into two contributions is in general *unjustified*, even when the most favorable conditions do exist, namely, that relation (A.27) is valid.

Substituting (A.27) into (A.26) we obtain

$$\exp(-\beta \, \Delta\mu^{\circ}_{A-B}) = \langle \exp[-\beta U(\mathbf{X}^N \mid A-B)] \rangle_0$$
$$= \langle \exp[-\beta U(\mathbf{X}^N \mid A)] \exp[-\beta U(\mathbf{X}^N \mid B)] \rangle_0 \quad \text{(A.30)}$$

where we used the symbol $\langle \; \rangle_0$ to denote an average over all configurations of the solvent molecules.

In the second form on the right-hand side of (A.30) we have rewritten the average of the random variable $\exp[-\beta U(X^N \mid A-B)]$ as an average over a product of two random variables, corresponding to the groups A and B. This average may be factored into a product of two averages only if the two random variables are independent. Clearly, since the groups A and B are very close to each other in the molecule A—B, the above two random variables cannot be independent, and such a factorization is in general impossible.

If the groups A and B were very far apart, or if for any other reason the two random variables may be considered to be independent, then we could write

$$\exp(-\beta \, \Delta\mu^{\circ}_{A-B}) = \langle \exp[-\beta U(\mathbf{X}^N \mid A)] \rangle_0 \langle \exp[-\beta U(\mathbf{X}^N \mid B)] \rangle_0$$
$$= \exp(-\beta \, \Delta\mu_A{}^{\circ}) \exp(-\beta \, \Delta\mu_B{}^{\circ}) \quad \text{(A.31)}$$

or, equivalently

$$\Delta\mu^{\circ}_{A-B} = \Delta\mu_A{}^{\circ} + \Delta\mu_B{}^{\circ} \quad \text{(A.32)}$$

We thus conclude that if A and B are close to each other, and even when relation (A.27) is valid exactly, we cannot expect the additivity of $\Delta\mu^{\circ}_{A-B}$ as written in (A.32) to be valid. This is true for any dense solvent, including a solvent consisting of hard spheres.

From the molecular-theoretical point of view the quantity

$$\exp[-\beta(\Delta\mu_{A-B}^\circ - \Delta\mu_A^\circ - \Delta\mu_B^\circ)]$$

is exactly the function y_{AB} at the point where A and B form the molecule A—B, and this in general is larger than unity for short distances between A and B; for more details, see Ben-Naim (1974), Chapter 2.

However, since relations of the form (A.32) are often cited in the literature (see the example discussed in Section 2.3), it is interesting to examine what kind of additivity may be written that is similar in appearance to (A.32).

To this end let us rewrite (A.31) in more detail as

$$\exp(-\beta\,\Delta\mu_{A-B}^\circ) = \frac{\int \cdots \int d\mathbf{X}^N \exp[-\beta U_N - \beta U(N\,|\,B) - \beta U(N\,|\,A)]}{\int \cdots \int d\mathbf{X}^N \exp(-\beta U_N)}$$

$$= \frac{\int \cdots \int d\mathbf{X}^N \exp[-\beta U_{N+B} - \beta U(N\,|\,A)]}{\int \cdots \int d\mathbf{X}^N \exp(-\beta U_{N+B})}$$

$$\times \frac{\int \cdots \int d\mathbf{X}^N \exp(-\beta U_{N+B})}{\int \cdots \int d\mathbf{X}^N \exp(-\beta U_N)}$$

$$= \int \cdots \int d\mathbf{X}^N P(\mathbf{X}^N/B) \exp[-\beta U(N\,|\,A)]$$

$$\times \int \cdots \int d\mathbf{X}^N P(\mathbf{X}^N) \exp[-\beta U(N\,|\,B)]$$

$$= \langle \exp[-\beta U(N\,|\,A)]\rangle_B \langle \exp[-\beta U(N\,|\,B)]\rangle_0 \qquad (A.33)$$

In the first form on the right-hand side of (A.33) we have used an obvious simplified notation for relation (A.26), using also (A.27). In the second form we have expanded the quotient by the factor $\int \cdots \int d\mathbf{X}^N \exp(-\beta U_{N+B})$, where U_{N+B} stands for $U_N + U(N\,|\,B)$. In the third form we identify the two factors; one is an average over all configurations of the solvent molecules denoted by $\langle \rangle_0$, the second is a *conditional* average, denoted by $\langle \rangle_B$, which is an average over all the configurations of the solvent molecules given B at some fixed position.

We now define a *conditional* standard free energy of solution of A, given B, by

$$\exp(-\beta\,\Delta\mu_{A/B}^\circ) = \langle \exp[-\beta U(N\,|\,A)]\rangle_B \qquad (A.34)$$

and relation (A.33) is rewritten as

$$\Delta\mu_{A-B}^\circ = \Delta\mu_B^\circ + \Delta\mu_{A/B}^\circ \qquad (A.35)$$

Since all the arguments are symmetrical with respect to A and B, we could also write

$$\Delta\mu^{\circ}_{A-B} = \Delta\mu_A{}^{\circ} + \Delta\mu^{\circ}_{B/A} \tag{A.36}$$

Relation (A.35) is the exact, or the corrected, analog of (A.32) for the case when the two random variables in (A.30) are dependent [in both cases we assumed the validity of (A.27)].

The meaning of (A.35) [or (A.36)] is very simple. The work required to transfer the molecule A—B from a fixed position and orientation in the gas to a fixed position and orientation in the liquid is split into two parts. First we transfer B alone. Then we transfer A from a fixed position in the gas to a fixed position, but next to B, in the liquid. The two parts of the process are indicated in Figure A.1. Note that in this process we have "cut" a chemical bond at some arbitrary point. But the combination of the two steps has effectively transferred the *entire* molecule A—B from one phase to the other. Therefore we do not have to take the bond energy into consideration. Clearly, the fact that B is held at a fixed position before transferring A has a significant effect on the work of transferring A; this is the reason $\Delta\mu^{\circ}_{A/B}$ should be different from $\Delta\mu_A{}^{\circ}$ as written in (A.32), and a concrete experimental example is worked out in Section 2.3.

Before concluding this section it should be noted that although A and B are not free molecular entities, it is conceptually legitimate to cut a molecule, say CH_3—CH_3, into two parts CH_3 and visualize the transfer of each part separately. (Note that by a "CH_3" we simply mean "half" of the ethane molecule; the methyl group has the same configuration as in the original molecule. A *real* methyl radical CH_3 might have a planar configuration and therefore a different behavior than "half" an ethane molecule.)

Furthermore, it is reasonable to assume that $\Delta\mu^{\circ}_{CH_3}$ is approximately equal to $\Delta\mu^{\circ}_{CH_4}$; the latter is an experimentally measurable quantity. With this approximation we can make an estimate of the difference between $\Delta\mu_A{}^{\circ}$ and $\Delta\mu^{\circ}_{A/B}$. Consider the molecule CH_3—CH_3 for A—B (i.e., A and B are the same). Thus, (A.35) can be written as

$$\Delta\mu^{\circ}_{CH_3-CH_3} = \Delta\mu^{\circ}_{CH_3} + \Delta\mu^{\circ}_{CH_3/CH_3}$$
$$\approx \Delta\mu^{\circ}_{CH_4} + \Delta\mu^{\circ}_{CH_3/CH_3} \tag{A.37}$$

Using experimental values for ethane and methane at 20°C, we obtain (see data in Section 2.3)

$$\Delta\mu^{\circ}_{CH_3-CH_3} = 2.36 \text{ kcal/mol} \qquad \Delta\mu^{\circ}_{CH_4} = 1.57 \text{ kcal/mol} \tag{A.38}$$

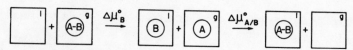

Figure A.1. A molecule A—B is transferred from a gaseous phase (g) into a liquid (l) in two steps. First the radical B is transferred alone, then A is transferred to a position next to B in the liquid.

hence

$$\Delta\mu^{\circ}_{\text{CH}_3/\text{CH}_3} = 2.36 - 1.57 = 0.79 \text{ kcal/mol} \qquad (A.39)$$

This is quite different from $\Delta\mu^{\circ}_{\text{CH}_4}$. We believe that the major reason for this difference is the *condition* imposed in the very definition of $\Delta\mu^{\circ}_{\text{CH}_3/\text{CH}_3}$, and partially due to the replacement of CH_3 by CH_4. [Note that had we taken larger hydrocarbons such as CH_3—$(\text{CH}_2)_n$—CH_3, the difference between CH_3—$(\text{CH}_2)_n$— and CH_3—$(\text{CH}_2)_n\text{H}$ would be negligible for large n.]

In conclusion, we list here three possible situations:

(1) If additivity of the form (A.27) is valid, and if the assumption of independence of the two random variables in (A.30) may be assumed, then the split of $\Delta\mu^{\circ}_{A-B}$ into two terms as in (A.32) is valid.

(2) If (A.27) is valid (say for solute molecules having a weak interaction with the solvent), but the independence of the two random variables in (A.30) is not presumed, then relations of the form (A.35) or (A.36) are valid. In this case, care must be exercised to distinguish between $\Delta\mu^{\circ}_{A/B}$ and $\Delta\mu_A^{\circ}$.

(3) If one of the groups A or B strongly interacts with the solvent, then the validity of (A.27) cannot be granted, and hence even the result (A.35) is incorrect. However, one may still use the formal split of $\Delta\mu^{\circ}_{A-B}$ in (A.35) with the understanding that $\Delta\mu^{\circ}_{A/B}$ is not defined by (A.34), but by

$$\exp(-\beta\,\Delta\mu^{\circ}_{A/B}) = \langle \exp[-\beta U^*(N\,|\,A)] \rangle_B \qquad (A.40)$$

where U^* is the modified potential that should be used in (A.29).

Thus the process of splitting $\Delta\mu^{\circ}_{A-B}$ into two parts as discussed in Section 2.3 is almost certainly invalid for molecules such as carboxylic acids, amino acids, etc. As we have seen in Section 2.3, $\Delta\mu^{\circ}_{A/B}$ may even have a different sign than $\Delta\mu_A^{\circ}$. With such an analysis we can rationalize the apparently contradictory results for the standard free energy of transfer between H_2O and D_2O (see Section 2.3).

A.3. DIMERIZATION FREE ENERGY OF MOLECULES OF THE TYPE R—Y

This appendix analyzes the validity of the split of $\Delta G_D{}^\circ$ into two contributions as discussed in Section 3.7. For simplicity of the mathematical presentation we assume that the molecules undergoing dimerization have the form R—Y, that the two groups R and Y do not involve internal rotational degrees of freedom, and that the configuration of the dimer is fixed as indicated below. The dimerization process is schematically represented by

$$R_1{-}Y_1 + R_2{-}Y_2 \rightleftarrows \quad \begin{array}{c} R_1{-}Y_1 \\ \vdots \\ R_2{-}Y_2 \end{array} \qquad (A.41)$$

A concrete example which we have in mind is the carboxylic acids, where in the dimer (D) a hydrogen bond is formed between the two Y groups (the carboxylic groups), and there is also hydrophobic interaction between the alkyl groups denoted by R (the numbers 1 and 2 serve to distinguish between the two specific monomers, which are chemically identical to each other).

The question under examination is to what extent one can split the standard free energy of dimerization $\Delta G_D{}^\circ$ into two contributions, as we have done in Section 3.7, namely,

$$\Delta G_D{}^\circ = \Delta G_{YY}^\circ + \Delta G_{HI}^\circ \qquad (A.42)$$

and whether one can really extract information on hydrophobic interaction from measurements of association constants.

The following treatment is partially dependent on the analysis made in Appendix A.2, which should be reviewed before reading this appendix. Using the notation of Section 1.2 we write the chemical potentials of the monomer and of the dimer in the form

$$\mu_M = kT \ln(\varrho_M \Lambda_M{}^3 q_{rot,M}^{-1} q_{vib,M}^{-1}) + W(M \mid W) \qquad (A.43)$$

$$\mu_D = kT \ln(\varrho_D \Lambda_D{}^3 q_{rot,D}^{-1} q_{vib,D}^{-1}) + U(D) + W(D \mid W) \qquad (A.44)$$

where q_{rot} and q_{vib} stand for the rotational and vibrational partition functions of a single molecule (monomer or dimer) and $U(D)$ is the direct interaction between the two monomers at the configuration of the dimer. In more detail, it may be represented as

$$U(D) = U(R_1, R_2) + U(R_1, Y_2) + U(R_2, Y_1) + U(Y_1, Y_2) \qquad (A.45)$$

where the configurations of R_1—Y_1 and R_2—Y_2 are as in Equation (A.41). We may combine the three factors $\Lambda^3 \cdot q_{rot}^{-1}q_{vib}^{-1}$ into a single one, denoted by q_{int}^{-1}, and write the condition of chemical equilibrium $2\mu_M = \mu_D$ as

$$2kT \ln(\varrho_M q_{int,M}^{-1}) + 2W(M \mid W)$$
$$= kT \ln(\varrho_D q_{int,D}^{-1}) + U(D) + W(D \mid W) \qquad (A.46)$$

Then using the conventional notation for the standard chemical potentials of M and D defined by

$$\mu_M{}^{\circ l} = W(M \mid W) + kT \ln q_{int,M}^{-1} \qquad (A.47)$$

$$\mu_D{}^{\circ l} = W(D \mid W) + U(D) + kT \ln q_{int,D}^{-1} \qquad (A.48)$$

we get the standard free energy of dimerization in the form

$$\Delta G_D{}^{\circ l} = \mu_D{}^{\circ l} - 2\mu_M{}^{\circ l} = W(D \mid W) + U(D) - 2W(M \mid W)$$
$$+ kT \ln(q_{int,M}^2/q_{int,D}) \qquad (A.49)$$

Clearly, if exactly the same dimerization process occurs in the absence of the solvent W, the corresponding standard free energy of dimerization would be

$$\Delta G_D{}^{\circ g} = U(D) + kT \ln(q_{int,M}^2/q_{int,D}) \qquad (A.50)$$

The statistical mechanical expression for $\Delta G_D{}^{\circ l}$ is

$$\Delta G_D{}^{\circ l} = U(D) - kT \ln\left\{\frac{\langle\exp[-\beta B(D)]\rangle_0}{\langle\exp[-\beta B(M)]\rangle_0{}^2}\right\} + kT \ln(q_{int,M}^2/q_{int,D})$$
$$(A.51)$$

where $B(D)$ and $B(M)$ are the binding energies of D and M to the solvent.

The main question posed here is whether there exist any "natural" splits of this quantity into two terms in such a way that the two terms may be assigned the meaning of the hydrogen bond (HB) and the HI contributions as in (A.42).

Clearly, even without having the solvent we run into difficulties if we try to carry out such a split. For instance, consider the interaction energy $U(D)$ in (A.45). $U(R_1, R_2)$ is the direct interaction between the alkyl groups, and $U(Y_1, Y_2)$ may be assigned the meaning of the contribution due to HB formation between the two carboxylic groups. The cross terms $U(R_1, Y_2)$ and $U(R_2, Y_1)$ cannot be classified in an obvious way as "belonging" to either of the two terms in (A.42). A similar difficulty arises in trying to split the last term on the right-hand side of (A.51) into two contributions as in (A.42). This term involves essentially the loss of some internal degrees of freedom of the monomers and the gain of some new ones in the dimer. At this stage we may conclude that a split of the form (A.42), though very appealing on intuitive grounds, cannot be justified in statistical mechanical terms. However, we propose a way of extracting information on a *conditional* HI from the same data. To do that we look at the difference between $\Delta G_D^{\circ l}$ and $\Delta G_D^{\circ g}$, which is

$$\delta G^{HI} = \Delta G_D^{\circ l} - \Delta G_D^{\circ g} = -kT \ln\left\{\frac{\langle \exp[-\beta B(D)]\rangle_0}{\langle \exp[-\beta B(M)]\rangle_0^2}\right\} \quad (A.52)$$

where we have eliminated the first and last terms on the right-hand side of (A.51). We have used the notation δG^{HI} since this has exactly the same form and the same meaning as the δG^{HI} introduced in Section 1.3, i.e., this is the indirect part of the work required to bring two monomers from fixed positions and orientations at infinite separation to the final configuration. The internal partition function does not appear in this expression; the very fact that we have separated the internal partition functions $q_{int,M}$ and $q_{int,D}$ from the coupling work means that on the average the binding energy of the molecule does not depend on the precise molecular states that are counted in q_{int}. However, if internal rotations do exist, then this assumption is no longer valid, since the binding energy of a solute changes with the change of an internal rotational angle. In such cases all of the above treatment is still true for a *fixed* set of internal rotational angles—i.e. for a fixed conformation of the solute. At the end one must perform an average over all conformations to get the proper expression for the standard free energy of dimerization. (See also Appendix A.5.)

The quantity δG^{HI} is the *total indirect* work as discussed above. We shall now try to extract from δG^{HI} a quantity which may be called the conditional HI between the R groups. To do that we use the idea of the conditional standard free energy of transfer that has been defined in Appendix A.2.

As in equation (A.33) we can write, using notation similar to that in Appendix A.2,

$$\langle \exp[-\beta B(M)]\rangle_0 = \frac{\int \cdots \int d\mathbf{X}^N \exp[-\beta U_N - \beta U(N\,|\,R) - \beta U(N\,|\,Y)]}{\int \cdots \int d\mathbf{X}^N \exp(-\beta U_N)}$$

$$= \langle \exp[-\beta U(N\,|\,R)]\rangle_Y \langle \exp[-\beta U(N\,|\,Y)]\rangle_0$$

$$= \exp(-\beta\,\Delta\mu_{R/Y}^\circ) \exp(-\beta\,\Delta\mu_Y^\circ) \quad (A.53)$$

where $\Delta\mu_Y^\circ$ is the standard free energy of solution (i.e., transfer from the gas to the liquid) of the group Y, and $\Delta\mu^\circ_{R/Y}$ is the *conditional* standard free energy of transferring R given Y (for more details see Appendix A.2).

The corresponding expression for the dimer may be written in a similar fashion:

$$\langle\exp[-\beta B(D)]\rangle_0$$

$$= \frac{\int \cdots \int d\mathbf{X}^N \exp[-\beta U_N - \beta U(N \mid R_1 R_2) - \beta U(N \mid Y_1 Y_2)]}{\int \cdots \int d\mathbf{X}^N \exp(-\beta U_N)}$$

$$= \langle\exp[-\beta U(N \mid R_1 R_2)]\rangle_{Y_1 Y_2} \langle\exp[-\beta U(N \mid Y_1 Y_2)]\rangle_0$$

$$= \exp(-\beta \, \Delta\mu^\circ_{R_1 R_2/Y_1 Y_2}) \exp(-\beta \, \Delta\mu^\circ_{Y_1 Y_2}) \qquad (A.54)$$

In (A.54) we have used a shorthand notation for $B(D)$, the binding energy of the dimers,

$$B(D) = U(N \mid R_1 R_2) + U(N \mid Y_1 Y_2) \qquad (A.55)$$

where

$$U(N \mid R_1 R_2) = U(N \mid R_1) + U(N \mid R_2)$$

and

$$U(N \mid Y_1 Y_2) = U(N \mid Y_1) + U(N \mid Y_2) \qquad (A.56)$$

Clearly, since there are four terms in $B(D)$ there are several possibilities for choosing the *condition* in the transformations carried out in (A.54) [similar to the two possible cases in (A.35) and (A.36) of Appendix A.2]. We have chosen one possibility that is most suitable for the end toward which we are striving.

Combining the results of (A.53) and (A.54) with (A.52) we can write

$$\delta G^{\mathrm{HI}} = \Delta\mu^\circ_{R_1 R_2/Y_1 Y_2} + \Delta\mu^\circ_{Y_1 Y_2} - 2(\Delta\mu^\circ_{R/Y} + \Delta\mu_Y{}^\circ) \qquad (A.57)$$

We also use the notation

$$\delta G^{\mathrm{HI}}_{R_1 R_2/Y_1 Y_2} = \Delta\mu^\circ_{R_1 R_2/Y_1 Y_2} - 2\Delta\mu^\circ_{R/Y} \qquad (A.58)$$

$$\delta G^{\mathrm{HI}}_{Y_1 Y_2} = \Delta\mu^\circ_{Y_1 Y_2} - 2\Delta\mu_Y{}^\circ \qquad (A.59)$$

Perhaps the quantity that is closest to the concept of the HI between the two alkyl groups R_1 and R_2 in the dimer is $\delta G^{\mathrm{HI}}_{R_1 R_2/Y_1 Y_2}$. This is a *conditional* HI, however, i.e., the indirect part of the work required to transfer two R groups from fixed positions next to Y groups, but at infinite separation, to two new fixed positions next to a pair of Y groups at a fixed configuration of the dimer as in (A.41). The process is schematically shown in Figure A.2.

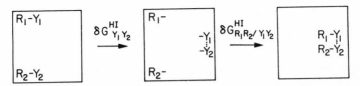

Figure A.2. A two-step formation of a dimer from two monomers of the type R—Y. The corresponding indirect parts of the free energy changes are defined in Equation (A.60).

Combining (A.58) and (A.59) with (A.51) we obtain

$$\Delta G_D{}^{ol} = U(D) + \delta G^{HI} + kT \ln(q^2_{int,M}/q_{int,D})$$
$$= \delta G^{HI}_{R_1R_2/Y_1Y_2} + \delta G^{HI}_{Y_1Y_2} + U(D) + kT \ln(q^2_{int,M}/q_{int,D}) \qquad (A.60)$$

By comparing this equation with (A.42) one can choose $\delta G^{HI}_{R_1R_2/Y_1Y}$ as the most faithful representative of the HI between the two R groups of the monomers [and perhaps one may also add the direct term $U(R_1, R_2)$ from (A.45) that is included in $U(D)$]. Clearly, all the remaining terms cannot be easily assigned the meaning of the contribution due to the Y_1-Y_2 interaction or to the formation of a HB.

To conclude this appendix we would like to propose that instead of looking at $\Delta G_D{}^{ol}$, one should preferably study the differences of $\Delta G_D{}^{\circ}$ in two phases, as defined in (A.52). This would lead to a meaningful measure of the HI between the two solute monomers, provided the same dimer is involved in the two phases. From δG^{HI} one can proceed to extract the quantity $\delta G^{HI}_{R_1R_2/Y_1Y_2}$ by approximating $\Delta\mu_Y{}^{\circ}$ and $\Delta\mu^{\circ}_{Y_1Y_2}$ by the corresponding standard free energies of solution of the lowest homologue molecules, say formic acid for —COOH, or glycine for —HC(NH₂)COOH. Adopting such an approximation, which we write symbolically as

$$\Delta\mu^{\circ}_{HY} \sim \Delta\mu_Y{}^{\circ}, \qquad \Delta\mu^{\circ}_{Y_1Y_2} \approx \Delta\mu^{\circ}_{HY_1HY_2} \qquad (A.61)$$

we can use (A.59) to estimate $\delta G^{HI}_{Y_1Y_2}$, and from the measurement of δG^{HI} through (A.52) one can extract $\delta G^{HI}_{R_1R_2/Y_1Y_2}$ from

$$\delta G^{HI}_{R_1R_2/Y_1Y_2} = \delta G^{HI} - \delta G^{HI}_{Y_1Y_2} \qquad (A.62)$$

This quantity would reflect the HI between two alkyl groups next to the two Y groups. Perhaps it should be noted that such a HI is expected to differ from the (unconditional) HI between two alkyl groups in a pure solvent. This should not necessarily be a disadvantage—in fact it may be beneficial, since after all we are interested in HI between groups attached to some kind of backbone as in proteins.

A.4. RUDIMENTS OF THE SCALED-PARTICLE THEORY (SPT)

In connection with the general problem of HI, the SPT has been mentioned in this book twice. The first place was in Section 2.5, where we were concerned with the problem of the solubility, or the standard free energy of transfer from the gas to the liquid. The second place was in Section 4.6, where we discussed the application of the SPT to the problem of aggregation of many simple solute molecules, forming a sort of large spherical "droplet" in the solvent.

In this appendix we present a very brief survey of this theory so that the reader can get an idea of the assumptions involved and of the applicability of the theory to aqueous fluids.

Basically the SPT has been devised and used for the study of a hard-sphere (HS) fluid. It was later also found useful and successful for simple fluids such as the inert gases in the liquid state. More recently the SPT has also been applied to aqueous solutions.

The basic ingredients of the SPT and the nature of the approximation involved are quite simple. Of course, there are some details in the theory that are quite sophisticated. These are of interest to the practitioners in this field and will not concern us in this short presentation.

The starting point of the SPT is the consideration of the work of creating a cavity at some *fixed* position in the fluid.

In a fluid consisting of HS particles of diameter a, a cavity of radius r at $\mathbf{R}_0$ is nothing but a stipulation that no centers of particles may be found in the sphere of radius r centered at $\mathbf{R}_0$. In this sense, creation of a cavity of radius r at $\mathbf{R}_0$ is equivalent to placing at $\mathbf{R}_0$ a HS solute of diameter b such that $r = (a + b)/2$. Hence the work required to create such a cavity is equivalent to the work required to introduce a HS solute at $\mathbf{R}_0$. This work is computed by using a continuous process of "building up" the particle in the solvent. This is the origin of the name "scaled-particle theory."

In a fluid of HS particles, the sole molecular parameter that fully describes the particles is the diameter a. It is important to bear this fact in mind when the theory is applied to real fluids, in which case one needs at least two molecular parameters to describe the molecules, and more than two parameters for complex molecules such as water. It is a unique feature of the HS fluid that only one molecular parameter is sufficient for its characterization.

The fundamental distribution function in the SPT is $P_0(r)$, the probability that no molecule has its center within the spherical region of radius r centered at some fixed point R_0 in the fluid. (In a homogeneous fluid all the points of the system are presumed to be equivalent, except for a small region near the boundaries, which is negligible in macroscopic systems.) This function was originally introduced by Hill (1958) and later formed the cornerstone of the SPT, as developed by Reiss *et al.* (1959) and Reiss (1966).

It is important to note that a cavity is considered "empty" if no centers of particles are found within it; such a cavity may well be "filled" in the ordinary sense of the word.

At this point it is also worth noting that a HS of zero diameter (or a point HS) produces a cavity of radius $a/2$ in the system. Alternatively, a cavity of radius zero is equivalent to placing a HS of *negative* diameter $b = -a$.

Let $P_0(r + dr)$ be the probability that a cavity of radius $r + dr$ be empty. (In all the following, a cavity is always assumed to be centered at some fixed point R_0, but this will not be mentioned explicitly.) This probability may be written as

$$P_0(r + dr) = P_0(r)P_0(dr/r) \tag{A.63}$$

where on the right-hand side of (A.63) we have introduced the symbol $P_0(dr/r)$ for the *conditional* probability of finding the spherical shell of width dr empty, given that the sphere of radius r is empty. The equality (A.63) is nothing but the well-known definition of a conditional probability in terms of the joint probability.

We now *define* an auxiliary function $G(r)$ by the relation

$$4\pi r^2 \varrho G(r)\, dr = 1 - P_0(dr/r) \tag{A.64}$$

Clearly, since $P_0(dr/r)$ is the conditional probability of finding the spherical shell empty, given that the sphere of radius r is empty, the right-hand side of (A.64) is the conditional probability of finding the center of *at least* one particle in this spherical shell, given that the sphere of radius r is empty.

Expanding $P_0(r + dr)$ to first order in dr, we get

$$P_0(r + dr) = P_0(r) + \frac{\partial P_0}{\partial r}\, dr + \cdots \tag{A.65}$$

Hence, from (A.63), (A.64), and (A.65) we obtain

$$\frac{\partial \ln P_0(r)}{\partial r} = -4\pi r^2 \varrho G(r) \tag{A.66}$$

Thus, the function $G(r)$ may be defined either through (A.64) or through (A.66); the latter may also be rewritten in the integral form

$$\ln P_0(r) - \ln P_0(r = 0) = -\varrho \int_0^r 4\pi \lambda^2 G(\lambda)\, d\lambda \qquad (A.67)$$

Since the probability of finding a cavity of radius zero is unity, we get the relation

$$\ln P_0(r) = -\varrho \int_0^r 4\pi \lambda^2 G(\lambda)\, d\lambda \qquad (A.68)$$

Next we turn to an important relation between the function $P_0(r)$, or $G(\lambda)$, and a thermodynamic quantity.

We shall derive this relation in the T, V, N ensemble. The probability density of finding a specific configuration $\mathbf{R}^N = \mathbf{R}_1, \ldots, \mathbf{R}_N$ is given by [see, for example, Hill (1956) or Ben-Naim (1974)]

$$P(\mathbf{R}^N) = \frac{\exp[-\beta U(\mathbf{R}^N)]}{\int \underset{V}{\cdots} \int \exp[-\beta U(\mathbf{R}^N)]\, d\mathbf{R}^N} \qquad (A.69)$$

where $\beta = (kT)^{-1}$ and $U(\mathbf{R}^N)$ is the interaction energy of the N particles at the configuration $\mathbf{R}^N$. Thus, the probability of finding an empty spherical region of radius r centered at $\mathbf{R}_0$ may be obtained from (A.69) by integrating over all of the complementary region $V - v(r)$, where $v(r)$ denotes the spherical region of radius r:

$$P_0(r) = \int \underset{V-v(r)}{\cdots} \int P(\mathbf{R}^N)\, d\mathbf{R}^N \qquad (A.70)$$

The restriction $V - v(r)$ under the integral sign means that the integration over each $\mathbf{R}_i$ extends through the region $V - v(r)$ only.

On the other hand we have the following basic relation between the Helmholtz free energy of a system and the corresponding partition function in the T, V, N ensemble:

$$\exp[-\beta A(T, V, N)] = \frac{1}{N!\, \varLambda^{3N}} \int \underset{V}{\cdots} \int \exp[-\beta U(\mathbf{R}^N)]\, d\mathbf{R}^N \qquad (A.71)$$

where $\varLambda^3$ is the momentum partition function, and no internal degrees of freedom are ascribed to the particles.

Similarly the free energy of a system with a cavity of radius r at $\mathbf{R}_0$ is given by

$$\exp[-\beta A(T, V, N; r)] = \frac{1}{N!\, \varLambda^{3N}} \int \underset{V-v(r)}{\cdots} \int \exp[-\beta U(\mathbf{R}^N)]\, d\mathbf{R}^N \qquad (A.72)$$

Thus the ratio of (A.71) and (A.72) gives

$$\exp\{-\beta[A(T, V, N; r) - A(T, V, N)]\}$$

$$= \left\{\int_{V-v(r)} \cdots \int \exp[-\beta U(\mathbf{R}^N)]\, d\mathbf{R}^N\right\} \Big/ \left\{\int_V \cdots \int \exp[-\beta U(\mathbf{R}^N)]\, d\mathbf{R}^N\right\}$$

$$= P_0(r) \tag{A.73}$$

where in the last equality on the right-hand side of (A.73) we have used the result (A.70).

Relation (A.73) is an important connection between the work (at given T, V, N) of creating a cavity of radius r and the probability of finding such a cavity in the system. We rewrite this relation in the form

$$W(r) = A(r) - A = -kT \ln P_0(r)$$

$$= kT\varrho \int_0^r 4\pi\lambda^2 G(\lambda)\, d\lambda \tag{A.74}$$

Since the work required to create a cavity of radius r is the same as the work required to place a hard sphere of diameter $b = 2r - a$ at $\mathbf{R}_0$, we can write, using the notation of Appendix A.1, the pseudo-chemical-potential of the added "solute" as

$$\tilde{\mu}_b = W(r) = kT\varrho \int_0^{(a+b)/2} 4\pi\lambda^2 G(\lambda)\, d\lambda \tag{A.75}$$

In order to get the chemical potential of the solute (having diameter b) we have to add to (A.75) the liberation free energy, namely,

$$\mu_b = \tilde{\mu}_b + kT \ln \varrho_b \Lambda_b^3 \tag{A.76}$$

Note that $\varrho_b = 1/V$ is the "solute" density, whereas $\varrho = N/V$ is the "solvent" density.

A particular case of (A.75) is obtained when we place a "solute" having a diameter $b = a$, i.e., a solute which is indistinguishable from other particles in the system. In this case we have

$$\tilde{\mu}_a = W(r = a) = kT\varrho \int_0^a 4\pi\lambda^2 G(\lambda)\, d\lambda \tag{A.77}$$

The function $G(\lambda)$ was introduced here through relations (A.64) and (A.66). It is now interesting to identify it with another function which plays a central role in the theory of liquids. This is the pair correlation function $g(r)$. To do this we cite, without proof, a relation equivalent to (A.77) which reads

$$\tilde{\mu}_a = kT\varrho \int_0^a 4\pi\lambda^2 g(\lambda, \lambda) \, d\lambda \tag{A.78}$$

where $\varrho g(x, y)$ is the local density of "solvent" particles at a distance x from the center of a "solute" particle that produces a cavity of radius y. Thus if we take the derivatives of (A.77) and (A.78) with respect to a we obtain the equality

$$G(\lambda) = g(\lambda, \lambda) \tag{A.79}$$

and, in particular, for $\lambda = a$, the quantity $\varrho G(a)$ is the density of "solvent" particles at the distance a from the center of any given particle (the diameter of which is also a). In other words, this is the local density of HS's at the boundary of a cavity produced by placing another HS of the same diameter at some fixed position. The integral on the right-hand side of (A.78) describes the work of "coupling" a new particle to the system using a continuous "charging" parameter λ. This process is very well known in the derivation of the expression for the chemical potential (Hill, 1956).

At this stage it is interesting to present the equation of state for a system of hard spheres of diameter a, namely,

$$P/kT = \varrho + \tfrac{2}{3}\pi a^3 \varrho^2 G(a) \tag{A.80}$$

i.e., the equation of state is determined by the function $G(r)$ at a single point $r = a$. Note that for the chemical potential one needs the entire *function* $G(\lambda)$, and not just its value at a single point.

The SPT provides an approximate expression for $P_0(r)$ or equivalently for $G(\lambda)$. Before presenting this expression we note that an exact expression is available for $P_0(r)$ at very small r. If the diameter of the HS particles is a, then in a sphere of radius $r < a/2$ there can be at most one center of a particle at any given time. Thus, for such a small r, the probability of finding the sphere occupied is $4\pi r^3 \varrho/3$. Since this sphere may be occupied by at most one center of a HS, the probability of finding it empty is simply

$$P_0(r) = 1 - \frac{4\pi r^3}{3}\varrho \quad \left(\text{for } r \le \frac{a}{2}\right) \tag{A.81}$$

For spheres with a slightly larger radius, namely, for $r \le a/3^{1/2}$, there can be at most *two* centers of HS's in it; the corresponding expression for $P_0(r)$ is

$$P_0(r) = 1 - \frac{4\pi r^3}{3}\varrho + \frac{\varrho^2}{2}\iint\limits_{v(r)} g(\mathbf{R}_1, \mathbf{R}_2)\, d\mathbf{R}_1\, d\mathbf{R}_2 \qquad (A.82)$$

where $g(\mathbf{R}_1, \mathbf{R}_2)$ is the pair correlation function, and the integration is carried out over the region defined by the sphere of radius r. The last equation is valid for a radius smaller than $a/3^{1/2}$. In a formal fashion one can write expressions similar to (A.82) for larger cavities, but these involve higher-order molecular distribution functions, and therefore are not useful in practical applications.

Clearly the information on $P_0(r)$ in (A.81) may be converted to information on $G(r)$. Using relation (A.66) we obtain

$$G(r) = \left(1 - \frac{4\pi r^3}{3}\varrho\right)^{-1} \qquad \text{for } r \leq \frac{a}{2} \qquad (A.83)$$

and for $W(r)$ we have

$$W(r) = -kT\ln\left(1 - \frac{4\pi r^3}{3}\varrho\right) \qquad \text{for } r \leq \frac{a}{2} \qquad (A.84)$$

We now turn to the other extreme, namely, to very large cavities. In this case the cavity becomes macroscopic and the work required to create a very large cavity is simply

$$W(r) = Pv(r) \qquad \text{(for } r \to \infty) \qquad (A.85)$$

where P is the pressure and $v(r)$ is the volume of the cavity.

Another way of obtaining (A.85) is to use the basic probability in the grand canonical ensemble [see Münster (1974), p. 435]. For a very large cavity, we can treat the volume $v(r)$ as the volume of a macroscopic system, in the T, v, μ ensemble. The probability of finding the system empty is

$$P_0(r) = \Xi(T, v(r), \mu)^{-1} = \exp[-Pv(r)/kT] \qquad (A.86)$$

where Ξ is the grand partition function and the last equality holds for mascroscopic systems. Using relation (A.74) we obtain

$$W(r) = -kT\ln P_0(r) = Pv(r) \qquad (r \to \infty) \qquad (A.87)$$

which is the same as (A.85) obtained from purely thermodynamic considerations.

Also from relation (A.66) we may obtain the limit of $G(r)$ at $r \to \infty$, namely,

$$G(r \to \infty) = P/kT\varrho \qquad (A.88)$$

At this stage we have two exact results for $G(r)$; one for very small r, (A.83), and one for very large r, (A.88). This information suggests that we may try to bridge the two ends by a smooth function of r. In fact this was precisely the procedure taken by Reiss *et al.* (1959). The arguments used by the authors to make a particular choice of such a smooth function are quite lengthy and involved. They assumed that $G(r)$ is a monotonic function of r in the entire range of r. They suggested a trial function of the form

$$G(r) = A + Br^{-1} + Cr^{-2} \qquad (A.89)$$

The coefficients A, B, and C were determined by the use of all the available information on the behavior of the function $G(r)$ for a fluid of hard spheres.

The final expression obtained for $G(r)$, after being translated into $W(r)$ [i.e., integration of relation (A.74)] is the following:

$$W(r) = K_0 + K_1 r + K_2 r^2 + K_3 r^3 \qquad (A.90)$$

with the coefficients given by

$$
\begin{aligned}
K_0 &= kT[-\ln(1 - y) + 4.5z^2] - \tfrac{1}{6}\pi Pa^3 \\
K_1 &= -(kT/a)(6z + 18z^2) + \pi Pa^2 \\
K_2 &= (kT/a^2)(12z + 18z^2) - 2\pi Pa \\
K_3 &= 4\pi P/3
\end{aligned}
\qquad (A.91)
$$

where a is the diameter of the hard spheres, and y and z are defined by

$$y = \pi\varrho a^3/6, \qquad z = y/(1 - y) \qquad (A.92)$$

Thus, in essence, what we have obtained is an approximate expression for the work required to create a cavity of radius r in a fluid of HS's characterized by the diameter a.

Using this expression for the particular choice of $r = a$ we obtain an expression for the chemical potential [see A.76)]:

$$\mu = kT \ln \varrho \Lambda^3 + W(r = a) \qquad (A.93)$$

With this we end our presentation of the SPT. So far we have derived our relations for a fluid consisting of HS particles. We should now examine the applicability of this theory to real fluids such as liquid water.

Before doing that, there is one important comment regarding the SPT that should be borne in mind. The computation of the chemical potential through (A.90) and (A.93) differs from the traditional process in statistical mechanics. Namely, if we work in the T, V, N ensemble, then a purely molecular theory of a fluid should, in principle, provide all the thermodynamic quantities of the system as a function of T, V, and N and of the molecular diameter a (in the case of HS's). In particular, one should be able to compute *both* the chemical potential μ and the pressure P in terms of these variables. However, in the above procedure we have expressed the chemical potential μ in terms of T, ϱ, and a (ϱ replaces V and N, since we are concerned with intensive quantities). In *addition* we need to use the pressure [in (A.90)] as an *input* quantity. This quantity should, in principle be an output of the theory in the T, V, N ensemble, i.e., through the relation $P = -\partial A/\partial V$.

Because of the importance of this comment we repeat it in the language of the T, P, N ensemble. Here one should be able to compute both the chemical potential (through the derivative $\mu = \partial G/\partial N$) and the average density [through the relation $\varrho^{-1} = (\partial\mu/\partial P)_T$]. But in (A.90) and (A.93) we have expressed μ as a function of T and P, and *in addition* we need to use ϱ as an *input*, rather than an *output*, parameter.

In the above sense the SPT is not a purely statistical mechanical theory of fluids. This comment should be borne in mind when the theory is applied to complex fluids. In any real case, and certainly for water, we need a few molecular parameters to characterize the molecules, say ε and a in a Lennard-Jones fluid, or, in general, a set of molecular parameters $a, b, c, \ldots$. Thus, a proper statistical mechanical theory of water should provide us with the Gibbs free energy as a function of T, P, N and the molecular parameters, $a, b, c, \ldots$, i.e., a function of the form $G(T, P, N; a, b, c, \ldots)$. Instead the SPT makes use of *only* one molecular parameter, the diameter a. No provision for incorporating other molecular parameters is offered by the theory. This deficiency in the characterization of the molecules is partially compensated for by the use of the measurable density ϱ as an input parameter.

Water is currently referred to as being a "structural liquid" (in whichever sense we choose to use this term; for one particular definition, see, however, Section 5.7). One of the main goals in the study of liquid water is the understanding of the role of this peculiar "structure" in the determination of the outstanding properties of this liquid.

Clearly this goal may not be achieved through the use of the SPT. The very fact that we "inject" the density of the liquid is equivalent to introducing "structural" information into the theory. Therefore, even if we find that the SPT is successful in predicting some thermodynamic quantities, it cannot be used to *explain* them on a molecular level. Moreover, the apparent success of the SPT, even when applied to complex fluids, is not entirely surprising. After all, injecting one macroscopic quantity into the theory is likely to produce other thermodynamic quantities that are at least consistent with the input information.

We have also noted in Section 4.6 that for the computation of entropies, enthalpies, etc., one must also use the temperature dependence of the molecular diameter of the solvent. This quantity is also determined in such a way that the results are consistent with some measurable macroscopic quantity. In this way we further supply the theory with parameters which carry "structural" information on our particular system.

A.5. CONFORMATIONAL CHANGE OF A BIOPOLYMER AND STANDARD FREE ENERGY OF TRANSFER

In Chapter 2 we discussed the problem of "hydrophobic hydration." The standard free energy of transferring a simple solute from a nonpolar solvent into water has been traditionally used as a measure of the "hydrophobic" contribution to the driving force for a conformational change of a biopolymer. In Chapters 1 and 2 we noted that the use of $\Delta\mu_{tr}^{\circ}$ as a measure of the hydrophobic interaction involves several approximations: e.g., replacing a —CH_3 group by a methane molecule, replacing the environment of a methyl group in the open-coiled form of a polymer by pure water, and at the same time replacing the medium in the interior of the compact form of the polymer by a nonpolar solvent. These assumptions are indeed quite appealing on intuitive grounds. In this appendix we address ourselves to a more fundamental question. Suppose we have an idealized polymer, which consists of an almost massless backbone along which *methane* molecules are hung on by very thin connecting lines. For simplicity we also assume that this polymer attains only two possible conformations, an open form (O) and a compact form (C). These are schematically shown in Figure A.3. There are no other groups on the polymers. Thus in the process of a conformational change from the O form to the C form, we are essentially transferring, say, m methane groups from almost pure water

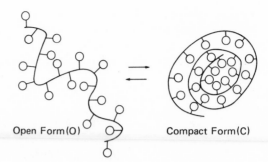

Figure A.3. Schematic representa-   Open Form(O)   Compact Form(C)
tion of a conformational change
of a polymer.

W to an almost nonpolar environment E (consisting of methane molecules).
We now pose the following question.

To what extent is it justifiable to use $\Delta\mu_{\mathrm{tr}}^{\circ}(W \to E)$ to represent the
driving force for such a conformational change? Note that we are asking
this question for the most favorable case, where we have deliberately
removed all other complicating factors that should be taken into account
when dealing with a real biopolymer.

The standard free energy change for the conformational transformation
is obtained from the equilibrium condition

$$\mu_C = \mu_O \qquad\qquad (A.94)$$

or

$$\Delta\mu^{\circ} = \mu_C^{\circ\varrho} - \mu_O^{\circ\varrho} = -kT \ln(\varrho_C/\varrho_O)_{\mathrm{eq}} \qquad (A.95)$$

where ϱ_C and ϱ_O are the equilibrium number densities of the two forms in
a very dilute solution of the polymer in a solvent W. The statistical mechanical expression for $\Delta\mu^{\circ}$ is

$$\Delta\mu^{\circ} = W(C \mid W) + U(C) - W(O \mid W) - U(O) + kT\ln(q_O/q_C) \quad (A.96)$$

Here $W(C \mid W)$ and $W(O \mid W)$ are the coupling work of C and O, respectively, to the solvent; $U(C)$ and $U(O)$ include all the intramolecular interactions of the two forms; and q_O, and q_C are essentially the rotational partition
functions of the two forms. The momentum partition functions of the two
forms are equal to each other, and we have neglected all other internal
partition functions.

From (A.96) we can see immediately that the term that includes the
rotational partition functions cannot be represented by the standard free
energy of transferring a *single* group from water into a nonpolar solvent.
The reason is that q_O and q_C depend on the distribution of mass along the
polymer in the two forms, whereas $\Delta\mu_{\mathrm{tr}}^{\circ}(W \to E)$ of a simple solute carries

no information of this kind. Next we write in some further detail the remaining terms in (A.96):

$$W(C \mid W) = -kT \ln \langle \exp(-\beta B_C) \rangle_0 \qquad (A.97)$$

$$W(O \mid W) = -kT \ln \langle \exp(-\beta B_O) \rangle_0 \qquad (A.98)$$

$$U(C) = U_b(C) + \frac{1}{2} \sum_{i=1}^{m} \sum_{j=1}^{m} U_{ij} \qquad (A.99)$$

$$U(O) = U_b(O) \qquad (A.100)$$

Here B_C and B_O are the total binding energies of the C and O forms to the solvent. Since we have assumed that the backbone is extremely thin and massless, we may neglect its interaction with the solvent; hence we may further factor $W(O \mid W)$ as

$$\begin{aligned}
W(O \mid W) &= -kT \ln \langle \exp(-\beta m B_s) \rangle_0 \\
&= -kTm \ln \langle \exp(-\beta B_s) \rangle_0 \\
&= m \, \Delta \mu_{\text{tr}}^{\circ}(G \rightarrow W) \qquad (A.101)
\end{aligned}$$

where we have assumed that all the groups s in the open form are very far apart, so that the total coupling *work* of O to the solvent is approximately equal to m times the coupling *work* of one solute molecule. No similar factorization of $W(C \mid W)$ is possible. In (A.99) and in (A.100) we have included in $U_b(O)$ and $U_b(C)$ all the interactions within the backbone, and between the side chains and the backbone, but not the interactions of the side chains with each other; the latter interactions are represented by the second term on the right-hand side of (A.99). Assuming that $U_b(O) = U_b(C)$ we may write

$$\begin{aligned}
U(C) - U(O) &= \frac{1}{2} \sum_{i=1}^{m} \sum_{j=1}^{m} U_{ij} \\
&= \frac{1}{2} \sum_{i=1}^{m} B_i \qquad (A.102)
\end{aligned}$$

where B_i is the binding energy of the ith group to the rest of the side chains in the compact conformation. On the other hand, the standard free energy of transferring s from the gas into a pure liquid s, the density of which is roughly the same as in the compact form C, is

$$\Delta \mu_{\text{tr}}^{\circ}(G \rightarrow S) = -kT \ln \langle \exp(-\beta B_s) \rangle \qquad (A.103)$$

We may further simplify (A.103) by assuming that the binding energy of s to all the other molecules is approximately independent of the configuration of these molecules; hence

$$\Delta\mu_{\text{tr}}^{\circ}(G \to S) \approx -kT \ln[\exp(-\beta B_s)] = B_s \qquad (A.104)$$

where B_s may be identified with those of the B_i in (A.102) which belong to the groups i in the interior of the polymer.

We now collect the various simplifying results to rewrite (A.96) as

$$\Delta\mu^{\circ} = W(C \mid W) - m\,\Delta\mu_{\text{tr}}^{\circ}(G \to W) + \frac{1}{2}\sum_{i=1}^{m} B_i + kT\ln(q_0/q_C) \qquad (A.105)$$

Now if the polymer becomes very large, so that $W(C \mid W)$ may be neglected compared with $W(O \mid W)$, and if the sum over B_i may be replaced by approximately m times the typical B_s of (A.104), we can rewrite (A.105) as

$$\begin{aligned}
\Delta\mu^{\circ} &= -m\,\Delta\mu_{\text{tr}}^{\circ}(G \to W) + \tfrac{1}{2}mB_s + kT\ln(q_0/q_C)\\
&= m[\Delta\mu_{\text{tr}}^{\circ}(G \to S) - \Delta\mu_{\text{tr}}^{\circ}(G \to W)] - \tfrac{1}{2}mB_s + kT\ln(q_0/q_C)\\
&= m\,\Delta\mu_{\text{tr}}^{\circ}(W \to S) - \tfrac{1}{2}m\,\Delta\mu_{\text{tr}}^{\circ}(G \to S) + kT\ln(q_0/q_C) \qquad (A.106)
\end{aligned}$$

Thus we see that even in the most favorable conditions the interpretation of $\Delta\mu^{\circ}$ associated with the conformational change of the polymer in terms of m times $\Delta\mu_{\text{tr}}^{\circ}$ is not fully justifiable. Besides the term that includes the ratio of the rotational partition functions, we also have the term $-\tfrac{1}{2}m\,\Delta\mu_{\text{tr}}^{\circ}(G \to S)$. This term arises from the inclusion in $\Delta\mu_{\text{tr}}^{\circ}(W \to S)$ of m times the binding of s with its surroundings, whereas in (A.105) we have only half of this quantity.

A.6. ON THE NATURE OF THE APPROXIMATION INVOLVED IN THE EXPRESSION FOR $\delta G^{\text{HI}}(\sigma_1)$ OF SECTION 3.3

In Section 3.3 we derived a relation between $\delta G^{\text{HI}}(\sigma_1)$ and the standard free energies of solution of ethane and methane. The basis for this approximation is the general expression for the indirect part of the free energy change for the HI process, i.e., the process of bringing two *methane* molecules from infinite separation to some small separation R,

$$\delta G^{\text{HI}}(R) = -kT \ln\left\{\frac{\langle\exp[-\beta B_{ss}(R)]\rangle_0}{\langle\exp[-\beta B_{ss}(\infty)]\rangle_0}\right\} \qquad (A.107)$$

The notation here is as in Section 3.3 with $\beta = (kT)^{-1}$ and $B_{ss}(R)$ the binding energy of the pair of *methane* molecules at a distance R from each other. We have noticed that in this expression the direct, solute–solute interaction does not appear, hence we could define $\delta G^{HI}(R)$ for any R, including $R \leq \sigma$, where σ is the molecular diameter of the solutes. In particular, for $R = \sigma_1 = 1.53$ Å, the C–C distance in ethane, we have approximately identified

$$-kT \ln\langle \exp[-\beta B_{ss}(\sigma_1)]\rangle_0 \approx \Delta\mu^{\circ}_{Et} \qquad (A.108)$$

This approximation essentially involves the neglect of the binding energy of the two hydrogen atoms that are removed in the process of replacing the pair of *methane* molecules at $R = \sigma_1$ by two *methyl* groups, and forming one ethane molecule. A more detailed discussion of this approximation is provided in Ben-Naim (1974). Here, we present a slightly different point of view of the same quantity $\delta G^{HI}(\sigma_1)$.

Consider the process depicted in two steps in Figure A.4. Two methane molecules are placed at fixed positions but at infinite separation from each other. As we did in Section 4.7, we now "cut" one hydrogen atom from each methane molecule and remove them to some fixed positions at infinite separation. The indirect free energy change for this process is

$$\delta G^{HI}(I) = -2\delta G^{HI}(R, H) \qquad (A.109)$$

where $\delta G^{HI}(R, H)$ is the indirect part of the free energy change for the process of bringing a hydrogen and a methyl radical to the configuration of methane. The exact location of the "cut" is of no importance since eventually we shall deal with the entire molecule. The radical is used only as an intermediate in our treatment.

Next we combine the two methyl radicals and the two hydrogen atoms to form an ethane and a hydrogen molecule. The indirect free energy change

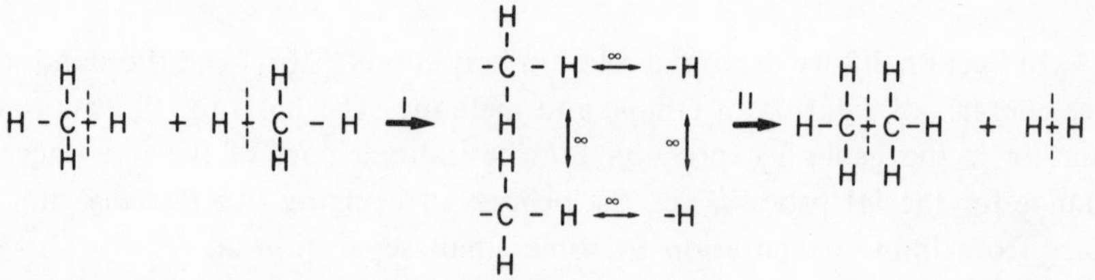

Figure A.4. A schematic "reaction" in which two methane molecules are transformed into ethane and hydrogen molecules.

for this process is

$$\delta G^{HI}(II) = \delta G^{HI}(R, R) + \delta G^{HI}(H, H) \tag{A.110}$$

Note that R in (A.109) and (A.110) stands for the methyl radical and not for the distance R as in (A.107). Thus for the entire process we have

$$\delta G^{HI}(I) + \delta G^{HI}(II) = \delta G^{HI}(R, R) + \delta G^{HI}(H, H) - 2\delta G^{HI}(R, H) \tag{A.111}$$

The overall "reaction" is

$$2CH_4 \rightarrow CH_3CH_3 + H_2 \tag{A.112}$$

Note, however, that this is not a conventional reaction since the molecules possess no translational and rotational degrees of freedom. In terms of standard free energies of solution, the indirect free energy for the "reaction" (A.112) may be rewritten as

$$\delta G^{HI}(R, R) + \delta G^{HI}(H, H) - 2\delta G^{HI}(R, H) = \Delta\mu_{Et}^{\circ} + \Delta\mu_{H_2}^{\circ} - 2\Delta\mu_{Me}^{\circ} \tag{A.113}$$

We now identify our previous measure

$$\delta G^{HI}(\sigma_1) \cong \Delta\mu_{Et}^{\circ} - 2\Delta\mu_{Me}^{\circ} \tag{A.114}$$

which, as noted above, involves an approximation. This quantity measures the strength of the HI between two *methane* molecules. We can now rewrite the exact relation (A.113) as

$$\Delta\mu_{Et}^{\circ} - 2\Delta\mu_{Me}^{\circ} = \delta G^{HI}(R, R) + [\delta G^{HI}(H, H) - 2\delta G^{HI}(R, H) - \Delta\mu_{H_2}^{\circ}] \tag{A.115}$$

We see here that the approximate measure of the HI between two *methane* molecules, on the left-hand side of (A.115), is reinterpreted in terms of the HI between two *methyl* radicals and a remainder which is approximately independent of the size of the alkyl group.

A.7. STANDARD FREE ENERGY OF SOLUTION OF SOLUTE MOLECULES HAVING INTERNAL ROTATIONS

In Appendix A.1 and in most of the book we have discussed the chemical potential and the standard thermodynamics of transfer of a *simple* solute S. For dilute solution of S in a solvent W we have written the chemical potential as

$$\mu_S = W(S \mid W) + kT \ln \varrho_S \Lambda_S^3 = \mu_S^{\circ \varrho W} + kT \ln \varrho_S \qquad (A.116)$$

where $W(S \mid W)$ denotes the coupling work of the solute S to the solvent. If S is a more complex solute, having internal relations, the expression for the chemical potential should be modified since the coupling work of S with the solvent depends on the conformation of the solute.

We generalize the expression for the chemical potential of S in two stages. First we consider a case where S may attain only two discrete conformations, say A and B as in Figure A.5. Then we shall proceed to treat the more general case in which continuous rotations about one or more bonds is possible. Consider a system of N solvent molecules at a given volume V and temperature T. The chemical potential of a solute S at very dilute solution may be obtained from

$$\mu_S(1) = A(T, V, N; S) - A(T, V, N) = -kT \ln[Q(S)/Q(O)] \qquad (A.117)$$

Here $\mu_S(1)$, the chemical potential of *one* S in the solvent, is expressed as a difference in the Helmholtz free energies of the system with and without the solute S; $Q(S)$ and $Q(O)$ are the corresponding canonical partition functions of the system with and without the solute S (for simplicity we use here the T, V, N ensemble, but all the results of this appendix are valid in the T, P, N ensemble as well).

From the definition of the partition function, we may write

$$Q(S) = Q(A) + Q(B) \qquad (A.118)$$

where we split the sum over all states of the system into two partial sums: $Q(A)$ is the "sum over all states of the system for which S is in state A," and $Q(B)$ is the "sum over all states of the system for which S is in the state B."

Thus from (A.117) and (A.118) we may write

$$\exp[-\beta \mu_S(1)] = \frac{Q(S)}{Q(O)} = \frac{Q(A)}{Q(O)} + \frac{Q(B)}{Q(O)}$$
$$= \exp[-\beta \mu_A(1)] + \exp[-\beta \mu_B(1)] \qquad (A.119)$$

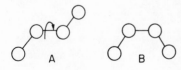

Figure A.5. Two conformations of a molecule. A and B are obtained from each other by rotation about a single bond.

where $\beta = (kT)^{-1}$, and $\mu_A(1)$ is the chemical potential of a system with *one* A solute in the solvent; a similar meaning is assigned to $\mu_B(1)$.

The statistical mechanical expression for $\mu_A(1)$ is (see also Appendix A.1)

$$\exp[-\beta\mu_A(1)] = \frac{Q(A)}{Q(O)} = \frac{q_A \int d\mathbf{X}^N \, d\mathbf{X}^A \exp[-\beta U_N - \beta B_A + U(A)]}{(8\pi^2)\Lambda_A{}^3 \int d\mathbf{X}^N \exp(-\beta U_N)}$$

$$= \frac{q_A V}{\Lambda_A{}^3} \exp[-\beta U(A)]\langle\exp(-\beta B_A)\rangle_0$$

$$= \frac{q_A V}{\Lambda_A{}^3} \exp[-\beta U(A) - \beta W(A \mid W)]$$

$$= \frac{q_A V}{\Lambda_A{}^3} \exp[-\beta U(A)] \exp[-\beta(\mu_A{}^{\circ\varrho W} - \mu_A{}^{\circ\varrho\varrho})] \quad \text{(A.120)}$$

$\Lambda_A{}^3$ is the momentum partition function of A, and is equal to $\Lambda_B{}^3$. q_A includes all the internal partition functions of the species A; in our particular example we may view q_A as essentially the rotational partition function of A. U_N is the total interaction energy among the N solvent molecules, B_A is the "binding energy" of A to the solvent, and $U(A)$ is the intramolecular interaction of the four groups in the conformation A. In the second form on the right-hand side of (A.120) we performed the integration over all locations and orientations of the molecule A. We further identify the coupling work of A and the standard free energy of solution of A in the last two forms on the right-hand side of (A.120). Similar expressions may be written for B.

Substituting (A.120) into (A.119) we obtain

$$\exp[-\beta\mu_S(1)] = \frac{V}{\Lambda_A{}^3} \{q_A \exp[-\beta U(A) - \beta \, \Delta\mu_A{}^\circ]$$
$$+ q_B \exp[-\beta U(B) - \beta \, \Delta\mu_B{}^\circ]\} \quad \text{(A.121)}$$

Using (A.116) we may identify the standard chemical potential of S in the liquid as

$$\exp(-\beta\mu_S{}^{\circ\varrho W}) = \frac{1}{\Lambda_A{}^3} \{q_A \exp[-\beta U(A) - \beta \, \Delta\mu_A{}^\circ]$$
$$+ q_B \exp[-\beta U(B) - \beta \, \Delta\mu_B{}^\circ]\} \quad \text{(A.122)}$$

Similarly for the gaseous phase we write

$$\exp(-\beta\mu_S{}^{\circ\varrho\varrho}) = \frac{1}{\Lambda_A{}^3} \{q_A \exp[-\beta U(A)] + q_B \exp[-\beta U(B)]\} \quad \text{(A.123)}$$

Hence the standard free energy of solution of S from the gas into the liquid is

$$\exp(-\beta\, \Delta\mu_S{}^\circ)$$

$$= \frac{q_A \exp[-\beta U(A) - \beta\, \Delta\mu_A{}^\circ] + q_B \exp[-\beta U(B) - \beta\, \Delta\mu_B{}^\circ]}{q_A \exp[-\beta U(A)] + q_B \exp[-\beta U(B)]}$$

$$= y_A \exp(-\beta\, \Delta\mu_A{}^\circ) + y_B \exp(-\beta\, \Delta\mu_B{}^\circ) \tag{A.124}$$

where y_A and y_B may be interpreted as the equilibrium mole fractions of A and B in the gaseous phase. We see that $\exp(-\beta\, \Delta\mu_S{}^\circ)$ is expressed in (A.124) as an average of the corresponding quantities of A and B, using the weights y_A and y_B that are determined by the equilibrium condition in the gaseous phase, namely,

$$\mu_A{}^g = U(A) + kT \ln(\varLambda_A{}^3 q_A{}^{-1}\varrho_A) = \mu_B{}^g = U(B) + kT \ln(\varLambda_B{}^3 q_B{}^{-1}\varrho_B) \tag{A.125}$$

or

$$\left(\frac{\varrho_A}{\varrho_B}\right)_{eq} = \left(\frac{y_A}{y_B}\right)_{eq} = \frac{q_A}{q_B} \exp[\beta U(B) - \beta U(A)] \tag{A.126}$$

We now generalize relation (A.124) to the case in which a continuous change of the conformation is possible, for example rotation about the second bond as indicated in Figure A.5. In such a case we select a specific conformation $\mathbf{P}$ for which the standard free energy of solution is written as

$$\exp[-\beta\, \Delta\mu^\circ(\mathbf{P})] = \frac{\int d\mathbf{X}^N \exp[-\beta U_N - \beta U(N\,|\,\mathbf{P})]}{\int d\mathbf{X}^N \exp(-\beta U_N)} \tag{A.127}$$

Here $\Delta\mu^\circ(\mathbf{P})$ is the free energy change for transferring the solute S, which is at a specific conformation $\mathbf{P}$, from a fixed position and orientation in the gas to a fixed position and orientation in the liquid.

The standard free energy of solution of S, in analogy with (A.124), is given by

$$\exp(-\beta\, \Delta\mu_S{}^\circ) = \frac{\int d\mathbf{P} q(\mathbf{P}) \exp[-\beta U(\mathbf{P})] \exp[-\beta\, \Delta\mu^\circ(\mathbf{P})]}{\int d\mathbf{P} q(\mathbf{P}) \exp[-\beta U(\mathbf{P})]}$$

$$= \int d\mathbf{P} y(\mathbf{P}) \exp[-\beta\, \Delta\mu^\circ(\mathbf{P})] \tag{A.128}$$

Here $q(\mathbf{P})$ is essentially the rotational partition function of S in the specific conformation $\mathbf{P}$, and $y(\mathbf{P})\, d\mathbf{P}$ denotes the mole fraction, or the probability of finding a conformation between $\mathbf{P}$ and $\mathbf{P} + d\mathbf{P}$ of S in the gaseous phase.

We now turn to some alternative expressions for the chemical potential of S having only two states A and B in the liquid. Let x_A and x_B be the equilibrium mole fractions of A and B in the liquid phase; these are determined by

$$\left(\frac{x_A}{x_B}\right)_{eq} = \frac{Q(A)}{Q(B)} = \frac{q_A \exp[-\beta U(A) - \beta \Delta\mu_A{}^\circ]}{q_B \exp[-\beta U(B) - \beta \Delta\mu_B{}^\circ]} \quad (A.129)$$

Eliminating x_A from (A.129) and using (A.119) and (A.120) we obtain

$$\exp[-\beta\mu_S(1)] = \frac{V}{\Lambda_A{}^3 x_A}\{q_A \exp[-\beta U(A) - \beta \Delta\mu_A{}^\circ]\} \quad (A.130)$$

Since in this particular case we have *one* S in the system, the density of A is

$$\varrho_A = \frac{N_A}{V} = \frac{x_A}{V} \quad (A.131)$$

Hence, from (A.130) we write

$$\begin{aligned}\mu_S(1) &= kT\ln(\varrho_A \Lambda_A{}^3 q_A{}^{-1}) + U(A) + \Delta\mu_A{}^\circ \\ &= kT\ln(\varrho_A \Lambda_A{}^3 q_A{}^{-1}) + U(A) + W(A \mid W) = \mu_A(\varrho_A) \quad (A.132)\end{aligned}$$

Similar relations could have been obtained for B, so

$$\mu_S(1) = \mu_A(\varrho_A) = \mu_B(\varrho_B) \quad (A.133)$$

which simply means that the chemical potential of the solute S is equal to the chemical potentials of A and of B, provided that these are evaluated at the equilibrium concentrations ϱ_A and ϱ_B, respectively. One may also rewrite $\mu_S(1)$ in a more symmetrical way in terms of $\mu_A(\varrho_A)$ and $\mu_B(\varrho_B)$ as follows:

$$\begin{aligned}\mu_S(1) &= x_A\mu_A(\varrho_A) + x_B\mu_B(\varrho_B) \\ &= (x_A\mu_A{}^{\circ\varrho} + x_B\mu_B{}^{\circ\varrho} + kTx_A \ln x_A + kTx_B \ln x_B) + kT\ln V^{-1} \\ &= \mu_S{}^{\circ\varrho} + kT\ln\varrho_S \quad (A.134)\end{aligned}$$

where $\varrho_S = V^{-1}$, and we have identified $\mu_S{}^{\circ\varrho}$ in terms of $\mu_A{}^{\circ\varrho}$ and $\mu_B{}^{\circ\varrho}$ and a mixing term of A and B. Both of the expressions (A.132) and (A.134) may be interpreted in terms of the free energy change of introducing S into the system along different routes, e.g., we may first introduce S, in the form of A, into a fixed position and orientation and then release it in a solution, the concentration of which is ϱ_A.

A.8. ON THE SOLVENT CONTRIBUTION TO THE STANDARD FREE ENERGY OF MICELLE FORMATION

We consider an equilibrium between monomers M and micelles A_n composed of n monomers. For simplicity we assume that the monomers are rigid and that the micelle has a compact spherical shape, as indicated in Figure A.6.

The standard free energy of micellization is given by

$$\Delta G^{\circ}(A_n) = \mu_{A_n}^{\circ\varrho} - n\mu_M^{\circ\varrho} = [U(A_n) + kT \ln(q_M{}^n/q_{A_n})] + \delta G(A_n) \quad \text{(A.135)}$$

where

$$\delta G(A_n) = -kT \ln\left\{ \frac{\langle \exp[-\beta B(A_n)]\rangle_0}{\langle \exp[-\beta B(M)]\rangle_0{}^n} \right\} \quad \text{(A.136)}$$

and where the square brackets on the right-hand side of (A.135) include all the internal contributions to the standard free energy of micellization. Any contribution to $\Delta G^{\circ}(A_n)$ from the solvent is included in $\delta G(A_n)$. (In q_α we included both translational and internal partition functions of the species α.)

We have discussed in Section 4.8 the possibility of splitting $U(A_n)$, the direct interaction energy, into head–head and alkyl–alkyl interactions:

$$U(A_n) \approx U_{HH} + U_{RR} \quad \text{(A.137)}$$

This may be a good approximation if we can neglect the head–alkyl interactions. U_{RR} is frequently referred to as the "hydrophobic interaction" in the micelle. This seems to be implied in the literature when $\Delta G^{\circ}(A_n)$ is presented as

$$\Delta G^{\circ}(A_n) = \Delta G_{RR}^{\circ} + \Delta G_{HH}^{\circ} \quad \text{(A.138)}$$

and the "hydrophobic" part is discussed in terms of the van der Waals interaction between the alkyl groups. This procedure is unsatisfactory for two reasons. In the first place, if we substitute (A.137) into (A.135) and identify ΔG_{RR}° as U_{RR}, then all the remaining terms in (A.135) may not be assigned the meaning that is ascribed to ΔG_{HH}°, i.e., the contribution of the head–head interaction to $\Delta G^{\circ}(A_n)$. Furthermore, if U_{RR} is identified with ΔG_{RR}°, then the very term "hydrophobic" becomes inappropriate. The reason is that U_{RR} is a property of the solutes and, to a good approximation, this is unaffected by the properties of the solvent. Thus if we are

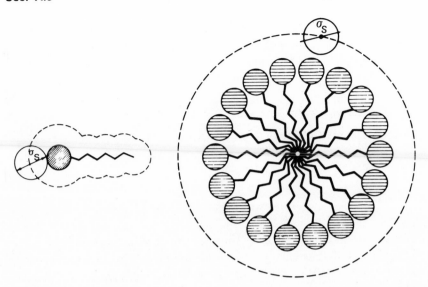

Figure A.6. A micelle and a surfactant monomer. The dashed curves indicate the excluded volumes for a solvent molecule with a diameter σ_s.

searching for any peculiarities of a specific solvent, such as water, we should examine the properties of the indirect interaction $\delta G(A_n)$, which is expressed in statistical mechanical terms in (A.136).

We shall now demonstrate that $\delta G(A_n)$ may not be split, in any obvious manner, into two terms corresponding to head–head and alkyl–alkyl interactions. To do this, we consider the "simplest" solvent, which consists of a *single* hard-sphere molecule, of diameter σ_S.

For this particular solvent, and following the assumptions made above, we can write explicitly, in the T, V, N ensemble, the quantity $\delta G(A_n)$. Thus

$$\frac{\langle \exp[-\beta B(A_n)]\rangle_0}{\langle \exp[-\beta B(M)]\rangle_0^n} = \frac{\{\int d\mathbf{R}_{HS} \exp[-\beta B(A_n)]\}(\int d\mathbf{R}_{HS})^n}{(\int d\mathbf{R}_{HS})\{\int d\mathbf{R}_{HS} \exp[-\beta B(M)]\}^n}$$

$$= \frac{[V - V_{ex}(A_n)]V^n}{V[V - V_{ex}(M)]^n}$$

$$= \frac{[1 - V_{ex}(A_n)/V]}{[1 - V_{ex}(M)/V]^n} \approx \frac{1 - V_{ex}(A_n)/V}{1 - nV_{ex}(M)/V} \quad \text{(A.139)}$$

In (A.139) the average over all the configurations of the solvent molecules reduces to an integration over all possible locations of the hard sphere. We have denoted by $V_{ex}(A_n)$ and $V_{ex}(M)$ the volume from which

the hard sphere is excluded by the micelle and by the monomer, respectively
—see Figure A.6. In this particular example we find an approximate expression for $\delta G(A_n)$ in terms of these two excluded volumes, namely,

$$\delta G(A_n) = -kT \ln \left[\frac{1 - V_{ex}(A_n)/V}{1 - nV_{ex}(M)/V} \right] \qquad (A.140)$$

Thus, for a compact aggregate as in Figure A.6 we expect that

$$V_{ex}(A_n) < nV_{ex}(M) \qquad (A.141)$$

Hence

$$\delta G(A_n) < 0 \qquad (A.142)$$

which means that this particular "solvent" will make a negative contribution
to the standard free energy of micellization. The important point that
should be emphasized now is not the particular value of $\delta G(A_n)$ as given
in (A.140), but the fact that this quantity may not be viewed as consisting
of two additive contributions due to head–head and alkyl–alkyl interactions.

In our particular example $\delta G(A_n)$ depends essentially on the volumes
of the monomers and the micelles. Clearly, in a real solvent, many molecules
contribute to the ratio of the two average quantities in (A.136). In particular, water may produce some outstandingly large ratio, leading to a
large negative value of $\delta G(A_n)$. It is impossible to say, at present, how the
peculiar properties of water might affect the quantity $\delta G(A_n)$. In general
one can always express this quantity as a difference between the free energy
of interaction (or coupling work) of A_n with the solvent and n times the
free energy of interaction of the monomer with the solvent, i.e.,

$$\delta G(A_n) = W(A_n \mid \text{solvent}) - nW(M \mid \text{solvent}) \qquad (A.143)$$

Again, we see that this quantity may not be split into separate contributions due to head–head and alkyl–alkyl interactions.

A.9. STANDARD FREE ENERGY OF TRANSFERRING A SOLUTE FROM THE SOLVENT INTO A MICELLE

In the study of solubilization phenomena, one encounters the concept of
the standard free energy of transferring the solute from the solvent into
the micelle. This quantity cannot be measured directly since we do not

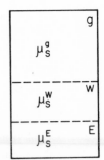

Figure A.7. Three phases g, W, and E in which a solute S is distributed. The two phases W and E are considered as a combined phase in the discussion of this section.

have a direct means of measuring the equilibrium concentration of the solute in the micelle. In fact, there is also some ambiguity in the very definition of the "concentration of the solute in the micelles." In order to clarify the nature of these difficulties, we first construct a simple experimental setup in which the standard free energy of transfer may be computed without measuring directly the partition coefficient of the solute in the two phases. Then we proceed to discuss a similar situation that actually occurs in micellar solutions.

Consider the three-phase system shown in Figure A.7. A solute s is distributed among the three phases, and at equilibrium we have

$$\mu_s^g = \mu_s^W = \mu_s^E \tag{A.144}$$

We are interested in the standard free energy of transferring s from W into E, which may be computed directly from the relation

$$\Delta\mu_{\mathrm{tr}}^\circ(W \to E) = \mu_s^{\circ\varrho E} - \mu_s^{\circ\varrho W} = kT\ln(\varrho_s^W/\varrho_s^E) \tag{A.145}$$

where ϱ_s^W and ϱ_s^E are the equilibrium (number) densities of s in the two phases W and E, respectively.

Now suppose that we cannot measure the densities of s in each phase E and W separately. Instead we can only measure the partition coefficient of s between the gaseous phase g and the liquid phase l, which consists of the two phases E and W. Also we assume, in anticipation of the application of this method to micelles, that we cannot construct a pure phase E (but a pure phase W does exist).

For simplicity we assume that the two phases E and W are immiscible liquids and s forms an ideal dilute solution in each of these phases. At equilibrium we have

$$\mu_s^g = \mu_s^l$$

from which we may obtain the standard free energy of transfer

$$\Delta\mu_{tr}^\circ(g \to l) = \mu_s^{\circ\varrho l} - \mu_s^{\circ\varrho g} = kT\ln(\varrho_s^{\,g}/\varrho_s^{\,l}) \qquad (A.146)$$

where

$$\varrho_s^{\,l} = \frac{N_s^{\,l}}{V^l} = \frac{N_s^{\,E} + N_s^{\,W}}{V^E + V^W} \qquad (A.147)$$

is the overall (number) density of s in the combined liquid phases. The measurable quantity is the overall partition coefficient $\varrho_s^{\,g}/\varrho_s^{\,l}$ at equilibrium. V^E and V^W are the volumes of the E and the W phases, respectively.

The questions we pose now are essentially two: First, how is $\Delta\mu_{tr}^\circ(g \to l)$ related to $\Delta\mu_{tr}^\circ(g \to E)$ and $\Delta\mu_{tr}^\circ(g \to W)$? Second, how can one compute $\Delta\mu_{tr}^\circ(W \to E)$ from measurements of $\Delta\mu_{tr}^\circ(g \to l)$?

The chemical potential of the (simple) solute s in the phases E and W are written as (see Appendix A.1)

$$\mu_s^{\,E} = W(S \mid E) + kT\ln\varrho_s^{\,E}\Lambda_s^{\,3} \qquad (A.148)$$

$$\mu_s^{\,W} = W(S \mid W) + kT\ln\varrho_s^{\,W}\Lambda_s^{\,3} \qquad (A.149)$$

From the definition of $\varrho_s^{\,l}$ in (A.147) we can relate $\varrho_s^{\,l}$ to $\varrho_s^{\,E}$ and $\varrho_s^{\,W}$ through

$$\begin{aligned}
\varrho_s^{\,l} &= \frac{N_s^{\,E} + N_s^{\,W}}{V^E + V^W} = \frac{V^E(N_s^{\,E}/V^E) + V^W(N_s^{\,W}/V^W)}{V^E + V^W} \\
&= x^E\varrho_s^{\,E} + x^W\varrho_s^{\,W}
\end{aligned} \qquad (A.150)$$

where x^E and x^W are the volume fractions of the phases E and W, respectively:

$$x^E = V^E/V^l, \qquad x^W = V^W/V^l \qquad (A.151)$$

In (A.150) we expressed $\varrho_s^{\,l}$ as an average of $\varrho_s^{\,E}$ and $\varrho_s^{\,W}$.

We denote the partition coefficient of s between the phases E and W by

$$K = (\varrho_s^{\,E}/\varrho_s^{\,W}) = \exp[-\Delta\mu_{tr}^\circ(W \to E)/kT] \qquad (A.152)$$

Hence from (A.150) we have

$$\varrho_s^{\,l} = x^E K\varrho_s^{\,W} + x^W\varrho_s^{\,W} \qquad (A.153)$$

or

$$\varrho_s^{\,W} = \frac{\varrho_s^{\,l}}{Kx^E + x^W} \qquad (A.154)$$

At equilibrium we may express $\mu_s{}^l$ in either one of the following forms:

$$\mu_s{}^l = W(S \mid E) + kT \ln \varrho_s{}^E \Lambda_s{}^3$$

$$= \left[W(S \mid E) + kT \ln\left(\frac{K}{Kx^E + x^W}\right) \right] + kT \ln \varrho_s{}^l \Lambda_s{}^3$$

$$= y_s{}^E W(S \mid E) + y_s{}^W W(S \mid W) + kT y_s{}^E \ln \varrho_s{}^E \Lambda_s{}^3 + kT y_s{}^W \ln \varrho_s{}^W \Lambda_s{}^3$$

$$\text{(A.155)}$$

In the last form on the right-hand side of (A.155) we have used the "mole fractions"

$$y_s{}^E = \frac{N_s{}^E}{N_s{}^E + N_s{}^W}, \qquad y_s{}^W = \frac{N_s{}^W}{N_s{}^E + N_s{}^W} \qquad \text{(A.156)}$$

to express $\mu_s{}^l$ in a more symmetrical form as a combination of $\mu_s{}^E$ and $\mu_s{}^W$. Each of the quantities on the right-hand side of (A.155) may be interpreted in terms of the free energy change of adding one s to the liquid. The first corresponds to placing the solute particle at a fixed position in E and then releasing it to wander in V^E. The second corresponds to placing s in l, then releasing it into the entire volume $V^E + V^W$. The third corresponds to placing a "fraction" $y_s{}^E$ of a molecule in E and a "fraction" $y_s{}^W$ in W and then releasing the entire molecule into the entire volume (of course in this case it is better to introduce one mole of solute molecules rather than a single molecule).

Clearly by comparing (A.155) with (A.148) and (A.149) we can express $\mu_s{}^{\circ\varrho l}$ in terms of $\mu_s{}^{\circ\varrho E}$ and $\mu_s{}^{\circ\varrho W}$.

We now turn to the question of the computability of $\Delta\mu_{\text{tr}}^{\circ}(W \to E)$ from measurements of $\Delta\mu_{\text{tr}}^{\circ}(g \to l)$. To do that we express the measurable quantity $(\varrho_s{}^l/\varrho_s{}^g)$, at equilibrium, in terms of $W(S \mid W)$ and K. Thus from the second form on the right-hand side of (A.155) we obtain

$$\gamma_s \equiv (\varrho_s{}^l/\varrho_s{}^g) = \exp[-(\mu_s{}^{\circ\varrho l} - \mu_s{}^{\circ\varrho g})/kT]$$

$$= \exp[-W(S \mid W)/kT](Kx^E + x^W) \qquad \text{(A.157)}$$

where γ_s is the measurable Ostwald absorption coefficient of s in the combined liquid phase. Taking the derivative of γ_s with respect to x^E we obtain

$$\frac{\partial \gamma_s}{\partial x^E} = (K - 1) \exp\left[-\frac{W(S \mid W)}{kT} \right] \qquad \text{(A.158)}$$

Thus from the two measurements

$$\gamma_s(x^E = 0) = \exp[-W(S \mid W)/kT] \qquad \text{(A.159)}$$

and

$$\frac{\partial \gamma_s}{\partial x^E} = (K - 1)\gamma_s(x^E = 0) \tag{A.160}$$

we may extract the required quantity K or, equivalently, $\Delta\mu_{\mathrm{tr}}^\circ(W \to E)$.

We recall that we have made the assumption that a pure phase E does not exist. Otherwise we could have measured $\gamma_s(x^E = 1)$ and $\gamma_s(x^E = 0)$ and from their ratio obtain K. Instead we measure γ_s in pure W and the slope of γ_s as a function of x^E. This example has demonstrated that in spite of the impossibility of measuring the concentration of s in E we can still determine $\Delta\mu_{\mathrm{tr}}^\circ(W \to E)$.

We now turn to the aqueous micellar solution. In order to apply the method outlined above we must make some further assumptions. First we view the micellar solution as a quasi-two-phase system. Clearly, we cannot measure the concentration of s in the micellar phase E. In fact there is an ambiguity in the very definition of the volume of this phase. From the experimental observation that below the CMC the solubility of s is almost independent of the total concentration of the surfactant, we assume that in this region the solubility of s is almost identical to the solubility in pure water. Above the CMC we assume that all the added surfactant is used to build up more micelles and for simplicity we assume that only one kind of micelle exists. Let N_M and N_T be the number of monomeric and the total number of surfactant molecules, respectively. The number of surfactant molecules that build up the micelles is

$$N_{\mathrm{mic}} = N_T - N_M \tag{A.161}$$

We now assume that each micelle has a characteristic volume V_{mic}. Since these micelles are of size m, the total "volume" of the phase E is taken to be

$$V^E = \frac{N_{\mathrm{mic}}V_{\mathrm{mic}}}{m} = \alpha N_{\mathrm{mic}} \tag{A.162}$$

In the more general case, when there is a distribution of micelle sizes, we can assume that the volume of the phase E is simply proportional to N_{mic}. Here α may be interpreted as the volume of the phase E per particle forming the micelles.

The volume fraction is thus

$$x^E = V^E/V = \alpha(\varrho_T - \varrho_M) \tag{A.163}$$

where ϱ_T is the total density of the surfactant, and ϱ_M is the density of the monomeric surfactant (and usually is taken to be equal to the CMC).

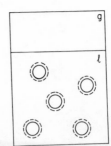

Figure A.8. Two phases, g and l. Micelles are formed in the liquid phase l. The dashed curves indicate the regions around the micelles in which the properties of the solvent differ appreciably from those of pure water.

Now for any surfactant concentration above the CMC, we assume that all the solute molecules s may be classified into two groups: those that are in the aqueous phase W, and all the others. By "all the others" we mean all those solute molecules with surroundings which are appreciably different from those of pure water. The total regions, shown schematically by the dashed circles in Figure A.8, will be referred to as the micellar phase E.

Following the assumptions made above, we may write the measurable quantity $\varrho_s{}^l/\varrho_s{}^g$ as in (A.157), using (A.163):

$$\gamma_s = (\varrho_s{}^l/\varrho_s{}^g) = \exp[-W(S \mid W)/kT]$$
$$\times [(K-1)\alpha(\varrho_T - \varrho_M) + 1] \qquad \text{(A.164)}$$

$$\gamma_s(\varrho_T = \varrho_M = \text{CMC}) = \exp[-W(S \mid W)/kT] \qquad \text{(A.165)}$$

$$\frac{\partial \gamma_s}{\partial \varrho_T} = \gamma_s(\text{CMC})(K-1)\alpha \qquad \text{(above the CMC)} \quad \text{(A.166)}$$

In (A.165) the assumption is made that the solubility of s in pure water is approximately equal to the solubility of s in the micellar solution at the CMC.

We see that from the measurement of γ_s in pure water and its slope above the CMC we may determine the quantity $\alpha(K-1)$. Clearly the meaning of $\Delta\mu_{\text{tr}}^{\circ}(W \to E)$ depends on the model we use to define the volume of the phase E in (A.163). The only determinable quantity here is $\alpha(K-1)$. For each choice of a *model* for α we can obtain the corresponding quantity K.

The fact that only the combination αK, rather than K itself, is the experimentally determinable quantity, reflects the ambiguity in the very definition of the micellar "volume" introduced in (A.162).

References

Ablett, S., M. D. Barrat, and F. Franks (1975), *J. Sol. Chem.* **4**, 797.

Abragam, A. (1961), *The Principles of Nuclear Magnetism*, Chapter 8, Oxford University Press, London.

Atherton, N. M. (1973), *Electron Spin Resonance*, John Wiley and Sons Inc., New York.

Barone, G., V. Crescenzi, B. Pispisa, and F. Quadrifoglio (1966), *J. Macromol. Chem.* **1**, 761.

Barone, G., V. Crescenzi, A. M. Liquori, and F. Quadrifoglio (1967), *J. Phys. Chem.* **71**, 2341.

Battino, R., and H. L. Clever (1966), *Chem. Rev.* **66**, 395.

Ben-Naim, A. (1968), *J. Phys. Chem.* **72**, 2998.

Ben-Naim, A. (1969), *J. Chem. Phys.* **50**, 404.

Ben-Naim, A. (1970), *Trans. Faraday Soc.* **66**, 2749.

Ben-Naim, A. (1971a), *Chem. Phys. Lett.* **11**, 389.

Ben-Naim, A. (1971b), *J. Chem. Phys.* **54**, 1387.

Ben-Naim, A. (1971c), *J. Chem. Phys.* **54**, 3696.

Ben-Naim, A. (1972a), *J. Chem. Phys.* **57**, 5257.

Ben-Naim, A. (1972b), *Mol. Phys.* **24**, 723.

Ben-Naim, A. (1972c), in *Water: A Comprehensive Treatise*, Vol. 1, edited by F. Franks, Plenum Press, New York.

Ben-Naim, A. (1972d), in *Water and Aqueous Solutions: Structure, Thermodynamics and Transport Processes*, edited by R. A. Horne, John Wiley, New York.

Ben-Naim, A. (1974), *Water and Aqueous Solutions: Introduction to a Molecular Theory*, Plenum Press, New York.

Ben-Naim, A. (1975), *J. Phys. Chem.* **79**, 1268.

Ben-Naim, A. (1977), *J. Chem. Phys.* **67**, 4884.

Ben-Naim, A. (1978a), *J. Phys. Chem.* **82**, 792.

Ben-Naim, A. (1978b), *J. Phys. Chem.* **82**, 874.

Ben-Naim, A. (1978c), in *Progress in Liquid Physics*, edited by C. A. Croxton, John Wiley and Sons, London.

Ben-Naim, A., and R. Tenne (1977), *J. Chem. Phys.* **67**, 627.

Ben-Naim, A., and J. Wilf (1979), *J. Chem. Phys.* **70**, 771.

Ben-Naim, A., and M. Yaacobi (1974), *J. Phys. Chem.* **78**, 170.

Ben-Naim, A., and M. Yaacobi (1975), *J. Phys. Chem.* **79**, 1263.

Ben-Naim, A., J. Wilf, and M. Yaacobi (1973), *J. Phys. Chem.* **77**, 95.

Birdi, K. S. (1976), *Anal. Biochem.* **74**, 620.

Birdi, K. S. (1977), in *Micellization, Solubilization, and Microemulsions*, Vol. 1, edited by K. L. Mittal, Plenum Press, New York.

Blyth, C. A., and J. R. Knowles (1971), *J. Am. Chem. Soc.* **93**, 3017.

Cerami, A., and C. M. Peterson (1975), *Sci. Am.* **232**, 45.

Clark, A. H., F. Franks, M. D. Pedley, and D. S. Reid (1977), *J. Chem. Soc. Faraday Trans. 1* **73**, 290.

Clever, H. L., R. Battino, J. S. Saylor, and P. M. Gross (1954), *J. Phys. Chem.* **61**, 1078.

Dashevsky, V. G., and G. N. Sarkisov (1974), *Mol. Phys.* **27**, 1271.

Dickerson, R. (1972), *Sci. Am.* **226**, 58.

Dubin, P. L., and U. P. Strauss (1970), *J. Phys. Chem.* **74**, 2842.

Edwards, J. T., P. G. Farrell, and F. Shahidi (1977), *J. Chem. Soc., Faraday Trans. 1* **73**, 705.

Eisenberg, D., and W. Kauzmann (1969), *The Structure and Properties of Water*, Oxford University Press, New York.

Franks, F. (1975), in *Water: A Comprehensive Treatise*, Vol. 4, edited by F. Franks, Plenum Press, New York.

Friedman, H. L., and C. V. Krishnan (1973), *J. Sol. Chem.* **2**, 119.

Goldammer, E. V., and H. G. Hertz (1970), *J. Phys. Chem.* **74**, 3734.

Goodman, D. S. (1958), *J. Am. Chem. Soc.* **80**, 3887.

Hall, D. G., and B. A. Pethica (1967), in *Nonionic Surfactants*, Chap. 16, edited by M. J. Schick, Marcel Dekker, New York.

Hansen, J. P., and I. R. McDonald (1976), *Theory of Simple Liquids*, Academic Press, New York.

Henriksson, U., and L. Ödberg (1976), *Colloid Polym. Sci.* **254**, 35.

Hermann, R. B. (1971), *J. Phys. Chem.* **75**, 363.

Hermann, R. B. (1972), *J. Phys. Chem.* **76**, 2754.

Hermann, R. B. (1974), in *Molecular and Quantum Pharmacology*, edited by E. Bergmann and B. Pullman, D. Reidel Publishing Company, Dordrecht, Holland.

Hermann, R. B. (1975), *J. Phys. Chem.* **79**, 163.

Hertz, H. G. (1973a), *J. Sol. Chem.* **2**, 239.

Hertz, H. G. (1973b), in *Water: A Comprehensive Treatise*, Vol. 3, edited by F. Franks, Plenum Press, New York.

Hertz, H. G., and R. Tutsch (1976), *Ber. Bunsenges. Phys. Chem.* **80**, 1268.

Hertz, H. G., and M. D. Zeidler (1964), *Ber. Bunsenges. Phys. Chem.* **68**, 621.

Hill, T. L. (1956), *Statistical Mechanics*, McGraw-Hill, New York.

Hill, T. L. (1958), *J. Chem. Phys.* **28**, 1179.

Hill, T. L. (1960), *Introduction to Statistical Thermodynamics*, Addison-Wesley, Reading, Massachusetts.

Horiuti, J. (1931), *Sci. Papers, Inst. Phys. Chem. Res. (Tokyo)* **17**, 125.

Howarth, O. W. (1975), *J. Chem. Soc. Faraday Trans. 1* **12**, 2303.

Hvidt, A. (1975), *J. Theor. Biol.* **50**, 245.

Jolicoeur, C., and J. Boileau (1974), *J. Sol. Chem.* **3**, 889.

Jolicoeur, C., and H. L. Friedman (1971), *Ber. Bunsenges. Phys. Chem.* **75**, 248.

Jolicoeur, C., and H. L. Friedman (1974), *J. Sol. Chem.* **3**, 15.

Jolicoeur, C., and G. Lacroix (1976), *Can. J. Chem.* **54**, 624.

Jolicoeur, C., P. Bernier, E. Firkins, and J. K. Saunders (1976), *J. Phys. Chem.* **80**, 1908.

Kauzmann, W. (1954), in *A Symposium on the Mechanism of Enzyme Action*, edited by W. D. McElroy and B. Glass, John Hopkins University Press, Baltimore, Maryland.

Kauzmann, W. (1959), *Adv. Protein Chem.* **14**, 1.

Kertes, A. S. (1977), in *Micellization, Solubilization and Microemulsions*, Vol. 1, edited by K. L. Mittal, Plenum Press, New York.

Kingston, B., and M. C. R. Symons (1973), *J. Chem. Soc. Faraday Trans. 2* **69**, 978.

Kirkwood, J. G. (1935), *J. Chem. Phys.* **3**, 300.

Kirkwood, J. G. (1939), *J. Chem. Phys.* **7**, 919.

Kirkwood, J. G. (1954), in *A Symposium on the Mechanism of Enzyme Action*, edited by W. D. McElroy and B. Glass, John Hopkins University Press, Baltimore, Maryland.

Kirkwood, J. G., and F. P. Buff (1951), *J. Chem. Phys.* **19**, 774.

Klapper, M. H. (1973), *Progr. Bioorganic Chem.* **2**, 55.

Knowles, J. R., and C. A. Parsons (1967), *Chem. Commun.*, 755.

Knowles, J. R., and C. A. Parsons (1969), *Nature* **221**, 53.

Kozak, J. J., W. S. Knight, and W. Kauzmann (1968), *J. Chem. Phys.* **48**, 675.

Kresheck, G. C. (1975), in *Water: A Comprehensive Treatise*, Vol. 4, edited by F. Franks, Plenum Press, New York.

Kresheck, G. C., H. Schneider, and H. A. Scheraga (1965), *J. Phys. Chem.* **69**, 3132.

Lebowitz, J. L., E. Helfand, and E. Praestgaad (1965), *J. Chem. Phys.* **43**, 774.

Leonard, P. J., D. Henderson, and J. A. Barker (1970), *Trans. Faraday Soc.* **66**, 2469.

Liquori, A. M., G. Barone, V. Crescenzi, F. Quadrifoglio, and V. Vilagliano (1966), *J. Macromol. Chem.* **1**, 291.

McAuliffe, C. (1966), *J. Phys. Chem.* **70**, 1267.

McMillan, W. G., and J. E. Mayer (1945), *J. Chem. Phys.* **13**, 276.

Marčelja, S., D. J. Mitchell, B. W. Ninham, and M. J. Sculley (1977), *J. Chem. Soc. Faraday Trans. 2* **73**, 630.

Martin, D. L., and F. J. C. Rossotti (1959), *Proc. Chem. Soc. London*, 60.

Masterton, W. L. (1954), *J. Chem. Phys.* **22**, 1830.

Miller, T. A., R. N. Adams, and P. M. Richards (1966), *J. Chem. Phys.* **44**, 4022.

Mittal, K. L. (1977), editor, *Micellization, Solubilization, and Microemulsions*, Vols. 1 and 2, Plenum Press, New York.

Morrison, T. J., and E. Billet (1952), *J. Chem. Soc.*, 3819.

Mukerjee, P. (1974), *J. Pharmaceutical Sci.* **63**, 972.

Mukerjee, P. (1977), *J. Phys. Chem.* **76**, 565.

Mukerjee, P., and J. R. Cardinal (1976), *J. Pharmaceutical Sci.* **65**, 882.

Mukerjee, P., and K. J. Mysels (1971), *Critical Micelle Concentrations of Aqueous Surfactant Systems*, National Bureau of Standards, Washington, D.C., Feb.

Muller, N., and R. H. Birkhahn (1967), *J. Phys. Chem.* **71**, 957.

Münster, A. (1969), *Statistical Thermodynamics*, Springer-Verlag, Berlin.

Münster, A. (1974), *Statistical Thermodynamics*, Vol. 2, Springer-Verlag, Berlin.

Murayama, M. (1973), *CRC Critical Rev. Biochem.* **1**, 461.

Nash, G. R., and C. B. Monk (1957), *J. Chem. Soc.*, 4274.

Némethy, G. (1967), *Angew. Chem. Int. Ed. Engl.* **6**, 195.

Némethy, G., and H. A. Scheraga (1962), *J. Phys. Chem.* **66**, 1773.

Némethy, G., and H. A. Scheraga (1964), *J. Chem. Phys.* **41**, 680.

Oakenfull, D. G. (1973), *J. Chem. Soc., Perkin Trans. II*, 1006.

Oakenfull, D. G., and D. E. Fenwick (1973), *Aust. J. Chem.* **26**, 2646.

Oakenfull, D. G., and D. E. Fenwick (1974a), *Aust. J. Chem.* **27**, 2149.

Oakenfull, D. G., and D. E. Fenwick (1974b), *J. Phys. Chem.* **78**, 1759.

Oakenfull, D. G., and D. E. Fenwick (1975), *Aust. J. Chem.* **28**, 715.

Oakenfull, D. G., and D. E. Fenwick (1977), *Aust. J. Chem.* **30**, 741.

Ödberg, L., B. Svens, and I. Danielson (1972), *J. Colloid Interface Sci.* **41**, 298.

Owicki, J. C., and H. A. Scheraga (1977), *J. Am. Chem. Soc.* **99**, 7413.

Packter, A., and M. Donbrow (1962), *Proc. Chem. Soc.* 220.

Paquette, J., and C. Jolicoeur (1977), *J. Sol. Chem.* **6**, 403.

Philip, P., and C. Jolicoeur (1973), *J. Phys. Chem.* **77**, 3071.

Pierotti, R. A. (1963), *J. Phys. Chem.* **67**, 1840.

Pierotti, R. A. (1965), *J. Phys. Chem.* **69**, 281.

Pierotti, R. A. (1976), *Chem. Rev.* **76**, 717.

Pratt, L. R., and D. Chandler (1977), *J. Chem. Phys.* **67**, 3683.

Priel, Z., and A. Silberberg (1970), *J. Polym. Sci.* **8**, 689, 705, 713.

Ramanathan, P. S., C. V. Krishnan, and H. L. Friedman (1972), *J. Sol. Chem.* **1**, 237

Reiss, H. (1966), *Adv. Chem. Phys.* **9**, 1.

Reiss, H., H. L. Frisch, and J. L. Lebowitz (1959), *J. Chem. Phys.* **31**, 369.

Saylor, J. H., and R. Battino (1958), *J. Phys. Chem.* **62**, 1334.

Scheraga, H. L., G. Némethy, and I. Z. Steinberg (1962), *J. Biol. Chem.* **234**, 2506.

Schrier, E. E., M. Pottle, and H. A. Scheraga (1964), *J. Am. Chem. Soc.* **86**, 3444.

Stryer, L. (1975), *Biochemistry*, W. H. Freeman, San Francisco, California.

Suzuki, K., and M. Tsuchiya (1971), *Bull. Chem. Soc., Japan* **44**, 967.

Suzuki, K., Y. Taniguchi, and T. Watanabe (1973), *J. Phys. Chem.* **77**, 1918.

Tanford, C. (1973), *The Hydrophobic Effect: Formation of Micelles and Biological Membranes*, Wiley Interscience, New York.

Tanford, C. (1974), *J. Phys. Chem.* **78**, 2469.

Tenne, R., and A. Ben-Naim (1976), *J. Phys. Chem.* **80**, 1120.

Tenne, R., and A. Ben-Naim (1977), *J. Chem. Phys.* **67**, 4632.

Timasheff, S. N., and G. D. Fasman (1969), editors, *Structure and Stability of Biological Macromolecules*, Marcel Dekker, New York.

Wen, W. Y., and H. G. Hertz (1972), *J. Sol. Chem.* **1**, 17.

Wen, W. Y., and J. H. Hung (1970), *J. Phys. Chem.* **74**, 170.

Wenograd, J., and R. A. Spurr (1957), *J. Am. Chem. Soc.* **79**, 5844.

Wertz, J. E., and J. R. Bolton (1972), *Electron Spin Resonance*, Chap. 9, McGraw-Hill, New York.

Wetlaufer, D. B., S. K. Malik, L. Stoller, and R. L. Coffin (1964), *J. Am. Chem. Soc.* **86**, 508.

Wilf, J., and A. Ben-Naim (1979), *J. Chem. Phys.* **70**, 3079.

Wilhelm, E., and R. Battino (1971), *J. Chem. Phys.* **55**, 4012.

Wilhelm, E., and R. Battino (1972), *J. Chem. Phys.* **56**, 563.

Wilhelm, E., R. Battino, and R. J. Wilcock (1977), *Chem. Rev.* **77**, 219.

Winsor, P. A., (1954), *Solvent Properties of Amphiphilic Compounds*, Butterworths Scientific Publications, London.

Wishnia, A. (1962), *Proc. Natl. Acad. Sci. U.S.A.* **48**, 2200.

Wishnia, A., and T. W. Pinder (1964), *Biochemistry* **3**, 1377.

Wishnia, A., and T. W. Pinder (1966), *Biochemistry* **5**, 1534.

Yaacobi, M., and A. Ben-Naim (1973), *J. Sol. Chem.* **2**, 425.

Yaacobi, M., and A. Ben-Naim (1974), *J. Phys. Chem.* **78**, 175.

Zeidler, M. D. (1973), in *Water: A Comprehensive Treatise*, Vol. 2, edited by F. Franks, Plenum Press, New York, p. 529.

Index